Number Theory Related to Modular Curves Momose Memorial Volume

CONTEMPORARY MATHEMATICS

701

Number Theory Related to Modular Curves Momose Memorial Volume

Seminar in Memory of Fumiyuki Momose
Barcelona-Boston-Tokyo Number Theory
Barcelona

Joan-Carles Lario
V. Kumar Murty
Editors

AMS AMERICAN MATHEMATICAL SOCIETY

2010 *Mathematics Subject Classification*. Primary 11G18, 11G05, 14G50, 11F41, 11R37, 11M99; Secondary 14G35, 11G40, 11G30.

Library of Congress Cataloging-in-Publication Data

Names: Lario, Joan-Carles, 1963-editor. | Murty, Vijaya Kumar, 1956-editor.

Title: Number theory related to modular curves : Momose memorial volume : Barcelona-Boston-Tokyo Number Theory Seminar, in memory of Fumiyuki Momose, Universitat Politecnica de Catalunya, Barcelona, Spain / Joan-Carles Lario, V. Kumar Murty, editors.

Description: Providence, Rhode Island : American Mathematical Society, [2018]| Series: Contemporary mathematics ; number 701 | Seminar held May 21–23, 2012, in Barcelona, Spain | Includes bibliographical references.

Identifiers: LCCN 2017042713 | ISBN 9781470419912 (alk. paper)

Subjects: LCSH: Momose, Fumiyuki. | Modular curves. | Number theory. | Forms, Modular. | AMS: Number theory – Arithmetic algebraic geometry (Diophantine geometry) – Arithmetic aspects of modular and Shimura varieties. msc | Algebraic geometry – Arithmetic problems. Diophantine geometry – Rational points. msc | Algebraic geometry – Arithmetic problems. Diophantine geometry – Applications to coding theory and cryptography. msc | Number theory – Discontinuous groups and automorphic forms – Automorphic forms on. msc | Number theory – Algebraic number theory: global fields – Class field theory. msc | Number theory – Zeta and L-functions: analytic theory – None of the above, but in this section. msc | Algebraic geometry – Arithmetic problems. Diophantine geometry – Modular and Shimura varieties. msc | Number theory – Arithmetic algebraic geometry (Diophantine geometry) – L-functions of varieties over global fields; Birch-Swinnerton-Dyer conjecture. msc | Number theory – Arithmetic algebraic geometry (Diophantine geometry) – Curves of arbitrary genus or genus $\neq 1$ over global fields. msc

Classification: LCC QA567.2.M63 N86 2018 | DDC 512.7/4–dc23

LC record available at https://lccn.loc.gov/2017042713

DOI: http://dx.doi.org/10.1090/conm/701

∞ The paper used in this book is acid-free and falls within the guidelines established to ensure permanence and durability.

Visit the AMS home page at http://www.ams.org/

10 9 8 7 6 5 4 3 2 1 23 22 21 20 19 18

To the memory of Fumiyuki Momose

Photo courtesy of Math. Dept., Chuo University, Tokyo, Japan

Contents

Preface

This volume is dedicated to the memory of Fumiyuki Momose. It mostly contains articles based on talks given at a conference in his honour that was held in 2012 in Barcelona. This conference was organized by one of us (Lario) together with Jinhui Chao, Francesc Fité, Josep González Rovira and Tsutomu Sekiguchi. It was held during May 21-23 at the Facultat de Matemàtiques i Estadística of the Universitat Politècnica de Catalunya in Barcelona. There were 17 invited speakers who discussed a variety of topics in Arithmetic Geometry, and some of them have contributed articles to this volume. Many of the talks dealt with matters which were directly or indirectly inspired or motivated by the work of Momose. About 33 mathematicians from around the world participated in the conference. In addition, there were a number of friends and well-wishers of Momose who wanted to join in honouring his memory but who could not be present for the conference. They have also contributed articles for this volume.

The themes of the conference and of most of the papers in this volume are related to the arithmetic of modular curves and abelian varieties, two themes that were of great interest to Momose and which occupied most of his mathematical work. In particular, there are three articles on rational points on modular curves and on elliptic curves. The paper of Derickx, Kamienny and Mazur discusses rational families of elliptic curves over number fields that possess a rational point of order 17. The paper of Bars discusses points on modular curves that are defined over a quadratic extension of the rationals. And the paper by Arai gives an overview of his results, some joint with Momose, on points on Shimura curves defined over a number field.

There are two articles that deal with applications of arithmetic geometry to cryptography, a topic that Momose became interested in as Chuo University was home to a major Japanese Center of Excellence grant to support work in this area. One of the articles is in fact a previously unpublished work of Hashizume, Momose and Chao in which they showed how to implement the so-called Gaudry-Hess-Smart attack on a large class of elliptic curves defined over a cubic extension field of odd characteristic. The second paper by Burko shows how to do efficient point counting on certain singular superelliptic curves.

There are two articles on subjects related to automorphic forms, a topic that appeared largely indirectly in Momose's work. The paper by Gon and Oda describes explicit integral representations for Siegel-Whittaker functions on the group $Sp(2,\mathbb{R})$. The paper by Lario and Somoza discusses the Sato-Tate conjecture for a Picard curve with complex multiplication.

The paper by Castano-Bernard deals with the arithmetic of elliptic curves defined over the rational number field. He considers an elliptic curve whose L-function has a simple zero at the central critical point. For such a curve, one knows that the Heegner point construction produces a subgroup P of the Mordell-Weil group M of finite index. Castano-Bernard formulates a conjecture on the 2-part of the index $[M : P]$ in terms of the 2-part of the Shafarevich-Tate group of the elliptic curve (which in this case is known to be finite by the work of Kolyvagin and the theorems on modularity).

The paper by Ram Murty gives an overview of recent work on transcendence of special values of Dirichlet series. A second paper by Ram Murty in collaboration with Tapas Chatterjee and Siddhi Pathak discusses a criterion for Dirichlet series with periodic coefficients to vanish at the edge of the critical strip. This is in the spirit of a conjecture of Chowla.

Finally, the article of Kumar Murty deals with the construction of some abelian extensions of real quadratic fields in the spirit of work of Shimura. It uses ideas of Momose and Ribet to generalize Shimura's construction.

At the beginning of the collection, Takeshi Saito gives a brief description of Momose's mathematical work as well as a full list of Momose's publications.

We would like to thank a number of people and institutions for their effort and patience to make the conference and this volume possible. First, the authors for their generosity in contributing an article and the anonymous referees for their help in reviewing the submissions. Second, the staff of the Facultat de Matemàtiques i Estadística at UPC where the conference took place for their collaboration. Third, the staff at the AMS, especially Sergei Gelfand, Christine Thivierge, and Lauren Foster for their patience and guidance. Finally, we thank all the close friends and family of Fumiyuki Momose for their interest and help taking care of all details.

Fumiyuki Momose was a gentle and kind human being, and an excellent mathematician. We are honoured to be associated with a volume in his memory.

Joan-Carles Lario
V. Kumar Murty

The Barcelona Conference

I do not know much about Fumiyuki Momose's life. But what I do know are the true reasons why one day I decided to promote the organization of this seminar in his honour. And that is what I would like to tell you now.

I met Fumiyuki Momose in 1993 in Boston. That year I was visiting the Mathematics Department of Harvard University. After a few months there, I was asked to give a talk in the Number Theory Seminar that meets weekly every Wednesday afternoon. It seemed to me that the more natural thing to do was to explain what I had been doing during my stay in that formidable atmosphere.

When I finished my talk, there were a number of questions and comments. They were all positive and suggestive, especially one by Noam Elkies. When we went out of the room of the seminar, Fumiyuki Momose, who had been in the audience, approached me. At that time he was also visiting the department but so far we had not had any opportunity to meet. His first words were: "I did what you have told 7 years ago and it is already published. Come with me to have dinner."

That was a surprise to me! His positive and open reaction strongly contrasted with what I was used to. Instead of getting angry on his part, I saw joy and understanding. Fumiyuki Momose went even further. What I just told you happened in June 1993, shortly before the first announcement of Andrew Wiles of a proof of Fermat's Last Theorem. Three months later, back into our respective countries, I received an email where Momose invited me to spend a year at the Chuo University in Tokyo. I spent 3 months there, and I felt at home thanks to Momose.

Since that time, we had a very good relationship with various comings and goings between Barcelona and Tokyo for short stays, for his and my part. Finally, Momose had decided to come for a long sabbatical year in Barcelona. All preparations were ready; his accommodation was already set. And then he received the news about his illness and we had to abort his stay in Barcelona.

I have no doubt that his reaction to my talk in 1993 led to a deep change in the way I see things now, and in the way I understand science and life. For a long time, I had been wondering what is more important in number theory, either the theory or the numbers. Thanks to Momose, now I know the answer. The most important thing in number theory is the people.

Joan-Carles Lario

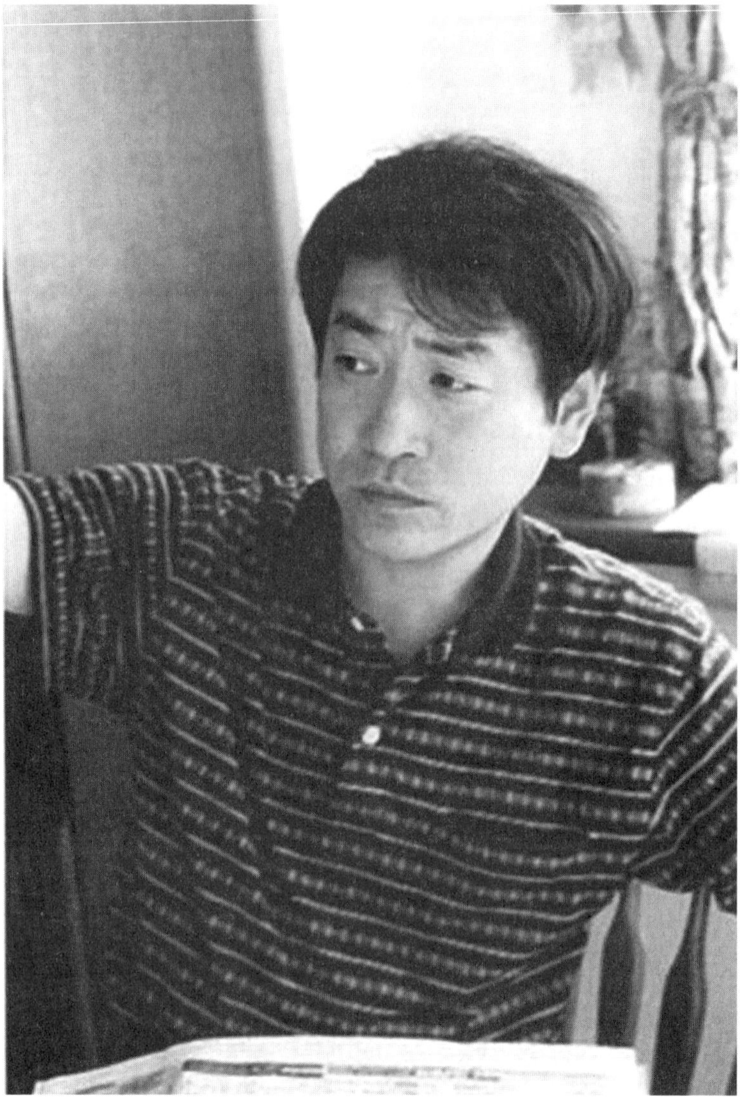

Photo circa 1995. Courtesy of Jinhui Chao

My friend, Fumiyuki Momose

I first heard of Fumiyuki Momose when I was a graduate student. At that time, I was thinking about the Tate conjecture for Abelian varieties. My advisor, John Tate, had just received a preprint from Ken Ribet in which he had proved the conjecture for Abelian varieties that are quotients of the Jacobians of modular curves. Tate gave me that paper to read and it turned out to be very important in the research that eventually formed my thesis. In that paper, Ribet mentioned that the same results had been obtained by Momose, who I later learnt was a student of Yasutaka Ihara.

That year, Takayuki Oda was visiting Harvard and was developing his theory of periods of Hilbert modular forms to prove the Tate conjecture for Hilbert modular surfaces. Oda, who I used to discuss mathematics with, told me that the following year, Momose would be visiting Harvard. I was delighted to hear this and was eagerly awaiting Momose's arrival.

When we met, I found Momose to be a very cheerful and modest human being, but with deep knowledge of arithmetic geometry. We became friends very quickly. After his thesis work, he had started to look at Mazur's work on rational points on modular curves. He was able to prove the analogue of Mazur's finiteness theorems for many curves that had not been covered in the earlier theory. He shared with me some of his preprints and the techniques I learnt there helped me in my own work. My contribution to this volume on the construction of some abelian extensions of real quadratic fields was directly influenced by Momose's work.

While Momose was very generous to everyone, he did not seem to be very attentive about his own health. He had an apendectomy soon after his arrival at Harvard. One day, I saw him in the Department limping around and he told me that he had 'escaped' from the hospital. I was amused at the time, but it was clear that he should have been resting and given himself more time for convalescing.

Over the years, we stayed in touch. When I visited Japan in 1990 for the International Congress of Mathematicians and for a satellite conference, Momose kindly took me around and showed me many wonderful places. During 2002-2006, we started meeting regularly again when Chuo University was involved with a Center of Excellence grant from the Japanese government to study number theory and cryptography. It was during this time that Momose proved some very interesting results on elliptic curves which were susceptible to the attack initially described by Frey and refined by Gaudry, Hess and Smart.

In May of 2009, our common friend Jinhui Chao of Chuo University informed me that Fumiyuki had been diagnosed with esophagus cancer and as a result had been hospitalized. He was operated on in June of that year, and he improved a little bit but much further treatment was required. Unfortunately, his condition deteriorated rapidly soon afterwards, and he passed away on April 24, 2010. This was devastating news to those who were near him and had not expected such a rapid negative turn of events.

Tragedy struck the family once again in 2011 when Chiaki, Fumiyuki's partner of 24 years also fell ill and passed away. The Momoses are survived by their son Takaaki, whom I met at the Barcelona conference. With the passing of his parents, Takaaki was being raised by Fumiyuki's sister Mrs. Suzuki. In my conversations with him at Barcelona, Takaaki expressed an interest in mathematics and perhaps he will pursue mathematics as a career.

Fumiyuki Momose had a very generous approach to life and to mathematics. Many mathematicians, both young and old, have benefitted from that generosity. It is an honour to present this volume of mathematics dedicated to his memory.

V. Kumar Murty

Photo of Chiaki, Takaaki, and Fumiyuki Momose. Courtesy of Jinhui Chao

Contemporary Mathematics
Volume **701**, 2018
http://dx.doi.org/10.1090/conm/701/14152

An overview of the mathematical work of Fumiyuki Momose

Takeshi Saito

1. Introduction

Momose san was a mathematician who had a deep understanding and insight on arithmetic of modular curves and modular forms. Since the seminal works by J-P. Serre and B. Mazur, the Galois action on the torsion points of an elliptic curve, or equivalently the rational points of modular curves, has been a central subject in number theory. Momose san made significant contributions on this subject. His early work on the image of the ℓ-adic representations attached to elliptic modular forms and the implications for the Tate conjecture are also significant, as is his later work on applications of arithmetic geometry to cryptography. In this article, we will give a very brief overview of his work.

2. ℓ-adic representations

In his first published article [1981] (see the list given in Section 6), which was based on his Master's thesis written in 1977, Momose studied the ℓ-adic representations associated to normalized Hecke eigen cusp forms. To such a cusp form f of weight k, level N and Nebentypus character ϵ, there is a family of ℓ-adic representations

$$\rho_{f,\ell} : \mathrm{Gal}(\overline{\mathbb{Q}}/\mathbb{Q}) \longrightarrow \mathrm{GL}_2(\mathcal{O}_f \otimes \mathbb{Z}_\ell)$$

constructed by Deligne. Here, $\mathcal{O}_f$ denotes the ring of integers of the number field generated by the Fourier coefficients of f. The representation $\rho_{f,\ell}$ has the property that it is unramified outside ℓN and for a prime p not dividing ℓN, the characteristic polynomial of any Frobenius automorphism $Frob_p$ at p is

$$X^2 - a_f(p)X + \epsilon(p)p^{k-1}.$$

where $a_f(p)$ is the p-th Fourier coefficient of f.

It is a very interesting question to ask what the image of $\rho_{f,\ell}$ is. A weaker question is to determine the ℓ-adic Lie algebra of the Zariski closure of the image (which is an ℓ-adic Lie group). In the case that all the Fourier coefficients of f are rational integers (in other words, $\mathcal{O}_f = \mathbb{Z}$), and the weight k of f is 2, there is an elliptic curve E_f associated to f which is defined over $\mathbb{Q}$. More generally, if we do not assume that the Fourier coefficients are rational integers, there is (by the work of Shimura) an Abelian variety A_f defined over $\mathbb{Q}$ associated to f. In this case the

Galois representation $\rho_{f,\ell}$ is just the action of $\mathrm{Gal}(\overline{\mathbb{Q}}/\mathbb{Q})$ on the ℓ-adic Tate module $T_\ell(A_f)$ of A_f.

Serre studied extensively the case of the Galois action on the Tate module of an elliptic curve over a number field. In particular, he showed that if the curve does not have complex multiplication, then for ℓ sufficiently large, the image is $\mathrm{GL}_2(\mathbb{Z}_\ell)$.

Following Ribet, we say that a modular form f has *complex multiplication* if there is a character χ (necessarily of order 2) with the property that $\rho_{f,\ell} \simeq \rho_{f,\ell} \otimes \chi$ for one (and hence, all) ℓ. From work of Ribet and Shimura, one knows that if the weight of f is 2, then f has complex multiplication (in the sense of Ribet) if and only if the corresponding Abelian variety A_f has complex multiplication.

In several papers, Ribet studied the question of the image of the Galois representations attached to modular forms. In particular, in [**10**], he realized that an important consideration is the possible existence of "inner twists", that is the existence of pairs (σ, χ) consisting of an automorphism σ and a character of finite order χ for which $f^\sigma = f \otimes \chi$. In the case that there are no such twists, Ribet determined the Lie algebra.

The paper [1981] of Momose determines the Lie algebra in general. Simultaneously and independently, Ribet [**11**] also determined it. In the case that f has weight 2, this actually allowed Ribet to prove the Tate conjecture on endomorphisms (equivalently, on divisors) of A_f. (Note that this was before Faltings proved the Tate conjecture for divisors on any Abelian variety defined over a number field.) In fact, as shown in Murty [**8**], it allows one to prove both the Hodge and Tate conjectures for A_f in all codimensions. Moreover, the algebra of inner twists can be used [**9**] to generalize Shimura's construction of some abelian extensions of real quadratic fields.

3. Rational Points

A striking result of Mazur [**6**] determined the complete list of prime numbers p such that the image of the representation of the absolute Galois group of $\mathbb{Q}$ associated to an elliptic curve lies inside the upper triangular subgroup of $\mathrm{GL}_2(\mathbb{F}_p)$, or equivalently, a cyclic subgroup of order p of an elliptic curve defined over $\mathbb{Q}$. It led naturally Momose san to investigate other maximal subgroups of $\mathrm{GL}_2(\mathbb{F}_p)$ or extensions of $\mathbb{Q}$. In the following, I consider some papers from the list of his publications given in Section 6 and give a brief account of the contents roughly in chronological order.

A series of papers on the rational points of modular curves is started in his doctor thesis in 1981, published as [1984A] in Compositio Mathematica. There he studies modular curves $X_{\mathrm{split}}(p)$ corresponding to the normalizer of a split Cartan subgroup. Via the modular interpretation, a rational point corresponds to an elliptic curve endowed with a pair of cyclic subgroups of order p, Galois conjugate to each other. The main result proved in [1984A] is the following. Let p be a prime number. Let $X_0(p)$ be the compact modular curve over $\mathbb{Q}$ classifying isogenies of elliptic curves of degree p and let $J_0(p)$ be its Jacobian. Let $J_0(p)^-$ denote the quotient of $J_0(p)$ where the Atkin-Lehner involution w_p acts as -1. Then, the modular curve $X_{\text{split}}(p)$ has no rational points other than cusps or those corresponding to elliptic curves with complex multiplication if p satisfies the following two conditions:

(i) $p = 11$ or $p \geqq 17$,
(ii) the group of $\mathbb{Q}$-rational points of $J_0(p)^-$ is finite.

The study initiated in [1984A] was later developed in many directions. Currently, by Bilu and Parent [**2**], the conditions (i) and (ii) are replaced by the condition that p is sufficiently large. In their proof, the integrality of the j-invariant of the elliptic curve corresponding to a non-cusp point partially established in [1984A] plays a crucial role.

In [1986], the above result on $X_{\text{split}}(p) = X_0^+(p^2)$ was generalized to the quotient $X_0^+(p^r)$ of $X_0(p^r)$ by the Atkin-Lerner involution w_{p^r}. In [1987A], Momose further considers $X_0^+(N)$ for a composite number N. He proves the non-existence of non-cusp and non-CM points assuming that N is divisible by a prime number p satisfying the two conditions (i) and (ii) above and the following condition:

(iii) $p \neq 37$.

The prime $p = 37$ is exceptional because it is the only prime for which the modular curve $X_0(37)$ is a hyperelliptic curve of genus $\geqq 2$ but the hyperelliptic involution is not given by the Atkin-Lehner involution w_p. This exceptional case was later studied in joint work [2010] and [2012] with K. Arai and published posthumously.

In a closely related but a different direction, he studied the automorphism groups of the modular curves $X_0(N)$ with Kenku in [1988A]. Generalizing an earlier work by Ogg for square free N, they proved that if the modular curve $X_0(N)$ has genus at least 2, its group of automorphisms is the image of the normalizer of $\Gamma_0(N)/\{\pm 1\}$ in $PGL_2^+(\mathbb{Q})$ except for $N = 37, 63$.

In his celebrated work, Mazur determined the complete list consisting of 12 primes $p \leqq 163$ such that there exists an isogeny of elliptic curves of degree p defined over $\mathbb{Q}$. In [1995], Momose studied this problem for imaginary quadratic fields k and showed that if k is not imaginary of class number one, there are only finitely many p such that there exists an isogeny of elliptic curves of degree p defined over k. This article appeared shortly before another celebrated work by Merel[**7**] proving the uniform boundedness of the torsion subgroup of elliptic curves over number fields of given degree.

Momose san enjoyed working with younger colleagues. It is quite impressive to find a number of joint papers as results of such fruitful collaborations. The works [2010], [2012] with K. Arai mentioned above is an example. In [1999A], he worked with K. Hashimoto and a younger colleague Y. Hasegawa to prove the modularity of some Abelian varieties of GL_2 type over $\mathbf{Q}$ using the method developed by Wiles and Taylor-Wiles. In [1999B], he worked with M. Shimura to give a lower bound for the absolute ramification index of a p-adic field having an elliptic curve with good supersingular reduction.

Photos of Momose with students courtesy of Math. Dept., Chuo University, Tokyo, Japan

4. Cryptography

Momose san was also interested in cryptosystems based on elliptic and hyperelliptic curves, especially the possibility to break such a system using Weil descent. He wrote about five papers on this topic including [2005], [2006], [2009], [2013] and [2017]. Note that the last two were actually preprints posted in 2008 and 2009. Strangely, the first three papers still only exist in preprint form.

The security of cryptosystems that are based on Abelian varieties over finite fields is connected to the presumed difficulty of solving the discrete logarithm problem in the group of points on such a variety. Given such an Abelian variety A over a finite field $\mathbb{F}$ of $q = p^a$ elements, we consider the group $A(\mathbb{F})$ of points on A rational over $\mathbb{F}$. We may also consider points on A defined over a finite extension of $\mathbb{F}$. Encryption and decryption is specified through arithmetic operations on $A(\mathbb{F})$ and this results in what is known as an asymmetric (or public-key) encryption scheme.

The security of the scheme rests on the presumed computational difficulty of solving the discrete logarithm problem. This is a computational problem which asks that given a point Q in $A(\mathbb{F})$ and another point R in the subgroup generated by Q, find an integer n so that $R = nQ$.

There is a generic way of solving this problem in any group G which runs in time proportional to the square root of the group order. This method does not use any specific information or structure about G. If one has additional information about G it is possible that there may be more efficient ways to solve the discrete logarithm problem. For example, the method of index calculus shows that for $G = \mathbb{F}_q^\times$, one has a subexponential algorithm, that is one that has running time proportional to $\exp\{c(\log q)^{1/3}(\log\log q)^{2/3}\}$ for some explicit and positive constant c.

Groups that arise from elliptic curves over finite fields are preferred as a basis of encryption schemes because there are efficient methods of performing arithmetic in such groups. On the other hand, though there has been a lot of study, there is in general no better way of solving the discrete logarithm problem than the generic attack, that is in time proportional to $q^{\frac{1}{2}}$ where the curve is defined over a finite field of q elements.

More generally, one may also use the group of points on an Abelian variety over a finite field. In this case, arithmetic is more complicated. The one case that has been studied extensively is that of Jacobians of hyperelliptic curves and one has now very efficient algorithms for group operations. However, as shown by Gaudry [**4**], Diem and others, as the genus of the curve increases, the discrete logarithm problem actually becomes easier.

G. Frey [**3**] showed that this has negative implications for elliptic curve cryptosystems as well. In particular, following his suggestion, Gaudry, Hess and Smart [**5**] formulated what is known as the Weil descent (or GHS) attack. Beginning with an elliptic curve E over a non-prime field $\mathbb{F}_q$, one considers the restriction of scalars of E to a subfield $\mathbb{F}_{q_0}$ of $\mathbb{F}_q$. This gives rise to an Abelian variety of dimension $[\mathbb{F}_q : \mathbb{F}_{q_0}]$ and one shows that within that Abelian variety, it is possible to locate a hyperelliptic curve C whose genus is in general large. One can then transfer the discrete logarithm problem from E to the Jacobian of C, where the attack of Gaudry is applicable.

Since this important discovery of Frey, many cryptographers were trying to identify which curves are thus compromised. Some felt that the curves thus compromised were few in number. However, Momose's works (with Hashizume, Iijima, and Chao), which include the classification of those weak curves over extensions of finite fields used in cryptosystems and the density analysis of such curves in [2009], [2013] and [2017] show that many curves are compromised by the GHS attack. In particular, the paper [2017] in this volume addresses the issue of curves C_0 over a field of p^3 elements (for a prime p) and uses non-hyperelliptic curves C which have the property that there is a covering map $C \longrightarrow C_0$ defined over $\mathbb{F}_{p^3}$ for which the composition

$$Jac(C) \longrightarrow Jac(C_0) \longrightarrow Res_{\mathbb{F}_{p^3}/\mathbb{F}_p} Jac(C_0)$$

defines an isogeny over $\mathbb{F}_p$. Here $Jac(C)$ (respectively $Jac(C_0)$ denotes the Jacobian of C (respectively C_0) and Res is the restriction of scalars functor. The earlier papers classify those curves C_0 of genus 1 which have a cover C as above which is non-hyperelliptic of genus 3. Using these, [2017] solves the discrete logarithm problem on C_0 in $\mathbf{O}(p)$ steps.

5. Concluding Remarks

Momose san made important contributions to several topics in arithmetic geometry including rational ponts on modular and Shimura curves, ℓ-adic representations associated to modular forms, modularity of elliptic, $\mathbb{Q}$ and QM-curves, and the discrete logarithm problem on certain elliptic and hyperelliptic curves over finite fields. The complete list of his papers is given in the next section. However, we note that Momose san was a very generous mathematician and he freely shared his ideas with others. His actual contributions, therefore, probably go beyond the formal list of publications.

We were very happy to share time with Momose san to work seriously and joyfully on mathematics. We regret that it was too early for us that we no longer have a chance to see him and discuss with him again.

6. Publications of Momose

1981 F. Momose, On the ℓ-adic representations attached to modular forms, *J. Fac. Sci. Univ. Tokyo Sect. IA Math.*, **28**(1981), 89-109.

1983 F. Momose, Galois action on some ideal section points of the abelian variety associated with a modular form and its application, *Nagoya Math. J.*, **91**(1983), 19-36.

1984A F. Momose, Rational points on the modular curves $X_{\text{split}}(p)$, *Compositio Math.*, **52**(1984), 115-137.

1984B F. Momose, p-torsion points on elliptic curves defined over quadratic fields, *Nagoya Math. J.*, **96**(1984), 139-165.

1986 Momose, Fumiyuki, Rational points on the modular curves $X_0^+(p^r)$, J. Fac. Sci. Univ. Tokyo Sect. IA Math. 33 (1986), no. 3, 441-466.

1987A F. Momose, Rational points on the modular curves $X_0^+(N)$, *J. Math. Soc. Japan*, **39**(1987), 269-286.

1987B I. Kuribayashi and F. Momose, On the action of automorphisms of a curve on the first l-adic cohomology, *Tsukuba J. Math.*, **11**(1987), 287-301.

1988A M. A. Kenku and F. Momose, Automorphism groups of the modular curves $X_0(N)$, *Compositio Math*, **65**(1988), 51-80.

1988B M. A. Kenku and F. Momose, Torsion points on elliptic curves defined over quadratic fields, *Nagoya Math. J.*, **109**(1988), 125-149.

1991 N. Ishii and F. Momose, Hyperelliptic modular curves, *Tsukuba J. Math.*, **15**(1991), 413-423.

1995 F. Momose, Isogenies of prime degree over number fields, *Compositio Math.*, **97**(1995), 329-348.

1996 Y. Hasegawa, K. Hashimoto, and F. Momose, QM-curves and $\mathbb{Q}$-curves, in: *Deformation of group schemes and applications to number theory* (Japanese) (Kyoto, 1995). Surikaisekikenkyusho Kokyuroku No. 942 (1996), 164-167.

1999A Y. Hasegawa, K. Hashimoto and F. Momose, Modularity conjecture for $\mathbb{Q}$-curves and QM-curves, *Internat. J. Math.*, **10**(1999), 1011-1036.

1999B F. Momose and M. Shimura, Moduli and modularity of $(\mathbb{Q}, F)$-abelian varieties of GL_2-type, in: *Algebraic number theory and related topics* (Japanese) (Kyoto, 1998). Surikaisekikenkyusho Kokyuroku No. 1097 (1999), 151-160.

2002 F. Momose and M. Shimura, Lifting of supersingular points on $X_0(p^r)$ and lower bound of ramification index, *Nagoya Math.J.*, **165**(2002), 159-178.

2005 F. Momose and J. Chao, Scholten forms and elliptic/hyperelliptic curves with weak Weil restrictions, http://eprint.iacr.org/2005/277.

2006 F. Momose and J. Chao, Classification of Weil restrictions obtained by $(2, ..., 2)$ coverings of $\mathbb{P}^1$, http://eprint.iacr.org/2006/347

2009 T. Iijima, F. Momose and J. Chao, Classification of elliptic/hyperelliptic curves with weak coverings against GHS attack without isogeny condition, http://eprint.iacr.org/2009/613.

2010 K. Arai, and F. Momose, Rational points on $X_0^+(37M)$, *J. Number Theory*, **130** (2010), 2272-2282.

2012 K. Arai and F. Momose, Points on $X_0^+(N)$ over quadratic fields, *Acta Arith.*, **152** (2012), 159-173.

2013 F. Momose and J. Chao, Elliptic curves with weak coverings over cubic extensions of finite fields with odd characteristics, *J. Ramanujan Math. Soc.*, **28**(2013), 299-357.

2014 K. Arai and F. Momose, Algebraic points on Shimura curves of $\Gamma_0(p)$-type, *J. Reine Angew. Math.*, **690** (2014), 179-202. (See also Errata, *J. Reine Angew. Math.*, **690**(2014), 203-205.)

2017 N. Hashizume, F. Momose and J. Chao, On the GHS Attack against Elliptic Curve Cryptosystems over Cubic Extension Fields of Odd Characteristics, this volume.

References

[1] Keisuke Arai, *Galois images and modular curves*, Algebraic number theory and related topics 2010, RIMS Kôkyûroku Bessatsu, B32, Res. Inst. Math. Sci. (RIMS), Kyoto, 2012, pp. 145–161. MR2986922

[2] Yuri Bilu and Pierre Parent, *Serre's uniformity problem in the split Cartan case*, Ann. of Math. (2) **173** (2011), no. 1, 569–584, DOI 10.4007/annals.2011.173.1.13. MR2753610

[3] G. Frey, How to disguise an elliptic curve, Talk at the Second Elliptic Curve Cryptography Workshop, 1998.

[4] Pierrick Gaudry, *An algorithm for solving the discrete log problem on hyperelliptic curves*, Advances in cryptology—EUROCRYPT 2000 (Bruges), Lecture Notes in Comput. Sci., vol. 1807, Springer, Berlin, 2000, pp. 19–34, DOI 10.1007/3-540-45539-6_2. MR1772021

[5] P. Gaudry, F. Hess, and N. P. Smart, *Constructive and destructive facets of Weil descent on elliptic curves*, J. Cryptology **15** (2002), no. 1, 19–46, DOI 10.1007/s00145-001-0011-x. MR1880933

[6] B. Mazur, *Rational isogenies of prime degree (with an appendix by D. Goldfeld)*, Invent. Math. **44** (1978), no. 2, 129–162, DOI 10.1007/BF01390348. MR482230

[7] Loïc Merel, *Bornes pour la torsion des courbes elliptiques sur les corps de nombres* (French), Invent. Math. **124** (1996), no. 1-3, 437–449, DOI 10.1007/s002220050059. MR1369424

[8] V. Kumar Murty, *Algebraic cycles on abelian varieties*, Duke Math. J. **50** (1983), no. 2, 487–504, DOI 10.1215/S0012-7094-83-05021-4. MR705036

[9] V. Kumar Murty, Arithmetic twists and abelian extensions, *this volume.*

[10] K. A. Ribet, Modular functions of one variable V (Bonn, 1976), pp. 17-51, Lecture Notes in Math. 601, Springer, Berlin, 1977.

[11] Kenneth A. Ribet, *Twists of modular forms and endomorphisms of abelian varieties*, Math. Ann. **253** (1980), no. 1, 43–62, DOI 10.1007/BF01457819. MR594532

Department of Mathematics, University of Tokyo
Email address: t-saito@ms.u-tokyo.ac.jp

Contemporary Mathematics
Volume **701**, 2018
http://dx.doi.org/10.1090/conm/701/14147

A note on algebraic points on Shimura curves

Keisuke Arai

To the memory of Fumiyuki Momose

ABSTRACT. In this note, we survey recent results of the author concerning algebraic points on Shimura curves, especially of $\Gamma_0(p)$-type. These results are based on a joint work of the author and Fumiyuki Momose. We also give several explicit examples.

1. Points on Shimura curves

For an integer $N \geq 1$, let $X_0(N)$ be the smooth compactification of the coarse moduli scheme over $\mathbb{Q}$ parameterizing isomorphism classes of pairs (E, C) where E is an elliptic curve and C is a cyclic subgroup of E of order N. We know that $X_0(N)$ is a proper smooth curve over $\mathbb{Q}$ (cf. [**14**, Théorème 3.4, p. 212]) and that $X_0(1)$ is isomorphic to the projective line $\mathbb{P}^1_{\mathbb{Q}}$ (cf. [**14**, Théorème 1.1, p. 267]). When N is a prime number p, Mazur and Momose studied points on the modular curve $X_0(p)$.

THEOREM 1.1. *(1)* [**18**, Theorem 7.1, p. 153] *For a prime number p, we have* $X_0(p)(\mathbb{Q}) = \{cusps\}$ *if and only if* $p \notin \{2, 3, 5, 7, 11, 13, 17, 19, 37, 43, 67, 163\}$.

(2) [**19**, Theorem B, p. 330] *Let k be a quadratic field which is not an imaginary quadratic field of class number one. Then there is a finite set* $\mathfrak{S}(k)$ *of prime numbers depending on k such that* $X_0(p)(k) = \{cusps\}$ *holds for every prime number* $p \notin \mathfrak{S}(k)$.

REMARK 1.2. In Theorem 1.1 (2), a quantity is estimated which bounds every prime number in $\mathfrak{S}(k)$ from above with at most one exception (see [**13**], [**18**, Theorem A, p. 160]). If such an exceptional prime number exists, it is concerned with a Siegel zero of Dirichlet L-functions. For related topics about modular curves, see [**2**].

We have obtained analogues of Theorem 1.1 for Shimura curves, as explained below. Let B be an indefinite quaternion division algebra over $\mathbb{Q}$. Let

$$d := \operatorname{disc} B$$

be the discriminant of B. Then d is the product of an even number of distinct prime numbers, and $d > 1$. Fix a maximal order $\mathcal{O}$ of B. For each prime number

2010 *Mathematics Subject Classification.* Primary 11G18, 14G05; Secondary 11G10, 11G15.
Key words and phrases. Rational points, Shimura curves, QM-abelian surfaces.

p not dividing d, fix an isomorphism

$$\mathcal{O} \otimes_{\mathbb{Z}} \mathbb{Z}_p \cong \mathrm{M}_2(\mathbb{Z}_p)$$

of $\mathbb{Z}_p$-algebras.

DEFINITION 1.3. (cf. [**12**, p. 591]) Let S be a scheme. A QM-abelian surface by $\mathcal{O}$ over S is a pair (A, i) where A is an abelian surface over S (i.e. A is an abelian scheme over S of relative dimension 2), and $i : \mathcal{O} \hookrightarrow \mathrm{End}_S(A)$ is an injective ring homomorphism (sending 1 to id). Here, $\mathrm{End}_S(A)$ is the ring of endomorphisms of A defined over S. We assume that A has a left $\mathcal{O}$-action.

Let M^B be the coarse moduli scheme over $\mathbb{Q}$ parameterizing isomorphism classes of QM-abelian surfaces by $\mathcal{O}$ (cf. [**15**, p. 93]). Then M^B is a proper smooth curve over $\mathbb{Q}$, called a Shimura curve. For a prime number p not dividing d, let $M_0^B(p)$ be the coarse moduli scheme over $\mathbb{Q}$ parameterizing isomorphism classes of triples (A, i, V) where (A, i) is a QM-abelian surface by $\mathcal{O}$ and V is a left $\mathcal{O}$-submodule of $A[p]$ with $\mathbb{F}_p$-dimension 2. Here, $A[p]$ is the kernel of multiplication by p in A. Then $M_0^B(p)$ is a proper smooth curve over $\mathbb{Q}$, which we call a Shimura curve of $\Gamma_0(p)$-type. We have a natural map

$$\pi^B(p) : M_0^B(p) \longrightarrow M^B$$

over $\mathbb{Q}$ defined by $(A, i, V) \longmapsto (A, i)$. Note that M^B (resp. $M_0^B(p)$) is an analogue of the modular curve $X_0(1)$ (resp. $X_0(p)$). First, we obtained the following result.

THEOREM 1.4 ([**8**, Theorem 1.3, p. 180], cf. [**9**]). *Let k be a quadratic field which is not an imaginary quadratic field of class number one. Then there is a finite set $\mathcal{N}(k)$ of prime numbers depending on k satisfying the following.*

(1) If $B \otimes_{\mathbb{Q}} k \cong \mathrm{M}_2(k)$, then $M_0^B(p)(k) = \emptyset$ holds for every prime number $p \notin \mathcal{N}(k)$ not dividing d.

(2) If $B \otimes_{\mathbb{Q}} k \not\cong \mathrm{M}_2(k)$, then $M_0^B(p)(k) \subseteq \{$elliptic points of order 2 or 3$\}$ holds for every prime number $p \notin \mathcal{N}(k)$ not dividing d.

Theorem 1.4 has been expanded to number fields of higher degree. Let k be a number field, and let $\mathcal{O}_k$ be the ring of integers of k. For a prime $\mathfrak{q}$ of k, put $\mathrm{N}(\mathfrak{q}) := \sharp(\mathcal{O}_k/\mathfrak{q})$. We say that a prime $\mathfrak{q}$ of k is of odd degree if $\mathrm{N}(\mathfrak{q})$ is an odd power of q, where q is the residue characteristic of $\mathfrak{q}$. Let h_k be the class number of k. Let $\mathcal{M}^{new}$ (resp. $\mathbf{Ram}(k)$) be the set of prime numbers which split completely in k (resp. which are ramified in k). Let $\mathcal{N}^{new}$ be the set of primes $\mathfrak{q}$ of k such that $\mathfrak{q}$ divides some prime number $q \in \mathcal{M}^{new}$. Take a finite subset $\emptyset \neq \mathcal{S}^{new} \subseteq \mathcal{N}^{new}$ which generates the ideal class group of k. For each prime $\mathfrak{q} \in \mathcal{S}^{new}$, fix an element $\alpha_{\mathfrak{q}} \in \mathcal{O}_k \setminus \{0\}$ satisfying $\mathfrak{q}^{h_k} = \alpha_{\mathfrak{q}}\mathcal{O}_k$. For an integer $n \geq 1$, define the set

$$\mathcal{FR}(n) := \left\{ \beta \in \mathbb{C} \;\middle|\; \beta^2 + a\beta + n = 0 \text{ for some integer } a \in \mathbb{Z} \text{ with } |a| \leq 2\sqrt{n} \right\}.$$

When k is Galois over $\mathbb{Q}$, define the sets

$$\mathcal{E}(k) := \left\{ \varepsilon_0 = \sum_{\sigma \in \mathrm{Gal}(k/\mathbb{Q})} a_\sigma \sigma \in \mathbb{Z}[\mathrm{Gal}(k/\mathbb{Q})] \;\middle|\; a_\sigma \in \{0, 8, 12, 16, 24\} \right\},$$

$$\mathcal{M}_1^{new}(k) := \{ (\mathfrak{q}, \varepsilon_0, \beta_{\mathfrak{q}}) \mid \mathfrak{q} \in \mathcal{S}^{new},\ \varepsilon_0 \in \mathcal{E}(k),\ \beta_{\mathfrak{q}} \in \mathcal{FR}(\mathrm{N}(\mathfrak{q})) \},$$
$$\mathcal{M}_2^{new}(k) := \{ \mathrm{Norm}_{k(\beta_{\mathfrak{q}})/\mathbb{Q}}(\alpha_{\mathfrak{q}}^{\varepsilon_0} - \beta_{\mathfrak{q}}^{24h_k}) \in \mathbb{Z} \mid (\mathfrak{q}, \varepsilon_0, \beta_{\mathfrak{q}}) \in \mathcal{M}_1^{new}(k) \} \setminus \{0\},$$
$$\mathcal{N}_0^{new}(k) := \{ l : \text{prime number} \mid l \text{ divides some integer } m \in \mathcal{M}_2^{new}(k) \},$$
$$\mathcal{T}^{new}(k) := \{ l' : \text{prime number} \mid l' \text{ is divisible by some prime } \mathfrak{q}' \in \mathcal{S}^{new} \} \cup \{2, 3\},$$

$\mathcal{N}_1^{new}(k) := \mathcal{N}_0^{new}(k) \cup \mathcal{T}^{new}(k) \cup \mathbf{Ram}(k)$.
Note that all the sets $\mathbf{Ram}(k)$, $\mathcal{FR}(n)$, $\mathcal{E}(k)$, $\mathcal{M}_1^{new}(k)$, $\mathcal{M}_2^{new}(k)$, $\mathcal{N}_0^{new}(k)$, $\mathcal{T}^{new}(k)$, and $\mathcal{N}_1^{new}(k)$ are finite.

THEOREM 1.5 ([**4**], [**5**], [**6**], [**7**, Theorem 2.3]). *Let k be a finite Galois extension of $\mathbb{Q}$ which does not contain the Hilbert class field of any imaginary quadratic field. Assume that there is a prime $\mathfrak{q}$ of k of odd degree whose residue characteristic q satisfies $B \otimes_{\mathbb{Q}} \mathbb{Q}(\sqrt{-q}) \not\cong \mathrm{M}_2(\mathbb{Q}(\sqrt{-q}))$. Let p be a prime number satisfying $p > 4\mathrm{N}(\mathfrak{q})$, $p \nmid 13d$ and $p \notin \mathcal{N}_1^{new}(k)$. Then $M_0^B(p)(k) = \emptyset$.*

REMARK 1.6. (1) In Theorem 1.4, the proof proceeds in a way similar to that of [**19**, Theorem B, p. 330]. It relies on an analytic method. A quantity can be estimated which bounds every prime number in $\mathcal{N}(k)$ from above with at most one exception. In the proof of Theorem 1.5, we make use of an algebraic method ([**5**], [**6**]) instead of an analytic method. Furthermore, we slightly modify the set $\mathcal{N}(k)$, so that we can estimate an upper bound of $\mathcal{N}_1^{new}(k)$ without exception (see §4 or [**4**]).

(2) We know $M^B(\mathbb{R}) = \emptyset$ by [**22**, Theorem 0, p. 136]. Since there is a map $\pi^B(p) : M_0^B(p) \longrightarrow M^B$ defined over $\mathbb{Q}$, we have $M_0^B(p)(\mathbb{R}) = \emptyset$ for any p.

The author is very sorry for the loss of Fumiyuki Momose, and dedicates this work to his memory.

Acknowledgments. The author would like to thank the organizers Jinhui Chao, Francesc Fité, Josep Gonzàlez Rovira, Joan-C. Lario and Tsutomu Sekiguchi for giving him an opportunity to talk at the conference. He would also like to thank the anonymous referee for helpful comments.

2. Automorphism groups and elliptic points

We consider the automorphism groups of a QM-abelian surface. Let (A, i) be a QM-abelian surface by $\mathcal{O}$ over a field k. Let $\mathrm{End}(A)$ (resp. $\mathrm{Aut}(A)$) be the ring of endomorphisms (resp. the group of automorphisms) of A defined over an algebraic closure $\overline{k}$ of k. Put

$$\mathrm{End}_{\mathcal{O}}(A) := \{f \in \mathrm{End}(A) \mid f \circ i(g) = i(g) \circ f \text{ for any } g \in \mathcal{O}\},$$

$$\mathrm{Aut}_{\mathcal{O}}(A) := \mathrm{Aut}(A) \cap \mathrm{End}_{\mathcal{O}}(A).$$

If k is of characteristic 0, then $\mathrm{Aut}_{\mathcal{O}}(A) \cong \mathbb{Z}/2\mathbb{Z}$, $\mathbb{Z}/4\mathbb{Z}$ or $\mathbb{Z}/6\mathbb{Z}$ (see [**8**, §3]). Let p be a prime number not dividing d. Let (A, i, V) be a triple where (A, i) is a QM-abelian surface by $\mathcal{O}$ over k and V is a left $\mathcal{O}$-submodule of $A[p](\overline{k})$ with $\mathbb{F}_p$-dimension 2. Define a subgroup $\mathrm{Aut}_{\mathcal{O}}(A, V)$ of $\mathrm{Aut}_{\mathcal{O}}(A)$ by

$$\mathrm{Aut}_{\mathcal{O}}(A, V) := \{f \in \mathrm{Aut}_{\mathcal{O}}(A) \mid f(V) = V\}.$$

If k is of characteristic 0, then $\mathrm{Aut}_{\mathcal{O}}(A, V) \cong \mathbb{Z}/2\mathbb{Z}$, $\mathbb{Z}/4\mathbb{Z}$ or $\mathbb{Z}/6\mathbb{Z}$.

We express the sets $M^B(\mathbb{C})$ and $M_0^B(p)(\mathbb{C})$ as quotients of the complex upper half-plane $\mathbb{H}$. Let $\mathcal{O}'$ be the order of B satisfying $\mathcal{O}' \subseteq \mathcal{O}$ and

$$\begin{cases} \mathcal{O}' \otimes_{\mathbb{Z}} \mathbb{Z}_l = \mathcal{O} \otimes_{\mathbb{Z}} \mathbb{Z}_l \quad (l \neq p), \\ \mathcal{O}' \otimes_{\mathbb{Z}} \mathbb{Z}_p = \left\{ \begin{pmatrix} s & t \\ u & v \end{pmatrix} \in \mathrm{M}_2(\mathbb{Z}_p) \cong \mathcal{O} \otimes_{\mathbb{Z}} \mathbb{Z}_p \,\middle|\, u \in p\mathbb{Z}_p \right\}. \end{cases}$$

Consider the groups

$$\Gamma := \{x \in \mathcal{O} \mid \mathrm{Nrd}(x) = 1\}, \qquad \Gamma' := \{x \in \mathcal{O}' \mid \mathrm{Nrd}(x) = 1\}$$

where Nrd is the reduced norm. Then Γ and Γ' are considered to be subgroups of $\mathrm{SL}_2(\mathbb{R})$ via $\mathrm{SL}_2(\mathbb{R}) \subseteq \mathrm{GL}_2(\mathbb{R}) \cong (B \otimes_{\mathbb{Q}} \mathbb{R})^{\times}$. We have isomorphisms of Riemann surfaces

$$M^B(\mathbb{C}) \cong \Gamma\backslash\mathbb{H}, \qquad M_0^B(p)(\mathbb{C}) \cong \Gamma'\backslash\mathbb{H}$$

(cf. [**10**, 4.3, p. 669]). A point of $M_0^B(p)(\mathbb{C})$ is an elliptic point of order 2 (resp. 3) if and only if the corresponding triple (A, i, V) over $\mathbb{C}$ satisfies $\mathrm{Aut}_{\mathcal{O}}(A, V) \cong \mathbb{Z}/4\mathbb{Z}$ (resp. $\mathrm{Aut}_{\mathcal{O}}(A, V) \cong \mathbb{Z}/6\mathbb{Z}$) (cf. [**21**, 1.2, p. 9]). Let $\nu_2(p)$ (resp. $\nu_3(p)$) be the number of elliptic points of order 2 (resp. 3) of $M_0^B(p)(\mathbb{C})$. Then we have

$$\nu_2(p) = \left(1 + \left(\frac{-1}{p}\right)\right) \prod_{l \mid d} \left(1 - \left(\frac{-1}{l}\right)\right),$$

$$\nu_3(p) = \left(1 + \left(\frac{-3}{p}\right)\right) \prod_{l \mid d} \left(1 - \left(\frac{-3}{l}\right)\right)$$

where l runs through all prime divisors of d, and

$$\left(\frac{-1}{q}\right) = \begin{cases} 1 & \text{if } q \equiv 1 \mod 4, \\ -1 & \text{if } q \equiv -1 \mod 4, \\ 0 & \text{if } q = 2, \end{cases}$$

$$\left(\frac{-3}{q}\right) = \begin{cases} 1 & \text{if } q \equiv 1 \mod 3, \\ -1 & \text{if } q \equiv -1 \mod 3, \\ 0 & \text{if } q = 3 \end{cases}$$

for a prime number q (see [**20**, Theorem 3.12, p. 111]).

3. Examples

The genus of the Shimura curve M^B is 0 if and only if $d \in \{6, 10, 22\}$ ([**1**, Lemma 3.1, p. 168]). By [**16**, Theorem 1-1, p. 279], the defining equations of such M^B are

$$\begin{cases} x^2 + y^2 + 3 = 0 & \text{if } d = 6, \\ x^2 + y^2 + 2 = 0 & \text{if } d = 10, \\ x^2 + y^2 + 11 = 0 & \text{if } d = 22. \end{cases}$$

In these cases, for a field k of characteristic 0 the condition $M^B(k) \neq \emptyset$ implies $M^B \otimes_{\mathbb{Q}} k \cong \mathbb{P}^1_k$, and so $\sharp M^B(k) = \infty$. We have the following example concerning $M^B(k)$.

LEMMA 3.1 (cf. [**3**, Lemma 4.4, p. 193]). *Suppose* $d \in \{6, 10, 22\}$ *and* $k \in \{\mathbb{Q}(\sqrt{3}, \sqrt{-5}), \mathbb{Q}(\zeta_5), \mathbb{Q}(\zeta_{17}), \mathbb{Q}(\zeta_{31})\}$. *If* $(d, k) = (22, \mathbb{Q}(\zeta_5))$ *or* $(10, \mathbb{Q}(\zeta_{31}))$, *then* $M^B(k) = \emptyset$. *Otherwise* $\sharp M^B(k) = \infty$.

We give examples concerning hypotheses of Theorem 1.5.

LEMMA 3.2 (cf. [**3**, Proof of Proposition 4.1, p. 194]). *The fields* $\mathbb{Q}(\sqrt{3}, \sqrt{-5})$, $\mathbb{Q}(\zeta_5)$, $\mathbb{Q}(\zeta_{17})$, $\mathbb{Q}(\zeta_{31})$ *do not contain the Hilbert class field of any imaginary quadratic field.*

LEMMA 3.3 ([**3**, Lemma 4.3, p. 193]). *Suppose* $d \in \{6, 10, 22\}$. *For a prime number* q, *the condition* $B \otimes_{\mathbb{Q}} \mathbb{Q}(\sqrt{-q}) \not\cong \mathrm{M}_2(\mathbb{Q}(\sqrt{-q}))$ *is equivalent to*

- $q \equiv 2, 5, 7, 11, 17, 23 \bmod 24$ *if* $d = 6$,
- $q \equiv 1, 7, 9, 11, 19, 21, 23, 29, 31, 39 \bmod 40$ *if* $d = 10$,
- $q \equiv 2, 7, 13, 15, 17, 19, 21, 23, 29, 31, 35, 39, 41, 43, 47, 51, 57, 61, 63, 65, 71, 73, 79, 83, 85, 87 \bmod 88$ *if* $d = 22$.

LEMMA 3.4. *Suppose* $d \in \{6, 10, 22\}$ *and* $k \in \{\mathbb{Q}(\sqrt{3}, \sqrt{-5}), \mathbb{Q}(\zeta_5), \mathbb{Q}(\zeta_{17}), \mathbb{Q}(\zeta_{31})\}$. *Then the least* $\mathrm{N}(\mathfrak{q})$ *for primes* $\mathfrak{q}$ *of* k, *of odd degree, with residue characteristic* q, *satisfying* $B \otimes_{\mathbb{Q}} \mathbb{Q}(\sqrt{-q}) \not\cong \mathrm{M}_2(\mathbb{Q}(\sqrt{-q}))$, *is as follows.*

$k \diagdown d$	6	10	22
$\mathbb{Q}(\sqrt{3}, \sqrt{-5})$	23	23	23
$\mathbb{Q}(\zeta_5)$	11	11	**31**
$\mathbb{Q}(\zeta_{17})$	103	103	103
$\mathbb{Q}(\zeta_{31})$	2^5	**311**	2^5

We then have the following examples of Theorem 1.5.

PROPOSITION 3.5. *(1) Suppose* $k = \mathbb{Q}(\sqrt{3}, \sqrt{-5})$ *and* $d \in \{6, 10, 22\}$. *Then* $\sharp M^B(k) = \infty$. *Furthermore, if* $p > 92$ *and* $p \notin \mathcal{N}_1^{new}(k)$, *then* $M_0^B(p)(k) = \emptyset$.

(2) Suppose $k = \mathbb{Q}(\zeta_5)$ *and* $d \in \{6, 10\}$. *Then* $\sharp M^B(k) = \infty$. *Furthermore, if* $p > 44$ *and* $p \notin \mathcal{N}_1^{new}(k)$, *then* $M_0^B(p)(k) = \emptyset$.

(3) Suppose $k = \mathbb{Q}(\zeta_{17})$ *and* $d \in \{6, 10, 22\}$. *Then* $\sharp M^B(k) = \infty$. *Furthermore, if* $p > 412$ *and* $p \notin \mathcal{N}_1^{new}(k)$, *then* $M_0^B(p)(k) = \emptyset$.

(4) Suppose $k = \mathbb{Q}(\zeta_{31})$ *and* $d \in \{6, 22\}$. *Then* $\sharp M^B(k) = \infty$. *Furthermore, if* $p > 128$ *and* $p \notin \mathcal{N}_1^{new}(k)$, *then* $M_0^B(p)(k) = \emptyset$.

4. Estimate of $\mathcal{N}_1^{new}(k)$

In this section, we estimate the set $\mathcal{N}_1^{new}(k)$ by using the method of [**13**]. Let k be a number field. Let d_k (resp. r_k, resp. R_k) be the absolute value of the discriminant of k (resp. the rank of the unit group $\mathcal{O}_k^\times$, resp. the regulator of k), and put $n_k := [k : \mathbb{Q}]$. For a place v of k and an element $\alpha \in k$, define $||\alpha||_v$ as follows.

- If v is finite, let $\mathfrak{q}$ be the prime of k corresponding to v, and let $||\alpha||_v := \mathrm{N}(\mathfrak{q})^{-\mathrm{ord}_{\mathfrak{q}}(\alpha)}$ where $\mathrm{ord}_{\mathfrak{q}}(\alpha)$ is the order of α at $\mathfrak{q}$. Here, we put $||\alpha||_v := 0$ if $\alpha = 0$.
- If v is real, let $\tau : k \hookrightarrow \mathbb{R}$ be the embedding corresponding to v, and let $||\alpha||_v := |\tau(\alpha)|$.
- If v is complex, let $\tau : k \hookrightarrow \mathbb{C}$ be one of the embeddings corresponding to v, and let $||\alpha||_v := |\tau(\alpha)|^2$.

For an element $\alpha \in k$, let $\mathrm{H}(\alpha)$ denote the absolute height of α defined by

$$\mathrm{H}(\alpha) := \left(\prod_v \max(1, ||\alpha||_v) \right)^{\frac{1}{n_k}},$$

where v runs through all places of k. We know that there is a positive constant δ_k, depending only on k, such that for every non-zero element $\alpha \in k$ that is not a root of unity we have $\log \mathrm{H}(\alpha) \geq \delta_k / n_k$ (cf. [**11**, p. 70]). We can take $\delta_k = \dfrac{\log 2}{r_k + 1}$ for

$n_k = 1, 2$. Both $\delta_k = \dfrac{1}{53n_k \log 6n_k}$ and $\delta_k = \dfrac{1}{1201}\left(\dfrac{\log\log n_k}{\log n_k}\right)^3$ are appropriate choices for $n_k \geq 3$. Fix such a constant δ_k. Define the constants

$$C_1(k) := \frac{r_k^{1+r_k}\delta_k^{1-r_k}}{2}, \quad C_2(k) := \exp(24n_kC_1(k)R_k),$$

$$C(k,a) := (a^{24h_k}C_2(k) + a^{12h_k})^{2n_k},$$

where $a > 0$. The following theorem is a key ingredient of the estimate.

THEOREM 4.1 ([**17**, Theorem 1.1, p. 272]). *There is an absolute, effectively computable constant $A_1 > 1$ such that for every finite extension k_1 of $\mathbb{Q}$, every finite Galois extension k_2 of k_1 and every conjugacy class C of* $\mathrm{Gal}(k_2/k_1)$, *there is a prime $\mathfrak{q}$ of k_1 which is unramified in k_2, for which* $\mathrm{Frob}_{\mathfrak{q}} = C$, *for which* $\mathrm{N}(\mathfrak{q})$ *is a prime number, and which satisfies the bound* $\mathrm{N}(\mathfrak{q}) \leq 2d_{k_2}^{A_1}$. *Here,* $\mathrm{Frob}_{\mathfrak{q}}$ *is the (arithmetic) Frobenius conjugacy class at $\mathfrak{q}$ in* $\mathrm{Gal}(k_2/k_1)$.

Then we have the following estimate of $\mathcal{N}_1^{new}(k)$.

THEOREM 4.2. *(1)* [**13**, Proposition 4.2] *Assume that k is Galois over $\mathbb{Q}$, and let A_1 be the constant in Theorem* 4.1. *Then we can take $\mathcal{S}^{new}$ so that every prime $\mathfrak{q} \in \mathcal{S}^{new}$ satisfies* $\mathrm{N}(\mathfrak{q}) \leq 2d_k^{A_1h_k}$.

(2) [**4**, Lemma 7.3, p. 351] *Let $\mathfrak{q}$ be a prime of k. Then there is a non-zero element $\alpha'_{\mathfrak{q}} \in \mathcal{O}_k$ satisfying $\mathfrak{q}^{h_k} = \alpha'_{\mathfrak{q}}\mathcal{O}_k$ and* $\mathrm{H}(\alpha'_{\mathfrak{q}}) \leq |\mathrm{Norm}_{k/\mathbb{Q}}(\alpha'_{\mathfrak{q}})|^{\frac{1}{n_k}} \exp(C_1(k)R_k)$.

(3) [**4**, Theorem 7.7, p. 352] *Assume that k is Galois over $\mathbb{Q}$. Take $\mathcal{S}^{new}$ as in (1), and take $\alpha_{\mathfrak{q}}$ to be $\alpha'_{\mathfrak{q}}$ in (2) for each $\mathfrak{q} \in \mathcal{S}^{new}$. Then for any $p \in \mathcal{N}_1^{new}(k)$, we have $p \leq C(k, 2d_k^{A_1h_k})$.*

References

[1] K. Arai, *On the Galois images associated to QM-abelian surfaces*, Proceedings of the Symposium on Algebraic Number Theory and Related Topics, RIMS Kôkyûroku Bessatsu, B4, Res. Inst. Math. Sci. (RIMS), Kyoto, 2007, pp. 165–187. MR2402009

[2] K. Arai, *Galois images and modular curves*, Algebraic number theory and related topics 2010, RIMS Kôkyûroku Bessatsu, B32, Res. Inst. Math. Sci. (RIMS), Kyoto, 2012, pp. 145–161. MR2986922

[3] K. Arai, *On the Rasmussen-Tamagawa conjecture for QM-abelian surfaces*, Algebraic number theory and related topics 2011, RIMS Kôkyûroku Bessatsu, B44, Res. Inst. Math. Sci. (RIMS), Kyoto, 2013, pp. 185–196. MR3221728

[4] K. Arai, *An effective bound of p for algebraic points on Shimura curves of $\Gamma_0(p)$-type*, Acta Arith. **164** (2014), no. 4, 343–354, DOI 10.4064/aa164-4-2. MR3244939

[5] K. Arai, *Algebraic points on Shimura curves of $\Gamma_0(p)$-type (II)*, Manuscripta Math. **149** (2016), no. 1-2, 63–70, DOI 10.1007/s00229-015-0770-6. MR3447140

[6] K. Arai, *Algebraic points on Shimura curves of $\Gamma_0(p)$-type (III)*, Ramanujan J. **43** (2017), no. 1, 15–28, DOI 10.1007/s11139-015-9766-9. MR3633820

[7] K. Arai, *Algebraic points on Shimura curves of $\Gamma_0(p)$-type (IV)*, Algebraic number theory and related topics 2013, RIMS Kôkyûroku Bessatsu, B53, Res. Inst. Math. Sci. (RIMS), Kyoto, 2015, pp. 3–11. MR3525161

[8] K. Arai and F. Momose, *Algebraic points on Shimura curves of $\Gamma_0(p)$-type*, J. Reine Angew. Math. **690** (2014), 179–202, DOI 10.1515/crelle-2012-0068. MR3200341

[9] K. Arai and F. Momose, *Errata to: Algebraic points on Shimura curves of $\Gamma_0(p)$-type [MR3200341]*, J. Reine Angew. Math. **690** (2014), 203–205, DOI 10.1515/crelle-2012-0109. MR3200342

[10] A. Besser, *CM cycles over Shimura curves*, J. Algebraic Geom. **4** (1995), no. 4, 659–691. MR1339843

[11] Y. Bugeaud and K. Győry, *Bounds for the solutions of unit equations*, Acta Arith. **74** (1996), no. 1, 67–80. MR1367579

[12] K. Buzzard, *Integral models of certain Shimura curves*, Duke Math. J. **87** (1997), no. 3, 591–612, DOI 10.1215/S0012-7094-97-08719-6. MR1446619

[13] *A. David*, Caractère d'isogénie et critères d'irréductibilité, preprint, available at the web page (http://arxiv.org/pdf/1103.3892.pdf).

[14] P. Deligne and M. Rapoport, *Les schémas de modules de courbes elliptiques* (French), Modular functions of one variable, II (Proc. Internat. Summer School, Univ. Antwerp, Antwerp, 1972), Springer, Berlin, 1973, pp. 143–316. Lecture Notes in Math., Vol. 349. MR0337993

[15] B. W. Jordan, *Points on Shimura curves rational over number fields*, J. Reine Angew. Math. **371** (1986), 92–114, DOI 10.1515/crll.1986.371.92. MR859321

[16] A. Kurihara, *On some examples of equations defining Shimura curves and the Mumford uniformization*, J. Fac. Sci. Univ. Tokyo Sect. IA Math. **25** (1979), no. 3, 277–300. MR523989

[17] J. C. Lagarias, H. L. Montgomery, and A. M. Odlyzko, *A bound for the least prime ideal in the Chebotarev density theorem*, Invent. Math. **54** (1979), no. 3, 271–296, DOI 10.1007/BF01390234. MR553223

[18] B. Mazur, *Rational isogenies of prime degree (with an appendix by D. Goldfeld)*, Invent. Math. **44** (1978), no. 2, 129–162, DOI 10.1007/BF01390348. MR482230

[19] F. Momose, *Isogenies of prime degree over number fields*, Compositio Math. **97** (1995), no. 3, 329–348. MR1353278

[20] H. Shimizu, *Hokei kansu. I–III* (Japanese), 2nd ed., Iwanami Shoten Kiso Sūgaku [Iwanami Lectures on Fundamental Mathematics], vol. 8, Iwanami Shoten, Tokyo, 1984. Daisū [Algebra], vii. MR857459

[21] G. Shimura, *Introduction to the arithmetic theory of automorphic functions*, Publications of the Mathematical Society of Japan, No. 11. Iwanami Shoten, Publishers, Tokyo; Princeton University Press, Princeton, N.J., 1971. Kanô Memorial Lectures, No. 1. MR0314766

[22] G. Shimura, *On the real points of an arithmetic quotient of a bounded symmetric domain*, Math. Ann. **215** (1975), 135–164, DOI 10.1007/BF01432692. MR0572971

Department of Mathematics, School of Science and Technology for Future Life, Tokyo Denki University, 5 Senju Asahi-cho, Adachi-ku, Tokyo 120-8551, Japan

Email address: araik@mail.dendai.ac.jp

Contemporary Mathematics
Volume **701**, 2018
http://dx.doi.org/10.1090/conm/701/14146

On quadratic points of classical modular curves

Francesc Bars

This paper is dedicated to the memory of F.Momose.

ABSTRACT. The infinitude of the set of quadratic points of a non-singular projective curve C defined over a number field k, (that is, the points on C defined over a quadratic extension of k), is related with the geometric property that C is an hyperelliptic or a bielliptic curve, and such geometric property may be tested, for example, if one knows the automorphism group of C.

In this survey paper we present the state of the art on finiteness or not for the set of quadratic points of classical modular curves. Moreover, we fix inaccuracies of the existing literature concerning automorphism groups of modular curves, and, we clarify the importance of k-points in the arithmetic statement between the relation of the curve C having an infinite set of quadratic points with C being an hyperelliptic or a bielliptic curve.

1. Introduction

Consider Γ a modular subgroup of integral 2×2 matrices $\mathrm{SL}_2(\mathbb{Z})$ which acts by linear transformation in the upper half-plane $\mathbb{H}$, defining the non-compact Riemann surface $Y_{\Gamma,\mathbb{C}} := \mathbb{H}/\Gamma$.

For this paper let us consider the following modular subgroups Γ of $\mathrm{SL}_2(\mathbb{Z})$ with N a positive integer and Δ a strict subgroup of $(\mathbb{Z}/N\mathbb{Z})^*$ with $-1\in\Delta$:

$$\Gamma(N)=\left\{\begin{pmatrix} a & b\\ c & d\end{pmatrix}\in \mathrm{SL}_2(\mathbb{Z})\,\middle|\,\begin{pmatrix} a & b\\ c & d\end{pmatrix}\equiv\begin{pmatrix} 1 & 0\\ 0 & 1\end{pmatrix}\pmod N\right\},$$

$$\Gamma_1(N)=\left\{\begin{pmatrix} a & b\\ c & d\end{pmatrix}\in \mathrm{SL}_2(\mathbb{Z})\,\middle|\,\begin{pmatrix} a & b\\ c & d\end{pmatrix}\equiv\begin{pmatrix} 1 & *\\ 0 & 1\end{pmatrix}\pmod N\right\},$$

$$\Gamma_0(N)=\left\{\begin{pmatrix} a & b\\ c & d\end{pmatrix}\in \mathrm{SL}_2(\mathbb{Z})\,\middle|\,\begin{pmatrix} a & b\\ c & d\end{pmatrix}\equiv\begin{pmatrix} * & *\\ 0 & *\end{pmatrix}\pmod N\right\},$$

$$\Gamma_\Delta(N)=\left\{\begin{pmatrix} a & b\\ c & d\end{pmatrix}\in \Gamma_0(N)\,|\,(a\pmod N)\in\Delta\right\}.$$

2010 *Mathematics Subject Classification.* Primary 11G18; Secondary 11G30 14G05 14H25 14K05.

Key words and phrases. Bielliptic curves, Hyperelliptic curves, Classical modular curves, Quadratic points.

Partially supported by grant MTM2016-75980-P.

The compactification of $Y_{\Gamma,\mathbb{C}}$ is an algebraic non-singular projective curve, named the modular curve associated to Γ, and we will denote such compact Riemann surface by $X_{\Gamma,\mathbb{C}}$. This compactification is obtained by adding a finite set of points, named cusps, and $X_{\Gamma,\mathbb{C}} = \mathbb{H}^*/\Gamma$ where $\mathbb{H}^*$ is the extended complex upper-half plane $\mathbb{H} \cup \mathbb{Q} \cup \{\infty\}$.

As usual let us denote by $X(N)_{\mathbb{C}} = X_{\Gamma(N),\mathbb{C}}$, $X_1(N)_{\mathbb{C}} = X_{\Gamma_1(N),\mathbb{C}}$, $X_\Delta(N)_{\mathbb{C}} = X_{\Gamma_\Delta(N),\mathbb{C}}$ and $X_0(N)_{\mathbb{C}} = X_{\Gamma_0(N),\mathbb{C}}$. Clearly $\Gamma(N) \leq \Gamma_1(N) \leq \Gamma_\Delta(N) \leq \Gamma_0(N)$, therefore we have natural maps:

$$X(N)_{\mathbb{C}} \to X_1(N)_{\mathbb{C}} \to X_\Delta(N)_{\mathbb{C}} \to X_0(N)_{\mathbb{C}}. \tag{1.1}$$

All the modular curves involved in (1.1) are algebraic and have as field of definition $\mathbb{Q}$. Over $\mathbb{C}$, they are curves associated to a moduli problem with a N-level structure. The moduli problem can be used to give a model of the curve over $\mathbb{Q}$ which is geometrically connected. This can be done directly for $X_1(N)_{\mathbb{C}}, X_\Delta(N)_{\mathbb{C}}$ and $X_0(N)_{\mathbb{C}}$ defining algebraic non-singular (modular) curves over $\mathbb{Q}$: $X_1(N), X_\Delta(N)$ and $X_0(N)$, respectively, and each of them have at least one point defined over $\mathbb{Q}$, see [**10**] or [**20**].

The usual complex moduli problem associated to $X(N)_{\mathbb{C}}$ is the coarse moduli space of the isomorphism classes of elliptic curves E over $\mathbb{C}$, together with two N-torsion points P_1, P_2 which satisfy: the points P_1, P_2 generate $E[N]$, the N-torsion of E, which is isomorphic to the group scheme $(\mathbb{Z}/N\mathbb{Z})^2$ and $e(P_1, P_2) = \zeta_N$ where e is the Weil pairing and ζ_N a fixed primitive N-root of unity. In order to define a curve over $\mathbb{Q}$ with some point over $\mathbb{Q}$ we need to modify the above moduli problem by (E, ϕ) with ϕ an isomorphism between $E[N]$ to the group scheme $\mathbb{Z}/N \times \mu_N$ such that it is equivariant by the Weil pairing where μ_N is the group scheme of the N-roots of unity. Denote by $X(N)$ the modular curve over $\mathbb{Q}$ by this moduli problem, which is isomorphic over $\mathbb{C}$ to the Riemann surface $X(N)_{\mathbb{C}}$, (see [**8**, §2] for more details concerning $X(N)$ or [**10**], [**20**]).

Denote by X_N any of the above geometrically connected modular curves over $\mathbb{Q}$ and assume that its genus ≥ 2.

By a great result of G. Faltings, we know that for any number field F the set of F-points of X_N, named $X_N(F)$, is always a finite set. We now consider the set of all quadratic points over $\mathbb{Q}$ of X_N:

$$\Gamma_2(X_N, \mathbb{Q}) := \cup_{[F:\mathbb{Q}]\leq 2, F\subset\overline{\mathbb{Q}}} X_N(F),$$

and ask if it is a finite set or not, where $\overline{\mathbb{Q}}$ denotes a fix algebraic closure of $\mathbb{Q}$. By the work of Harris and Silverman, we know that $\Gamma_2(X_N, \mathbb{Q})$ is an infinite set if and only if X_N has a degree two map defined over $\mathbb{Q}$ to a projective line (in particular is an hyperelliptic curve) or to an elliptic curve E over $\mathbb{Q}$ (in particular is a bielliptic curve) with positive rank, see §2 for the precise general statement. In this survey paper, we list all the N for which $\Gamma_2(X_N, \mathbb{Q})$ is not a finite set.

The contents of the paper are as follows. In §2 we recall Harris and Silverman results in [**13**], in particular, the relation between the infinitude of the set of quadratic points over some number field L of a nonsingular projective curve C and the property that C is an hyperelliptic or a bielliptic curve. We also detail this relation when one fixes the field L, fixing some inaccuracies in the literature. In §3 we present and fix inaccuracies of some results concerning the automorphism group of the above modular curves X_N. In §4 we present results and main ideas needed to obtain the exact list of N for which the set $\Gamma_2(X_0(N), \mathbb{Q})$ is a non-finite set. Finally in §5 we deal with the question on quadratic points over $\mathbb{Q}$ for the remaining X_N, observing in particular that any elliptic curve over $\mathbb{Q}$ parametrized by $X(N)$ satisfies that its conductor divides N^2. We finish the survey paper with a remark on the modular curves associated to the modular subgroup $\Gamma_1(M, N)$.

2. The set of quadratic points is a not finite set: Hyperelliptic and Bielliptic curves

Let k be a number field, and $\overline{k}$ a fix algebraic closure of k. Once and for all, write $C = C_{|k}$ for a non-singular projective curve defined over k and let $\overline{C}$ be $C \times_k \overline{k}$ the curve C over $\overline{k}$. Moreover, given $k \subset L \subset \overline{k}$ the curve C over L, named $C \times_k L$ will be also denoted by C if L/k is a finite extension by abuse of notation. We denote by g_C the genus of $\overline{C}$ and we assume that $g_C \geq 2$. We write $\mathrm{Aut}(C)$ for the automorphism group of $\overline{C}$.

As usual $C(L)$ denotes the set of points of C defined over L or L-points where L is a field extension of k inside $\overline{k}$.

THEOREM 2.1 (Faltings, 1983). *For L a number field, $C(L)$ is always a finite set.*

Consider the set of all quadratic points of C over a number field L:

$$\Gamma_2(C, L) := \cup_{[L':L] \leq 2} C(L'),$$

where L' runs over all the extensions of degree at most 2 of L inside $\overline{k}$ and we may ask concerning the finiteness or not for the set $\Gamma_2(C, L)$.

DEFINITION 2.2. The curve C is called hyperelliptic if $\overline{C}$ admits a degree two morphism to the projective line over $\overline{k}$.

We have the following well-known result.

PROPOSITION 2.3. *C is hyperelliptic if and only if there exists a $w \in \mathrm{Aut}(C)$ of order 2 (i.e. an involution of $\mathrm{Aut}(C)$) with $2g_C+2$ fixed points. If C is hyperelliptic, then w is unique and is defined over k and w is called the hyperelliptic involution of C.*

From the definition, it follows easily

LEMMA 2.4. *If C is hyperelliptic then exists a number field L where $\Gamma_2(C, L)$ is a not finite set.*

Next, let us obtain an arithmetic result fixing the number field L.

LEMMA 2.5. *If C is hyperelliptic, and w denotes the hyperelliptic involution, then*

1. *there exists a (unique) degree two map to a conic, all defined over k,*
2. *if $C(k)$ is not $\emptyset$ (the empty set), or, more generally, $C/ < w > (k) \neq \emptyset$, then, there exists a (unique) degree two map defined over k to the projective line over k, $\mathbb{P}^1_{|k}$.*

PROOF. The curve $C/ < w >$ has genus zero, therefore corresponds to a conic, and because w and C are defined over k the conic is also defined over k.

Consider $\pi : C \to C/ < w >$ a degree two morphism defined over $\overline{k}$. For every $\delta \in \mathrm{Gal}(\overline{k}/k)$, in the absolute Galois group of k, we have the degree two morphism $\pi^\delta : C \to C/ < w >$. By the uniqueness of the hyperelliptic involution the morphisms π and π^δ differ by an element $\xi_\delta \in \mathrm{Aut}(C/ < w >) = \mathrm{PGL}_2(\overline{k})$ (the projective linear matrices 2×2 matrices over $\overline{k}$), because the conic is isomorphic to the projective line in the algebraic closure. Thus, we have an application:

$$\xi : \mathrm{Gal}(\overline{k}/k) \to \mathrm{PGL}_2(\overline{k})$$
$$\delta \mapsto \xi_\delta.$$

Observe that given $\sigma, \delta \in \mathrm{Gal}(\overline{k}/k)$ we have $\xi_{\sigma\delta} = \xi_\sigma^\delta \circ \xi_\delta$ thus $(\xi_\sigma)_\sigma$ defines a 1-cocycle, in particular, an element of the first Galois cohomology group $H^1(\mathrm{Gal}(\overline{k}/k), \mathrm{PGL}_2(\overline{k}))$. This element corresponds to a twist of $C/ < w >$ (or a twist of the projective line over k), $\varphi_1 : B \to C/ < w >$ (here B is a conic defined over k) such that $\xi_\sigma = \varphi_1^\sigma \circ \varphi_1^{-1}$ for all $\sigma \in \mathrm{Gal}(\overline{k}/k)$, see [**34**, III.1]. The morphism:

$$\varphi := \varphi_1^{-1} \circ \pi : C \to B$$

is defined over k.

For the second statement, if $\varphi(C(k)) \neq \emptyset$ or $C/ < w > (k) \neq \emptyset$ we obtain that the conic has a point in k, therefore isomorphic over k to the projective line over k: $\mathbb{P}^1_{|k}$. □

REMARK 2.6. Mestre proved in [**28**, p.322-324] that if g_C is even and C hyperelliptic (recall that always C is defined over k), then, there exists a map $\varphi : C \to \mathbb{P}^1_{|k}$ of degree 2 all defined over k.

From Lemma 2.5 we obtain:

COROLLARY 2.7. *If C is hyperelliptic with hyperelliptic involution w, and $(C/ < w >)(k) \neq \emptyset$ then $\Gamma_2(C, k)$ is always a set with an infinite number of elements.*

REMARK 2.8. Let C be a non-singular projective curve over k, hyperelliptic with hyperelliptic involution w and assume $(C/ < w >)(k) = \emptyset$, (in particular g_C is odd by Remark 2.6). Then, $\Gamma_2(C, k)$ may have only a finite number of elements, see Corollary 2.17 and Lemma 2.18.

DEFINITION 2.9. A curve C is called bielliptic if $\overline{C}$ admits a degree two morphism to an elliptic curve over $\overline{k}$.

PROPOSITION 2.10. *C is bielliptic if and only if there exists an involution $\tilde{w} \in \mathrm{Aut}(C)$ with $2g_C - 2$ fixed points, and in such case, we call $\tilde{w}$ a bielliptic involution. A bielliptic involution is unique if $g_C \geq 6$ and then it is defined over k and belongs to the center of* $\mathrm{Aut}(C)$.

See a proof of Proposition 2.10 in [**35**, p.706, Prop.1.2.a)] with [**35**, Lemma 1.3].

COROLLARY 2.11. *If C is bielliptic, then, there exists a number field L such that $\Gamma_2(C, L)$ is not a finite set.*

PROOF. Take $\varphi : C \to E$ a degree two morphism where φ and E are defined in some finite extension M of k, E a genus 1 curve over M. Then take L defined over some finite extension of M such that $E(L) \neq \emptyset$ (i.e. E is an elliptic curve over L), φ is defined over L, and $rank E(L) \geq 1$ to conclude. □

Harris and Silverman obtained in [**13**]:

THEOREM 2.12 (Harris-Silverman). *Take C with $g_C \geq 2$. Then:*
There exists a number field L such that the set $\Gamma_2(C, L)$ is not finite $\Leftrightarrow$ C is a hyperelliptic or a bielliptic curve.

Let us state an arithmetic statement of Theorem 2.12 fixing the number field L.

We first recall Harris-Silverman result [**13**, Lemma 5],

LEMMA 2.13. *Let C be a bielliptic curve with $g_C \geq 6$. Then, there exists a genus 1 curve E defined over k and a morphism $\varphi : C \to E$ of degree 2 all defined over k.*

The following result is well-known to specialists.

THEOREM 2.14. *Take C with $g_C \geq 2$. Then:*
$\Gamma_2(C, k)$ is an infinite set if and only if C is hyperelliptic with a degree two morphism $\varphi : C \to \mathbb{P}^1_{|k}$ defined over k to the projective line over k or C is bielliptic with a degree two morphism $\phi : C \to E$ all defined over k where E is an elliptic curve over k with $rank(E(k)) \geq 1$.

REMARK 2.15. Schweizer, in [**35**, proof of Theorem 5.1] gave different details of the proof of Theorem 2.14. We warn the reader that the statement of [**35**, Theorem 5.1] is weaker than Theorem 2.14.

PROOF (SKETCH). One implication of the statement of Theorem 2.14 is clear.
Let us suppose that $\Gamma_2(C, k)$ is infinite.

Take $P \in \Gamma_2(C, k)$ a quadratic point and denote by P' its conjugate by the quadratic extension. Define then

$$\phi^{(2)} : S^2C \to Jac(C)$$

$$q_1 + q_2 \mapsto [q_1 + q_2 - P - P']$$

where S^2C corresponds to $(C \times C)/S_2$ with S_2 the permutation group and $Jac(C)$ is the Jacobian of C and denote by $proj$ the projection map $C \times C \to S^2C$. The map $\phi^{(2)}$ is defined over k.

Denote by $\phi^{(2)}(k) : S^2C(k) \to Jac(C)(k)$, the map on the k-points, observe $S^2C(k)$ is a not finite set because $\Gamma_2(C, k)$ is not finite.

LEMMA 2.16. *The map $\phi^{(2)}(k)$ is injective if and only if there does not exist a degree two map defined over k of C to the projective line over k. In particular, if C is hyperelliptic and the genus zero curve $C/ < w >$ has no k-point (where w denotes the hyperelliptic involution of C), then $\phi^{(2)}(k)$ is injective.*

PROOF. Consider $(q_1, q_2), (q'_1, q'_2) \in S^2C(k)$ and suppose

$$q_1 + q_2 - P - P' = q'_1 + q'_2 - P - P' \in Jac(C)(k),$$

and recall that $Jac(C)(k)$ coincides with the k-points of the Picard group of degree zero for the curve C. We have that is equivalent to $q_1 + q_2 - q'_1 - q'_2 = div(f)$ with $f \in k(C)$ of degree two. And this is equivalent to define a degree two map over k to the projective line over k.

In particular if C with $g_C \geq 2$ is hyperelliptic, the hyperelliptic involution is defined over k and we have a degree two morphism to C to a genus 0 curve all defined over k, and if the genus 0 curve has no points over k (this situation only happens with g_C odd by Remark 2.6) we have then that $\phi^{(2)}(k)$ is injective.

□

Now, we can suppose that C is not hyperelliptic or C is hyperelliptic but satisfies that $\phi^{(2)}(k)$ is injective. We will prove that under this assumption C is bielliptic with a degree two map φ defined over k of the shape $\varphi : C \to E$ with E an elliptic curve with $rank(E(k)) \geq 1$ proving theorem 2.14.

By Falting's Theorem [**12**] we have

$$Im(\phi^{(2)}(k)) = \cup P_i + B_i(k)$$

with P_i points of $Jac(C)$ and B_i abelian subvarieties of dimension lower or equal to 1 because $Im(\phi^{(2)}(k))$ is not an abelian variety. Therefore a B_i, say B_1 is an elliptic curve E where its k-points have positive rank, and E is defined over k (see for more details [**35**, proof Theorem 5.1]). Now Abramovich and Harris, in [**1**, Lemma 2], construct a degree two map φ from C to E as follows: take $\mathfrak{F} := \phi^{(2)} \circ proj : C \times C \to S^2C \to Im(\phi^{(2)})$, $\mathfrak{F}$ is defined over k. Then, $\mathfrak{F}^{-1}(E)$ is a union of irreducible projective varieties which have a degree 2 map to E via $\mathfrak{F}$. By [**1**, Lemma 2] there exists Z_1, an irreducible component of $\mathfrak{F}^{-1}(E)$, and $j : Z_1 \to C$ a rational map which is one-to-one where j corresponds to the projection of $C \times C$ to C in the first or the second component. In particular j is defined over k. Therefore j induces an isomorphism ι over k from the normalization of Z_1, named Z, to C, and the normalization map ψ of Z to Z_1 is defined over k. Therefore the degree two map $\varphi : C \to E$ defined in the proof of [**1**, Lemma 2] is $\varphi := \mathfrak{F}_{|Z_1} \circ \psi \circ \iota^{-1}$ and is defined over k. □

COROLLARY 2.17. *Let C be an hyperelliptic curve, w its hyperelliptic involution, with $(C/ < w >)(k) = \emptyset$ and $g_C \geq 2$ such that there does not exist a degree 2 morphism $C \to E$ all defined over k, where E an elliptic curve over k with $rank(E(k)) \geq 1$. Then $\Gamma_2(C,k)$ is a finite set.*

PROOF. Because C is hyperelliptic and $C/ < w > (k) = \emptyset$ we obtain that does not exist a degree two morphism of C to the projective line over k by the uniqueness of the hyperelliptic involution w. By the proof of Theorem 2.14, we conclude the result. □

LEMMA 2.18 (Xarles). *Consider the curve C in $\mathbb{P}^3$ given by the equations: $y^2 = -x^2 - t^2$ and $z^2t^4 = x^6 + x^4t^2 + x^2t^4 + t^6$.*

The curve C is defined over $\mathbb{Q}$, has genus 5, satisfies $C(\mathbb{Q}) = \emptyset$.

It is hyperelliptic with $C/ < w > (\mathbb{Q}) = \emptyset$ and there does not exist a degree 2 morphism $C \to E$ all defined over $\mathbb{Q}$ with E an elliptic curve over $\mathbb{Q}$ with $rank(E(\mathbb{Q})) \geq 1$. In particular $\Gamma_2(C,\mathbb{Q})$ is a finite set by Corollary 2.17.

PROOF. The quotient of C by the automorphism $w : z \mapsto -z$ induces a map of C to the conic $y^2 = -x^2 - t^2$ with no points on $\mathbb{Q}$, implying that C is hyperelliptic and $C/ < w > (\mathbb{Q}) = \emptyset$.

The Jacobian of C is isogenous to the product of the Jacobian of C_1 and C_2 where $C_1 : t^2 = x^6 + x^4 + x^2 + 1$ and $C_2 : t^2 = (-x^2 - 1)(x^6 + x^4 + x^2 + 1)$. By use of Magma the Jacobian of C_1 and C_2 are $\mathbb{Q}$-simple of dimension 2 and 3 respectively, justifying the genus and that there is no elliptic curve defined over $\mathbb{Q}$ in S^2C and in particular in the Jacobian of C. Thus, by the proof of Theorem 2.14, C does not admits a degree two morphism $C \to E$ to an elliptic curve E, all defined over $\mathbb{Q}$ with $rank(E(\mathbb{Q})) \geq 1$. □

To finish this section we state the following result ([**13**, Prop1], agreements in [**13**]), very useful for the study of bielliptic curves in a family of modular curves:

PROPOSITION 2.19 (Accola-Landman, Harris-Silverman). *Consider C a bielliptic curve, and $\overline{C'}$ a non-singular projective curve over $\overline{k}$ of genus ≥ 2 where C' is a non-singular projective curve over a field extension of k. If we have $\overline{C} \to \overline{C'}$ a finite map then the curve C' is either bielliptic or hyperelliptic.*

3. Automorphism group of classical modular curves

An important facet for the modular curves X_N with genus ≥ 2 is to compute the group $\mathrm{Aut}(X_N)$, which it is equal to the automorphism of the Riemann surface $(X_N)_{\mathbb{C}}$.

For the Riemann surface $X_{\Gamma,\mathbb{C}}$, with modular group $\Gamma \leq \mathrm{SL}_2(\mathbb{Z})$, we know that $Norm(\Gamma)/\pm\Gamma$ is a subgroup of $\mathrm{Aut}(X_{\Gamma,\mathbb{C}})$, where $Norm(\Gamma)$ denotes the normalizer of Γ in $\mathrm{PSL}_2(\mathbb{R})$, the projective linear group of $2{\times}2$ matrices over $\mathbb{R}$.

This normalizer can be computed explicitly for different classical modular curves.

For the modular curve $X_0(N)$, we have [**31**]:

PROPOSITION 3.1 (Newman). *Write $N = \sigma^2 q$ with $\sigma, q \in \mathbb{N}$ and q square-free. Let ϵ be the* gcd *of all integers of the form $a - d$ where a, d are integers such that $\begin{pmatrix} a & b \\ Nc & d \end{pmatrix} \in \Gamma_0(N)$. Denote by $v(N) := \mathrm{gcd}(\sigma, \epsilon)$. Then $M \in \mathrm{Norm}(\Gamma_0(N))/\pm\Gamma_0(N)$ if and only if M is represented in $\mathrm{PSL}_2(\mathbb{R})$ as a matrix of the form*

$$\sqrt{\delta}\begin{pmatrix} r\Delta & \frac{u}{v(N)\delta\Delta} \\ \frac{sN}{v(N)\delta\Delta} & l\Delta \end{pmatrix}$$

with $r, u, s, l \in \mathbb{Z}$ and $\delta | q$, $\Delta | \frac{\sigma}{v}$. Moreover $v(N) = 2^{\mu}3^{ss}$ with $\mu = min(3, [\frac{1}{2}v_2(N)])$ and $ss = min(1, [\frac{1}{2}v_3(N)])$ where $v_{p_i}(N)$ is the valuation at the prime p_i of the integer N.

For later convenience we define particular elements in Proposition 3.1 which induce elements of $\mathrm{Aut}(X_0(N))$.

DEFINITION 3.2. Let N be a fixed positive integer. For every Q, a Hall divisor of N (i.e. a positive divisor Q of N with $\mathrm{gcd}(Q, N/Q) = 1$), with $Q > 1$, the Atkin-Lehner involution w_Q is defined as follows,

$$\frac{1}{\sqrt{Q}}\begin{pmatrix} Qa & b \\ Nc & Qd \end{pmatrix} \in SL_2(\mathbb{R})$$

with a fix $a, b, c, d \in \mathbb{Z}$.

Always w_Q defines an involution of $\mathrm{Aut}(X_0(N))$ and $w_Q \cdot w_{Q'} = w_{QQ'} \in \mathrm{Aut}(X_0(N))$ with $(Q, Q') = 1$ (and another choice of $a, b, c, d \in \mathbb{Z}$ for w_Q in Definition 3.2, determines the same element in $\mathrm{Aut}(X_0(N))$).

Denote by $S_{v'} = \begin{pmatrix} 1 & \frac{1}{v'} \\ 0 & 1 \end{pmatrix}$ with $v' \in \mathbb{N} \setminus \{0\}$.

Atkin-Lehner wrongly claimed without proof in [**4**, Theorem 8] the group structure of $Norm(\Gamma_0(N))/\pm\Gamma_0(N))$. Akbas and Singerman [**3**] (rediscovered also by the author in [**7**]) obtained the correct statement. Here we only present the following intermediate result:

PROPOSITION 3.3 (Atkin-Lehner, Akbas-Singerman, Bars). *Any element* Υ *that belongs to* $\mathrm{Norm}(\Gamma_0(N))/\pm\Gamma_0(N)$ *has an expression of the form*

$$\Upsilon = w_Q\Omega,$$

where w_Q *is an Atkin-Lehner involution with* $(Q,6)=1$ *and* Ω *belongs to the subgroup generated by* $S_{v(N)}$ *and the Atkin Lehner involutions* $w_{2^{v_2(N)}}$, $w_{3^{v_3(N)}}$.

We have also an analog of Newman's result for the modular curves $X_1(N)$ in [**23**](or in [**25**]) (read the particular case $X_1(4)$ in [**23**]):

PROPOSITION 3.4 (Kim-Koo, Lang). *If* $N\neq 4$, *then* $\Xi\in Norm(\Gamma_1(N))/\pm\Gamma_1(N)$ *if and only if* Ξ *is represented in* $PSL_2(\mathbb{R})$ *as a matrix of determinant 1 of the form:*

$$\Xi = \frac{1}{\sqrt{Q}}\begin{pmatrix} Qa & b \\ Nc & Qd \end{pmatrix}$$

where Q *is a Hall divisor of* N *and* a,b,c,d *are all integers.*

Concerning elements in $\mathrm{Aut}(X_1(N))$ we observe (see also [**16**]):

COROLLARY 3.5. *Always* $w_Q\in\mathrm{Aut}(X_1(N))$ *for a Hall divisor* Q *of* N, *but* w_Q *is not necessarily an involution of* $X_1(N)$. *The full Atkin-Lehner involution* w_N *is always an involution of* $\mathrm{Aut}(X_1(N))$.

COROLLARY 3.6. *Consider* $\gamma\in\Gamma_0(N)$ *with* $\gamma\equiv\begin{pmatrix} a & * \\ 0 & * \end{pmatrix}$ *mod* N. *Then,* γ *represents an automorphism of* $X_1(N)$ *which depends only of* a *and we name it by* $[a]$. *In particular,* $Norm(\Gamma_1(N))/\pm\Gamma_1(N)$ *is generated by* w_Q *and* $[a]$; *i.e. generated by* $\Gamma_0(N)/\pm 1$ *and* w_Q *with* $Q|N$ *and* $(Q,N/Q)=1$.

For the classical modular curves $X(N)$ it follows easily following the proof for $X_1(N)$ in [**23**], (or see also [**8**]):

PROPOSITION 3.7. *If* $N\geq 5$ *the normalizer of* $\Gamma(N)$ *in* $PSL_2(\mathbb{R})$ *is* $PSL_2(\mathbb{Z})$ *and therefore* $Norm(\Gamma(N))/\pm\Gamma(N)\cong PSL_2(\mathbb{Z}/N\mathbb{Z})$.

REMARK 3.8. Lang in [**26**] gave an algorithm that, given a subgroup Γ of finite index in $PSL_2(\mathbb{Z})$, allows to determine the normalizer of Γ in $PSL_2(\mathbb{R})$.

A very deep question is to determine when $Norm(\Gamma)/\pm\Gamma$ coincides with the full group of automorphisms of the Riemann surface $X_{\Gamma,\mathbb{C}}$.

DEFINITION 3.9. Consider $g_{X_{\Gamma,\mathbb{C}}}\geq 2$. An automorphism

$$\vartheta\in\mathrm{Aut}(X_{\Gamma,\mathbb{C}})\setminus(Norm(\Gamma)/\pm\Gamma)$$

is called an exceptional automorphism for the Riemann surface $X_{\Gamma,\mathbb{C}}$.

It was known by A. Ogg [**32**]:

PROPOSITION 3.10 (Ogg). *If* p *is a prime* $p\neq 37$ *with* $g_{X_0(p)}\geq 2$, *then* $X_0(p)$ *has no exceptional automorphisms. For* $p=37$, $\mathrm{Aut}(X_0(37))$ *has index two with respect to the subgroup* $Norm(\Gamma_0(37))/\pm\Gamma_0(37)=\{id,w_{37}\}$ *an one of the exceptional automorphism corresponds to the hyperelliptic involution of* $X_0(37)$.

F. Momose contributed strongly to this question, [**21**]:

THEOREM 3.11 (Kenku-Momose). *For $N \neq 37, 63$ and 108 there are no exceptional automorphisms for $X_0(N)$ with $g_{X_0(N)} \geq 2$, and therefore*

$$\mathrm{Aut}(X_0(N)) = Norm(\Gamma_0(N))/\pm\Gamma_0(N).$$

REMARK 3.12. The case $\mathrm{Aut}(X_0(63))$ was completed by Elkies in [**11**] and the case $\mathrm{Aut}(X_0(108))$ was obtained by Harris in [**14**]. In both cases there appear exceptional automorphisms and the index of the full automorphism group with $Norm(\Gamma_0(N))/\pm\Gamma_0(N)$ is two. We point out to the reader that Kenku and Momose did not deal with the case N=108. This is dealt with in [**14**].

F. Momose also contributed to obtain the full automorphism group for the modular curves $X_1(N)$ and $X_\Delta(N)$, for N square-free. Unfortunately, his paper [**30**] is not available in the literature as far as I know. A particular statement of the general result in [**30**] reads as follows:

THEOREM 3.13 (Momose). *If N is square-free with $g_{X_1(N)} \geq 2$, then* $\mathrm{Aut}(X_1(N)) = Norm(\Gamma_1(N))/\pm\Gamma_1(N)$.

In [**30**] a similar conclusion is stated for $X_\Delta(N)$ with N square-free (where $\Delta = (\mathbb{Z}/N)^*$ is allowed), but A. Schweizer communicated us the following result which will appear in the work [**19**]:

LEMMA 3.14 (Jeon-Kim-Schweizer). *The curve $X_{\{\pm1,\pm6,\pm8,\pm10,\pm11,\pm14\}}(37)$, denoted by $X_{\Delta_3}(37)$, has exceptional automorphisms, and one of the exceptional automorphisms is a bielliptic involution for this curve.*

PROOF. $X_{\Delta_3}(37)$ has genus 4 and is an unramified Galois cover of the genus 2 curve $X_0(37)$. So by [**2**, Corollary 1, p. 346] it is bielliptic.

The not exceptional automorphisms of $X_{\Delta_3}(37)$ form a group S_3 generated by [2] (of order 3) and the involution w_{37}. The quotient by this group is the elliptic curve $X_0(37)/w_{37}$. Applying the Hurwitz formula to this S_3-covering, one sees that each of the 3 (conjugate) involutions has 2 fixed points. So the bielliptic involution must be exceptional. □

REMARK 3.15. The proof that $X_{\Delta_3}(37)$ is not hyperelliptic in the Tsukuba paper [**15**] uses the non-existence of exceptional automorphisms. To prove that $X_{\Delta_3}(37)$ is not hyperelliptic one can use that unramified Galois covers of degree 3 of a hyperelliptic curve are never hyperelliptic [**2**, Lemma 4].

One may wonder if a revision of [**15**] is needed concerning the list of hyperelliptic modular curves, but the work of Jeon and Kim in [**17**] [**18**] already did this work. See in particular [**18**] for a more computational proof that $X_{\Delta_3}(37)$ is not hyperelliptic.

Finally, we turn to the classical modular curves $X(N)$. It is known to the specialists that there are no exceptional automorphisms, but it was not an available proof in the literature until the recent work of [**8**]:

THEOREM 3.16. *With $g_{X(N)} \geq 2$, always* $\mathrm{Aut}(X(N)) = Norm(\Gamma(N))/\pm\Gamma(N)$.

4. Which curves $X_0(N)$ have an infinite set of quadratic points?

Fix N for which $g_{X_0(N)} \geq 2$. The question of the existence of a number field L for which $\Gamma_2(X_0(N), L)$ is not a finite set is equivalent to determining when the modular curve $X_0(N)$ is hyperelliptic or bielliptic, by Theorem 2.12.

A. Ogg in [**32**] completed the list of hyperelliptic curves:

THEOREM 4.1 (Ogg). *There are 19 values of N, such that $X_0(N)$ with $g_{X_0(N)} \geq 2$ is hyperelliptic. For $N = 37$, $N = 40$ and $N = 48$ the hyperelliptic involution is not an Atkin-Lehner involution. The rest, can be read off from the following table:*

N	v	N	v
22	w_{11}	35	w_{35}
23	w_{23}	39	w_{39}
26	w_{26}	41	w_{41}
28	w_7	46	w_{23}
29	w_{29}	47	w_{47}
30	w_{15}	50	w_{50}
31	w_{31}	59	w_{59}
33	w_{11}	71	w_{71}

In order to reduce to a finite list of N for which $X_0(N)$ could be a bielliptic curve: take N with $g_{X_0(N)} \geq 6$ and $X_0(N)$ bielliptic. Then, by Lemma 2.13 we have a degree two map $\varphi : X_0(N) \to E$ all defined over $\mathbb{Q}$. By [**20**], there exists $\mathfrak{X}_0(N)$ a non-singular proper and normal scheme over $Spec(\mathbb{Z}[1/N])$ and for any p prime with $p \nmid N$ the fiber at p defines a non-singular projective curve over $\mathbb{F}_p$ denoted by $X_0(N)_{\mathbb{F}_p}$. Now, by [**37**, theorem 6.1], E has a Neron model over $Spec(\mathbb{Z}[1/N])$ denoted by $\mathfrak{E}$, and by the theory of Neron models we lift φ to a map of $Spec(\mathbb{Z}[1/N])$-schemes $\hat{\varphi} : \mathfrak{X}_0(N) \to \mathfrak{E}$, and its fiber over the place p we obtain a degree two map from $X_0(N)_{|\mathbb{F}_p}$, to an elliptic curve over the finite field $\mathbb{F}_p$.

The curve $X_0(N)_{|\mathbb{F}_p}$ has a moduli interpretation, (not in the cusps), given by couples $(E', \mathfrak{C})$ where E' denotes an elliptic curve over $\overline{\mathbb{F}}_p$, the algebraic closure of $\mathbb{F}_p$, and $\mathfrak{C}$ a subgroup of E' of exact order N. Recall that the cusps of $X_0(N)_{|\mathbb{F}_p}$ are the ones coming from the cusps of $X_0(N)$, i.e. from the points $X_{\Gamma_0(N),\mathbb{C}} \setminus Y_{\Gamma_0(N),\mathbb{C}}$.

Ogg in [**32**] proved that any point $(E', \mathfrak{C}) \in X_0(N)_{|\mathbb{F}_p}$ with $p \nmid N$ and E' a supersingular curve defined over $\mathbb{F}_p$ satisfies that $(E', \mathfrak{C}) \in X_0(N)_{|\mathbb{F}_p}(\mathbb{F}_{p^2})$. This fact is the main tool to prove the following result in [**13**].

THEOREM 4.2 (Harris-Silverman). *If $X_0(N)$ is bielliptic with $g_{X_0(N)} \geq 6$ and p a prime with $p \nmid N$, then:*

$$\#cusps(\mathbb{F}_{p^2}) + 2n(p)\mu(N) \leq \#X_0(N)_{|\mathbb{F}_p}(\mathbb{F}_{p^2})$$
$$\leq min(2(p+1)^2, p^2 + pg_{X_0(N)} + 1)$$

where $\mu(N)$ is the index $(SL_2(\mathbb{Z}) : \Gamma_0(N))$, $cusps(\mathbb{F}_{p^2})$ are the cusps of $X_0(N)_{|\mathbb{F}_p}$ defined over $\mathbb{F}_{p^2}$, as usual $\#\Theta$ for a set Θ denotes the number of elements of the set and $n(p) = \sum \frac{1}{\#\mathrm{Aut}(E'')}$ where E'' runs over all the supersingular elliptic curves defined over $\mathbb{F}_p$.

We can control a lower bound for the cusps of $X_0(N)_{|\mathbb{F}_p}$ which are defined over $\mathbb{F}_{p^2}$ by [**33**, Prop.4.8] (or [**6**, Lemma 2.2]) and therefore we can deduce for example if $2 \nmid N$ then $N \leq 192$ ([**6**, Lemma 2.1]), and,

PROPOSITION 4.3. [**6**, Prop.2.3] *The curve* $X_0(N)$ *is not bielliptic for* $N > 210$.

Now it remains a case-by-case study if $X_0(N)$ is bielliptic or not with $N \leq 210$. By Proposition 2.10, it is enough to control the fixed points of all involutions of $\mathrm{Aut}(X_0(N))$.

* Because we know the number of fixed point on $X_0(N)$ of the Atkin-Lehner involutions $w_Q \in Norm(\Gamma_0(N))/\pm\Gamma_0(N)$ (see for example the formula in [**24**] or a table computing these numbers up to $N \leq 210$ in [**38**]):

LEMMA 4.4. *Assume that* $X_0(N)$ *is bielliptic with a bielliptic involution which it is an Atkin-Lehner involution. Then,* N *appears in the following list:*

22	*26*	*28*	*30*	*33*	*34*	*35*	*37*	*38*	*39*
40	*42*	*43*	*44*	*45*	*48*	*50*	*51*	*53*	*54*
55	*56*	*60*	*61*	*62*	*63*	*64*	*65*	*69*	*75*
79	*81*	*83*	*89*	*92*	*94*	*95*	*101*	*119*	*131*

* Now by Kenku-Momose Theorem 3.11 and the description of Newman in Proposition 3.1 we obtain that for $4 \nmid N$ and $9 \nmid N$ all the involutions of $\mathrm{Aut}(X_0(N))$ are Atkin-Lehner involutions.

* When $4|N$ or $9|N$ we have more involutions in $\mathrm{Aut}(X_0(N))$ than the Atkin-Lehner ones. For example when $4|N$, there is the involution S_2.

The author in [**6**] did a detailed study of the involutions which are not Atkin-Lehner involutions which appear in the remaining situations, in particular on the computation of the number of fixed points for these new involutions. We reproduce next a particular statement [**6**, p.159-160].

If $9||N$ and $4 \nmid N$, from Proposition 3.1, one obtains that all involutions which are not of Atkin-Lehner type are:

$$S_3 w_9 S_3^2, \quad S_3^2 w_9 S_3$$
$$w_r S_3 w_9 S_3^2, \quad w_r S_3^2 w_9 S_3 \ \ if\ r \equiv 1 (mod\ 3),$$
$$w_r S_3, \quad w_r S_3^2, \quad w_r w_9 S_3 w_9, \quad w_r w_9 S_3^2 w_9 \ \ if\ r \equiv 2 (mod\ 3),$$

PROPOSITION 4.5. [**6**, Prop.3.13] *The number of fixed points of* $S_3 w_9 S_3^2$ *and* $S_3^2 w_9 S_3$ *in the modular curve* $X_0(N)$ *is equal to the number of fixed points of* w_9 *in* $X_0(N)$.
For $r \equiv 1(mod\ 3)$ *the number of fixed points of* $w_r S_3 w_9 S_3^2, w_r S_3^2 w_9 S_3$ *in* $X_0(N)$ *is equal to the number of fixed points of* $w_r w_9$ *in* $X_0(N)$.
For $r \equiv 2(mod\ 3)$ *the number of fixed points of* $w_r S_3, w_r S_3^2, w_r w_9 S_3 w_9, w_r w_9 S_3^2 w_9$ *is bounded by 3 times the number of fixed points of* w_r *in* $X_0(N/3)$.

After all this study on involutions we obtained in [**6**, Theorem 3.15] all the bielliptic curves $X_0(N)$ with a list of bielliptic involutions, in particular:

THEOREM 4.6 (Bars). *There are 41 values of* N, *such that* $X_0(N)$ *is bielliptic. For each value,* $X_0(N)$ *has a bielliptic involuton which is an Atkin-Lehner involution, except for the bielliptic curve* $X_0(2^3 3^2)$. *The list of these* N, $N \neq 72$, *is given in Lemma* 4.4.

REMARK 4.7. Harrison noticed in 2011 that $\mathrm{Aut}(X_0(108))$ has an exceptional automorphism. This result does not affect the argument in [**6**] from 1999 because we discard $X_0(108)$ to become a bielliptic curve by arguments using Theorem 4.2, see for a precise proof [**5**, Corol.lari 4.11]. In particular, $X_1(108)$ or $X_\Delta(108)$ are not bielliptic by Proposition 2.19 with $C' = X_0(108)$.

From Theorem 4.1 and Theorem 4.6 we can list the modular curves $X_0(N)$ where the set of quadratic points over some number field is not finite by Theorem 2.12:

COROLLARY 4.8. *Assume $X_0(N)$ of genus greater than or equal to 2. Then*

$$\#\Gamma_2(X_0(N), L_N) = \infty$$

for some number field L_N if and only if N is in the following list:
22, 23, 26, 28, 30, 31, 33, 34,35, 37, 38, 39, 40, 41, 42,43, 44, 45, 46, 47, 48, 50, 51, 53,54, 55,56, 59, 60,61, 62, 63, 64, 65, 69, 71,72, 75, 79, 81, 83, 89, 92, 94, 95, 101, 119, 131.

If we want to determine $X_0(N)$ for which $\Gamma_2(X_0(N), \mathbb{Q})$ is not finite, we need to use Theorem 2.14. We obtained in [**6**]

THEOREM 4.9 (Bars). *Assume $g_{X_0(N)} \geq 2$. We have that $\Gamma_2(X_0(N), \mathbb{Q})$ is a finite set if and only if N does not appear in the following list*

22	*23*	*26*	*28*	*29*	*30*	*31*	*33*	*35*	*37*
39	*40*	*41*	*43*	*46*	*47*	*48*	*50*	*53*	*59*
61	*65*	*71*	*79*	*83*	*89*	*101*	*131*		

PROOF (SKETCH). Assume that $X_0(N)$ is bielliptic or hyperelliptic. We know that $X_0(N)(\mathbb{Q}) \neq \emptyset$ (some cusps are defined over $\mathbb{Q}$), thus we restrict to $X_0(N)$ bielliptic and no-hyperelliptic by Lemma 2.5. By Carayol's Theorem we discard N where the elliptic curves over $\mathbb{Q}$ with the conductor N and all of its divisors have rank zero. The remaining situations come from the study of the Weil strong modular parametrization studied in [**27**]. (We mention here that in the proof we use a weak form of Theorem 2.14 without worry if the degree two map from $X_0(N)$ to the elliptic curve is defined over $\mathbb{Q}$. This is so because in $\mathrm{Aut}(X_0(N))$ Atkin-Lehner involutions are defined over $\mathbb{Q}$ and for $4|N$ or $9|N$ we obtain the result only searching when $Jac(X_0(N))$ contains an elliptic curve over $\mathbb{Q}$ with positive rank and if it contains such elliptic curve then the theory of Weil strong modular parametrization gives the morphism). □

5. Other classical modular curves

For simplicity, we denote in the following X'_N, once and for all, any of the modular curves defined over $\mathbb{Q}$:

$$X(N),\ X_1(N),\ or\ X_\Delta(N)\ (with\ \{\pm 1\} \subset \Delta \subset (\mathbb{Z}/N)^*)$$

corresponding to modular groups $\Gamma(N), \Gamma_1(N) = \Gamma_{\{\pm 1\}}(N)$, or $\Gamma_\Delta(N)$. And recall once and for all we assume $g_{X'_n} \geq 2$ and we have the natural finite maps:

$$X(N)_\mathbb{C} \to X_1(N)_\mathbb{C} \to X_\Delta(N)_\mathbb{C} \to X_0(N)_\mathbb{C}.$$

5.1. Hyperelliptic modular curves.

Consider the natural map $X'_{N,\mathbb{C}} \to X_0(N)_\mathbb{C}$. If X'_N is hyperelliptic and $g_{X_0(N)} \geq 2$, then $X_0(N)$ is a hyperelliptic curve; therefore by Theorem 4.1 we are reduced to study when X'_N is hyperelliptic for a finite list of N.

THEOREM 5.1 (Mestre, Ishii-Momose). *$X_1(N)$ is hyperelliptic only for $N = 13, 16$ and 18.*

THEOREM 5.2 (Ishii-Momose, Jeon-Kim). *If $\{\pm 1\} \subsetneq \Delta \subsetneq (\mathbb{Z}/N)^*$ the only hyperelliptic curve for the family $X_\Delta(N)$ is $X_{\{\pm 1, \pm 8\}}(21)$.*

REMARK 5.3. Mestre listed the hyperelliptic modular curves $X_1(N)$ in [**29**]. Ishii and Momose studied again this problem for the modular curves $X_\Delta(N)$ in [**15**], reproving the result of Mestre because $X_{\{\pm 1\}}(N) = X_1(N)$. Ishii and Momose claimed in [**15**] that always $X_\Delta(N)$ is not a hyperelliptic curve when $\Delta \neq \{\pm 1\}$. Jeon and Kim proved that $X_{\{\pm 1, \pm 8\}}(21)$ is hyperelliptic in [**18**], clarifying the gap in [**15**], see also Remark 3.15.

REMARK 5.4. The Atkin-Lehner element w_Q is not always in $Norm(\Gamma_\Delta(N))$, and therefore is not an automorphism of the curve $X_\Delta(N)$. The study when w_Q belongs to $\mathrm{Aut}(X_\Delta(N))$ began already in [**15**], as far as I know, and is deeper developed in [**19**].

By the work of Ishii and Momose in [**15**] (or Bars, Kontogerogis and Xarles in [**8**]):

THEOREM 5.5. *There are no N for which $X(N)$ is hyperelliptic.*

5.2. Bielliptic modular curves.

By Proposition 2.19 with the natural map from X'_N to $X_0(N)$ and with Theorems 4.1 and 4.6:

COROLLARY 5.6. *Take N where $X_0(N)$ is not bielliptic and is not hyperelliptic. Then X'_N is not bielliptic, in particular X'_N is not a bielliptic curve for $N \geq 132$.*

For the families $X_1(N), X_\Delta(N), X(N)$ a detailed case-by-case analysis of the involutions is developed in order to obtain the exact list of N for which X'_N is a bielliptic curve, following Proposition 2.10.

THEOREM 5.7 (Jeon-Kim). *Take N with $g_{X_1(N)} \geq 2$, i.e. $N \geq 16$ or $N = 13$. We have that the curve $X_1(N)$ is bielliptic exactly when $2 \leq g_{X_1(N)} \leq 6$ (this corresponds to the values of N: 13, 16, 17, 18, 20, 21, 22, 24).*

Let us present some main ideas given by Jeon and Kim in [**16**] to obtain the complete list of the curves $X_1(N)$ which are bielliptic.

By proposition 3.4 and its corollaries, the non-exceptional automorphisms of $X_1(N)$ are of the form $[a]w_Q$, and we need to study involutions of this type. We recall again that w_Q is not necessarily an involution of $X_1(N)$ but it is an automorphism.

- For $2 \leq g_{X_1(N)} \leq 6$: Jeon and Kim prove that all $X_1(N)$ are bielliptic with an involution in $Norm(\Gamma_1(N))$. They observe that always w_N is an involution of $X_1(N)$ and gave some criteria for situations where the number of fixed points of w_N in $X_1(N)$ coincides with the number of fixed points of w_N in $X_0(N)$. Moreover they obtained some congruences between $[a]w_Q$ which allow the authors, (joint with the hyperelliptic involutions obtained by Ishii and Momose in [**15**] for $N = 13, 16, 18$ and properties of automorphism groups with hyperelliptic involution) to obtain a bielliptic involution belonging to $Norm(\Gamma_1(N))$.

- For $g_{X_1(N)} > 6$, the authors prove that all modular curves $X_1(N)$ are not bielliptic. If $X_1(N)$ is bielliptic, by proposition 2.10 the bielliptic involution is defined over $\mathbb{Q}$. The main argument is:
  Let $\tilde{w}$ be the bielliptic involution of $X_1(N)$ and consider the natural map to $X_0(N)$, thus we obtain an involution $\underline{\tilde{w}}$ of $X_0(N)$ and for $N \neq 37, 63$ is not an exceptional automorphism of $X_0(N)$, therefore the involution $\tilde{w}$ is not an exceptional automorphism of $X_1(N)$. By Theorem 3.4 $\underline{\tilde{w}}$ is necessarily of Atkin-Lehner type w_Q. Restrict now from [**6**, Theorem 3.15] to the pairs $[N, w_Q]$ where w_Q is a bielliptic involution of $X_0(N)$.

  The cusps of $X_0(N)$ (see [**5**, Prop.3.3.1] or [**36**]), $(\mathbb{Q} \cup \{\infty\})/\Gamma_0(N)$, are given by $\begin{pmatrix} x \\ d \end{pmatrix}$ where d runs the positive divisors of N and x is an natural representant modulo $t_d = (d, N/d)$ with $(x, d) = 1$.

  The argument in [**16**] if $Q \neq 2$, is as follows: $\underline{\tilde{w}}$ maps the cusp $\begin{pmatrix} 1 \\ 0 \end{pmatrix}$ to $\begin{pmatrix} y \\ Q \end{pmatrix}$ with $(y, Q) = 1$ for $Q \neq N$ and to $\begin{pmatrix} 0 \\ 1 \end{pmatrix}$ for $Q = N$, but the cusps over $\begin{pmatrix} 1 \\ 0 \end{pmatrix}$ are rational and the cusps over $\begin{pmatrix} y \\ Q \end{pmatrix}$ or $\begin{pmatrix} 0 \\ 1 \end{pmatrix}$ are not, therefore $\tilde{w}$ cannot exist.

  To conclude the remaining cases, the authors use properties of bielliptic involutions for example [**35**, Prop.1.2(b)]. We remark that the arguments in [**16**] only uses Theorem 3.13 for $N = 37$.

Now we turn to the the exact list of bielliptic modular curves $X(N)$:

THEOREM 5.8 (Jeon-Kim, Bars-Kontogeorgis-Xarles). *Take N with $g_{X(N)} \geq 2$, i.e. $N \geq 7$. Then $X(N)$ is bielliptic only for $X(7)$ or $X(8)$.*

We sketch the main ideas of this Theorem following the approach given in [**8**], see also remark 5.16.

Main points in getting the above result are:

- $X(7)$ corresponds to the Klein quartic, $X(8)_{\mathbb{C}}$ is the Wiman curve of genus 5 and both are known to be bielliptic curves.
- we have a morphism over $\mathbb{Q}$: $X(N) \to X_0(N^2)$ and from Theorem 4.6 and Theorem 4.2 we obtain that $X(N)$ is not bielliptic for $N \geq 10$. To deal with the case $N = 9$ one uses a Hurwitz formula argument for a convenient map with an explicit equation for $X(9)_{\mathbb{C}}$.

Recently Jeon, Kim and Schweizer completed the list of bielliptic intermediate curves $X_\Delta(N)$ in [**19**]:

THEOREM 5.9 (Jeon-Kim-Schweizer, [**19**]). *Except for $N \neq 37$, the list of bielliptic $X_\Delta(N)$ with $\{\pm 1\} \subsetneq \Delta \subsetneq (\mathbb{Z}/N)^*$ are the following:*
$X_{\{\pm1,\pm8\}}(21)$, $X_{\{\pm1,\pm5\}}(24)$, $X_{\{\pm1,\pm7\}}(24)$, $X_{\{\pm1,\pm5\}}(26)$,
$X_{\{\pm1,\pm3,\pm9\}}(26)$, $X_{\{\pm1,\pm13\}}(28)$, $X_{\{\pm1,\pm3,\pm9\}}(28)$,
$X_{\{\pm1,\pm4,\pm5,\pm6,\pm7,\pm9,\pm13\}}(29)$, $X_{\{\pm1,\pm11\}}(30)$, $X_{\{\pm1,\pm15\}}(32)$,
$X_{\{\pm1,\pm2,\pm4,\pm8,\pm16\}}(33)$, $X_{\{\pm1,\pm9,\pm13,\pm15\}}(34)$, $X_{\{\pm1,\pm6,\pm8,\pm13\}}(35)$,
$X_{\{\pm1,\pm4,\pm6,\pm9,\pm11,\pm16\}}(35)$, $X_{\{\pm1,\pm11,\pm13\}}(36)$,
$X_{\{\pm1,\pm4,\pm10,\pm14,\pm16,\pm17\}}(39)$, $X_{\{\pm1,\pm9,\pm11,\pm19\}}(40)$,
$X_{\{\pm1,\pm2,\pm4,\pm5,\pm8,\pm9,\pm10,\pm16,\pm18,\pm20\}}(41)$, $X_{\{\pm1,\pm4,\pm11,\pm14,\pm16,\pm19\}}(45)$,

$X_{\{\pm1,\pm11,\pm13,\pm23\}}(48)$, $X_{\{\pm1,\pm6,\pm8,\pm13,\pm15,\pm20,\pm22\}}(49)$, $X_{\{\pm1,\pm9,\pm11,\pm19,\pm21\}}(50)$,$X_{\{\pm1,\pm4,\pm6,\pm9,\pm14,\pm16,\pm19,\pm21,\pm24,\pm26\}}(55)$, $X_{\{\pm1,\pm7,\pm9,\pm15,\pm17,\pm23,\pm25,\pm31\}}(64)$.

Jeon, Kim and Schweizer, need new ideas to obtain the above Theorem. We list some of the main new items in [**19**]:

* A precise criteria to decide when w_Q is an automorphism of $X_\Delta(N)$ and when it defines an involution in $\mathrm{Aut}(X_\Delta(N))$, and in such case compute the number of fixed points. Mainly, they prove that the curves $X_\Delta(N)$ which are bielliptic have a bielliptic involution of the form w_Q or $[a]w_Q$.

* To discard and obtain few new bielliptic curves they study the natural morphism $X_\Delta(N)_\mathbb{C} \to X_0(N)_\mathbb{C}$ and how a bielliptic involution on $X_\Delta(N)$ should translate to $X_0(N)$ and compare with the known results on bielliptic involutions in $\mathrm{Aut}(X_0(N))$. In particular, they need to check for $X_\Delta(37)$ for different Δ, if these curves have exceptional automorphisms. They obtain, for example, that $X_{\Delta_3}(37)$ has exceptional automorphisms (this contradicts a result claimed in [**30**] see also Lemma 3.14) and obtain that this modular curve is bielliptic with an exceptional bielliptic involution. The authors also need a particular study for genus 4 curves to obtain that $X_\Delta(25)$ is not bielliptic.

5.3. On infiniteness of quadratic points over $\mathbb{Q}$ of modular curves.

Consider first the modular curves $X_1(N)$ that are defined over $\mathbb{Q}$ with moduli interpretation given by couples $(E, \mathfrak{C})$ with $\mathfrak{C}$ a N-torsion subgroup of the elliptic curve E.

Fix N. In order for $X_1(N)$ to have an infinite set of quadratic points, we need (from the moduli interpretation of $X_1(N)$) an infinite set of couples $(E, \mathfrak{C})$ defined over a degree 2 extension of $\mathbb{Q}$. In particular, $\mathfrak{C}$ should appear as torsion subgroup for elliptic curves over quadratic fields.

We have the following main result of Kenku and Momose [**22**]:

THEOREM 5.10 (Kenku-Momose). *Let $\mathfrak{q}$ be a fixed integer and E an elliptic curve over a quadratic field $\mathbb{Q}(\sqrt{\mathfrak{q}})$. Then the torsion subgroup of $E(\mathbb{Q}(\sqrt{\mathfrak{q}}))$ is isomorphic to one of the following groups:*
$\mathbb{Z}/m\mathbb{Z}$ with $m \leq 16$ or $m = 18$
$\mathbb{Z}/2 \times \mathbb{Z}/2k$ with $k \leq 6$
$\mathbb{Z}/3 \times \mathbb{Z}/3l$ with $l \leq 2$
$\mathbb{Z}/4 \times \mathbb{Z}/4$.

Now by Theorem 2.14 and because $X_1(13)$, $X_1(16)$ and $X_1(18)$ are hyperelliptic (Theorem 5.1) we have after Theorem 5.7 the following result in [**16**]:

COROLLARY 5.11. *Take N with $g_{X_1(N)} \geq 2$, i.e. $N \geq 16$ or $N = 13$. We have that $\Gamma_2(X_1(N), \mathbb{Q})$ is a finite set if and only if $N \notin \{13, 16, 18\}$.*

Consider now, the modular curves $X_\Delta(N)$ defined over $\mathbb{Q}$, and recall that we have natural morphisms,

$$X_1(N) \to X_\Delta(N) \to X_0(N).$$

COROLLARY 5.12. *Take $N \neq 37$ with $g_{X_\Delta(N)} \geq 2$, and $\{\pm1\} \subsetneq \Delta \subsetneq (\mathbb{Z}/N)^*$. Then the set $\Gamma_2(X_\Delta(N), \mathbb{Q})$ is not finite if and only if $X_\Delta(N) = X_{\{\pm1,\pm8\}}(21)$.*

PROOF. By Theorem 2.14 we only need to study $X_\Delta(N)$ bielliptic and not hyperelliptic. Take the list of $X_\Delta(N)$ bielliptic in Theorem 5.9, and suppose is given $\phi : X_\Delta(N) \to E$ all defined over $\mathbb{Q}$. In particular we have a morphism over $\mathbb{Q}$: $X_1(N) \to E$. By Carayol's result we have that the conductor of E should divide N. Now for all the $N \neq 37$ for which $X_\Delta(N)$ is bielliptic, we have $rankE(\mathbb{Q}) = 0$ for all elliptic curves over $\mathbb{Q}$ with conductor dividing N by Cremona tables [**9**]. □

REMARK 5.13. For $X_\Delta(37)$, we can not discard it from the arguments in the proof of Corollary 5.12 because we have the elliptic curve 37a $[0,0,1,-1,0]$ (following the notation in [**9**]) which has rank 1. Jeon, Kim and Schweizer in [**19**], with an ad-hoc argument, obtained that $\Gamma_2(X_\Delta(37),\mathbb{Q})$ is always a finite set for any Δ with $\pm 1 \subsetneq \Delta \subsetneq (\mathbb{Z}/N)^*$.

Now consider the modular curves $X(N)$.

First, we warn that one may take care of Carayol theorem for such modular curves, for example one can construct a map over $\mathbb{Q}$,

$$X(8) \to \{y^2 = x(x-1)(x+1)\} = E$$

and E has conductor 32 (and not a divisor of 8), where the rational model of $X(8)$ is the one given in [**39**] (see [**8**] to ensure that the model of $X(8)$ in [**39**] coincides with the one fixed in this paper).

PROPOSITION 5.14. *Assume that we have a morphism φ over $\mathbb{Q}$:*

$$\varphi : X(N) \to E$$

with E an elliptic curve defined also over $\mathbb{Q}$. Then the conductor of E divides N^2.

PROOF. By [**8**, Lemma1] we have natural morphisms defined over $\mathbb{Q}$: $X_1(N^2) \to X(N)$ and $X(N) \to X_0(N^2)$, therefore we have a morphism over $\mathbb{Q}$: $X_1(N^2) \to E$ and by the theorem of Carayol we obtain that the conductor of E divides N^2. □

We can prove by an ad-hoc method the following result in [**8**] (without use of Proposition 5.14):

THEOREM 5.15 (Bars-Kontogeorgis-Xarles). *For any $N \geq 7$, we have that the set $\Gamma_2(X(N) \times_{\mathbb{Q}} \mathbb{Q}(\zeta_N), \mathbb{Q}(\zeta_N))$ is always a finite set, in particular $\Gamma_2(X(N),\mathbb{Q})$ is always a finite set.*

REMARK 5.16 (The modular curves $X_1(N,M)$). For positive integers $M|N$ consider the congruence subgroup $\Gamma_1(M,N)$ of $SL_2(\mathbb{Z})$ defined by:

$$\Gamma_1(M,N) := \left\{ \begin{pmatrix} a & b \\ c & d \end{pmatrix} \in SL_2(\mathbb{Z}) | \begin{pmatrix} a & b \\ c & d \end{pmatrix} \equiv \begin{pmatrix} 1 & * \\ 0 & 1 \end{pmatrix} \ mod\ N,\ M|b \right\}.$$

Let $X_1(M,N)_{\mathbb{C}}$ be the Riemann surface associated to $\Gamma_1(M,N)$. It is the coarse moduli space of the isomorphism classes of elliptic curves E with a pair (P_M, P_N) of points P_M and P_N of E which generate a subgroup isomorphic to $\mathbb{Z}/M \oplus \mathbb{Z}/N$ and satisfy $e_E(P_M, \frac{N}{M}P_N) = \zeta_M$ where e_E is the Weil pairing and ζ_M a fixed M-th root of unity. It is known that the field of definition of the Riemann surface $X_1(M,N)_{\mathbb{C}}$ is $\mathbb{Q}$ and the above coarse moduli problem gives a model over $\mathbb{Q}(\zeta_M)$ and denote by $X_1(M,N)$ this model corresponding to a curve defined over $\mathbb{Q}(\zeta_M)$.

Observe that $X_1(M,N)$ is birational, over some number field, to $X_\Delta(MN)$ with $\Delta = \{\pm 1, \pm(N+1), \pm(2N+1), \ldots, \pm((M-1)N+1)\}$.

Jeon and Kim in [**17**] listed all $X_1(M,N)$ which are bielliptic (obtaining a lot of the results by working in $X_\Delta(MN)$ with arguments already explained in this survey paper), and they determined in [**17**, Theorem 4.5] a list of $X_1(M,N)$ with an infinite set of quartic points over $\mathbb{Q}$, i.e. points of degree 4 over $\mathbb{Q}$ (and because the model of $X(M,N)$ is over $\mathbb{Q}(\zeta_M)$ in [**17**], the result on quartic points [**17**, Theorem 4.5] assume that the number fields of degree at most 4 over $\mathbb{Q}$ contain $\mathbb{Q}(\zeta_M)$). Because $X_1(N,N) = X(N) \times_{\mathbb{Q}} \mathbb{Q}(\zeta_N)$ and $X_1(1,N) = X_1(N)$ their results cover results in [**16**] and [**8**] explained in the survey paper.

Acknowledgements

I am very happy to thank Andreas Schweizer for all his comments and suggestions, in particular for letting me know Lemma 3.14 with its proof and also the Remark 3.15 that I reproduced here. I am pleased to thank Xavier Xarles for several discussions about the related theory, in particular discussions concerning the proof of Theorem 2.14 and pointing out to me the example in Lemma 2.18 that I also reproduced here. I am also very pleased to thank Joseph H. Silverman on a remark concerning the works [**1**] and [**13**], and Kumar Murty for several suggestions.

References

[1] Dan Abramovich and Joe Harris, *Abelian varieties and curves in* $W_d(C)$, Compositio Math. **78** (1991), no. 2, 227–238. MR1104789

[2] Robert D. M. Accola, *On lifting the hyperelliptic involution*, Proc. Amer. Math. Soc. **122** (1994), no. 2, 341–347, DOI 10.2307/2161022. MR1197530

[3] M. Akbas and D. Singerman, *The normalizer of* $\Gamma_0(N)$ *in* $\mathrm{PSL}(2,\mathbf{R})$, Glasgow Math. J. **32** (1990), no. 3, 317–327, DOI 10.1017/S001708950000940X. MR1073672

[4] A. O. L. Atkin and J. Lehner, *Hecke operators on* $\Gamma_0(m)$, Math. Ann. **185** (1970), 134–160, DOI 10.1007/BF01359701. MR0268123

[5] F. Bars, Determinació de les corbes $X_0(N)$ biel.líptiques (1997). You find a copy in http://ddd.uab.es/record/75137?ln=es

[6] Francesc Bars, *Bielliptic modular curves*, J. Number Theory **76** (1999), no. 1, 154–165, DOI 10.1006/jnth.1998.2343. MR1688168

[7] Francesc Bars, *The group structure of the normalizer of* $\Gamma_0(N)$ *after Atkin-Lehner*, Comm. Algebra **36** (2008), no. 6, 2160–2170, DOI 10.1080/00927870801949682. MR2418382

[8] Francesc Bars, Aristides Kontogeorgis, and Xavier Xarles, *Bielliptic and hyperelliptic modular curves* $X(N)$ *and the group* $\mathrm{Aut}(X(N))$, Acta Arith. **161** (2013), no. 3, 283–299, DOI 10.4064/aa161-3-6. MR3145452

[9] Cremona tables. See http://homepages.warwick.ac.uk/ masgaj/ftp/data/count.00000-09999.gz

[10] P. Deligne and M. Rapoport, *Les schémas de modules de courbes elliptiques* (French), Modular functions of one variable, II (Proc. Internat. Summer School, Univ. Antwerp, Antwerp, 1972), Springer, Berlin, 1973, pp. 143–316. Lecture Notes in Math., Vol. 349. MR0337993

[11] Noam D. Elkies, *The automorphism group of the modular curve* $X_0(63)$, Compositio Math. **74** (1990), no. 2, 203–208. MR1047740

[12] Gerd Faltings, *Diophantine approximation on abelian varieties*, Ann. of Math. (2) **133** (1991), no. 3, 549–576, DOI 10.2307/2944319. MR1109353

[13] Joe Harris and Joe Silverman, *Bielliptic curves and symmetric products*, Proc. Amer. Math. Soc. **112** (1991), no. 2, 347–356, DOI 10.2307/2048726. MR1055774

[14] M. Harrison, A new automorphism of $X_0(108)$, http://arxiv.org/abs/1108.5595, (2011).

[15] N. Ishii and F. Momose, *Hyperelliptic modular curves*, Tsukuba J. Math. **15** (1991), no. 2, 413–423, DOI 10.21099/tkbjm/1496161667. MR1138196

[16] Daeyeol Jeon and Chang Heon Kim, *Bielliptic modular curves* $X_1(N)$, Acta Arith. **112** (2004), no. 1, 75–86, DOI 10.4064/aa112-1-5. MR2040593

[17] Daeyeol Jeon and Chang Heon Kim, *Bielliptic modular curves* $X_1(M,N)$, Manuscripta Math. **118** (2005), no. 4, 455–466, DOI 10.1007/s00229-005-0595-9. MR2190107

[18] Daeyeol Jeon and Chang Heon Kim, *On the arithmetic of certain modular curves*, Acta Arith. **130** (2007), no. 2, 181–193, DOI 10.4064/aa130-2-7. MR2357655

[19] D. Jeon, C.H. Kim and A. Schweizer, Bielliptic intermediate modular curves. Work in progress (2013).

[20] Nicholas M. Katz and Barry Mazur, *Arithmetic moduli of elliptic curves*, Annals of Mathematics Studies, vol. 108, Princeton University Press, Princeton, NJ, 1985. MR772569

[21] M. A. Kenku and Fumiyuki Momose, *Automorphism groups of the modular curves* $X_0(N)$, Compositio Math. **65** (1988), no. 1, 51–80. MR930147

[22] M. A. Kenku and F. Momose, *Torsion points on elliptic curves defined over quadratic fields*, Nagoya Math. J. **109** (1988), 125–149. MR931956

[23] Chang Heon Kim and Ja Kyung Koo, *The normalizer of* $\Gamma_1(N)$ *in* $\mathrm{PSL}_2(\mathbf{R})$, Comm. Algebra **28** (2000), no. 11, 5303–5310, DOI 10.1080/00927870008827156. MR1785501

[24] P. G. Kluit, *On the normalizer of* $\Gamma_0(N)$, Modular functions of one variable, V (Proc. Second Internat. Conf., Univ. Bonn, Bonn, 1976), Springer, Berlin, 1977, pp. 239–246. Lecture Notes in Math., Vol. 601. MR0480340

[25] Mong-Lung Lang, *Normalizer of* $\Gamma_1(m)$, J. Number Theory **86** (2001), no. 1, 50–60, DOI 10.1006/jnth.2000.2555. MR1813529

[26] Mong Lung Lang, *Normalisers of subgroups of the modular group*, J. Algebra **248** (2002), no. 1, 202–218, DOI 10.1006/jabr.2001.9024. MR1879013

[27] B. Mazur and P. Swinnerton-Dyer, *Arithmetic of Weil curves*, Invent. Math. **25** (1974), 1–61, DOI 10.1007/BF01389997. MR0354674

[28] Jean-François Mestre, *Construction de courbes de genre* 2 *à partir de leurs modules* (French), Effective methods in algebraic geometry (Castiglioncello, 1990), Progr. Math., vol. 94, Birkhäuser Boston, Boston, MA, 1991, pp. 313–334. MR1106431

[29] Jean-François Mestre, *Corps euclidiens, unités exceptionnelles et courbes élliptiques* (French, with English summary), J. Number Theory **13** (1981), no. 2, 123–137, DOI 10.1016/0022-314X(81)90001-9. MR612679

[30] F. Momose, Automorphism groups of the modular curves $X_1(N)$. Preprint.

[31] Morris Newman, *Structure theorems for modular subgroups*, Duke Math. J. **22** (1955), 25–32. MR0067936

[32] Andrew P. Ogg, *Hyperelliptic modular curves*, Bull. Soc. Math. France **102** (1974), 449–462. MR0364259

[33] A. P. Ogg, *Rational points on certain elliptic modular curves*, Analytic number theory (Proc. Sympos. Pure Math., Vol XXIV, St. Louis Univ., St. Louis, Mo., 1972), Amer. Math. Soc., Providence, R.I., 1973, pp. 221–231. MR0337974

[34] Jean-Pierre Serre, *Cohomologie galoisienne* (French), With a contribution by Jean-Louis Verdier. Lecture Notes in Mathematics, No. 5. Troisième édition, vol. 1965, Springer-Verlag, Berlin-New York, 1965. MR0201444

[35] Andreas Schweizer, *Bielliptic Drinfeld modular curves*, Asian J. Math. **5** (2001), no. 4, 705–720, DOI 10.4310/AJM.2001.v5.n4.a6. MR1913817

[36] Goro Shimura, *Introduction to the arithmetic theory of automorphic functions*, Publications of the Mathematical Society of Japan, No. 11. Iwanami Shoten, Publishers, Tokyo; Princeton University Press, Princeton, N.J., 1971. Kanô Memorial Lectures, No. 1. MR0314766

[37] Joseph H. Silverman, *Advanced topics in the arithmetic of elliptic curves*, Graduate Texts in Mathematics, vol. 151, Springer-Verlag, New York, 1994. MR1312368

[38] STNB 1992, Corbes modulars: Taules. Notes del seminari de Teoria de Nombres, UB-UAB-UPC, Barcelona 1992.

[39] Yifan Yang, *Defining equations of modular curves*, Adv. Math. **204** (2006), no. 2, 481–508, DOI 10.1016/j.aim.2005.05.019. MR2249621

DEPART. MATEMÀTIQUES, UNIVERSITAT AUTÒNOMA DE BARCELONA, 08193 BELLATERRA (BARCELONA), CATALONIA.

Email address: francesc@mat.uab.cat

Contemporary Mathematics
Volume **701**, 2018
http://dx.doi.org/10.1090/conm/701/14150

p-adic point counting on singular superelliptic curves

Robert M. Burko

ABSTRACT. Motivated by applications to cryptography, mathematicians have employed p-adic cohomological methods to compute the zeta functions of various classes of varieties defined over finite fields of order $q = p^a$ in an amount of time polynomial in a, assuming the characteristic p is fixed. The varieties that are considered generally have smooth representations in either affine or projective space. Extending the methods of Gaudry and Gürel, we present a polynomial-time algorithm which computes the zeta function of singular superelliptic curves, and provide the results of an implementation in MAGMA. Assuming the curve has geometric genus g, and that the characteristic p is fixed, the running time of the algorithm is $O(a^{3+\varepsilon}g^{5+\varepsilon})$.

1. Introduction

In 2001, K. Kedlaya introduced a new approach to counting the number of points on a hyperelliptic curve in characteristic p [**Ked01**]. His approach was groundbreaking in that it hinged on deep aspects of algebraic geometry that had previously only appealed to theoretical mathematicians, while being a significant contribution to the field of cryptography. It also distinguished itself from other methods by not relying on the group operation of the curve, but working directly with the equation of the curve itself. Subsequently, papers emerged extending his method to a broader range of varieties, including superelliptic curves [**GG03**], $C_{a,b}$ curves [**DV06b**], and higher dimensional projective hypersurfaces [**AKR09**]. Other extensions allowed for the $p = 2$ case [**DV06a**], and dealt more effectively with larger characteristic [**GG03**][**Har07**].

Kedlaya showed in odd characteristic how to explicitly construct the Monsky-Washnitzer cohomology groups of a curve C', obtained by removing points from a hyperelliptic curve $C/\mathbb{F}_q$ given by a nonsingular affine planar equation $y^2 = f(x)$. One lifts the coordinate ring of C' to the integer ring of a p-adic field and "weakly completes" it with respect to the p-adic norm, obtaining what is called a "dagger algebra". This defines an object with two fibres, one of which is isomorphic to C' (the special fibre), and the other defined over a field of characteristic 0 and possessing analytic properties (the generic fibre). One can then compute the trace of Frobenius from the cohomology of the generic fibre to obtain the zeta function of the curve as well as the number of $\mathbb{F}_q$-rational points.

2010 *Mathematics Subject Classification.* Primary 14H20, 11G20; Secondary 14Q05, 14G10.

A natural step forward in the development of Kedlaya's algorithm is to adapt it to singular varieties. In one sense this is a special case of the initial algorithm that can only be applied to specific situations. For instance, a homogeneous polynomial $f \in K[X,Y,Z]$ defines a singular curve in $\mathbb{P}^2_{\mathbb{Q}}$ if and only if f and its partial derivatives vanish simultaneously at some point $(X',Y',Z') \neq (0,0,0)$. From another viewpoint, it is sometimes convenient to introduce singularities into a curve in order to obtain a more useful representation, e.g. the planar representation $y^r = f(x)$ of a superelliptic curve has a singularity at infinity if $|\deg(f) - r| > 1$. In the algorithm of Gaudry and Gürel, this is the only singularity that is permitted. In this paper we take a step towards adapting Kedlaya-type algorithms to allow for additional singularities, that is, we permit $f(x)$ to have multiple roots.

In section 2 we introduce the relevant p-adic notation, the construction of Monsky-Washnitzer cohomology, and comparison theorems. In section 3 we explicitly describe these constructions for singular superelliptic curves, the method for reducing algebraic differential to a basis of cohomology, as well as precision-loss estimates incurred due to the reduction procedure. In this section we also compare the chosen basis to a basis which acts on Frobenius-stable lattice over the integer ring. Section 4 describes how to compute the action of Frobenius on cohomology, giving bounds on the number of terms to be computed as well as the required p-adic working precision. In section 5, we summarize the algorithm and show the overall complexity to be $O(pa^{3+\varepsilon}g^{5+\varepsilon})$, where $q = p^a$ and g is the genus of the curve (we henceforth will treat p as a constant when giving complexity estimates). In the final section, we give results of an implementation of the algorithm, including a discussion about what happens when we apply the algorithm to a particular equation with slightly more relaxed assumptions.

This work is an exposition of a part of the author's PhD thesis [**Bur14**, Chapter 3]. We note that since the publication of the thesis, J. Tuitman [**Tui**] has successfully employed Lauder's fibration method [**Lau06**] to extend Kedlaya-type algorithms to a broader class of curves, including the cases described in this paper. The running time of Tuitman's algorithm is $O(pa^{3+\varepsilon}g^{6+\varepsilon})$, slightly slower than the present setting, however his implementation does seem to be more effective.

2. Review of p-adic theory

Let $p > 0$ be a prime, $p \neq 2$. Fix a positive integer a, set $q = p^a$ and let $k = \mathbb{F}_q$. Let K denote the unique unramified extension of degree a of the field of p-adic numbers, and let $\mathscr{V}$ denote its integer ring (i.e. the ring of Witt vectors of k).

Definition 2.1. For a $\mathscr{V}$-algebra A, we will denote by $\hat{A}$ its completion with respect to the ideal pA, that is, the universal object of the inverse system $\cdots \to A/p^2A \to A/pA$.

Remark 2.2. If A is finitely generated by elements $\{x_1, ..., x_n\}$, then an element of $\hat{A}$ can be written as power series in multi-index notation

$$\sum c_\alpha x^\alpha$$

with $c_\alpha \in \mathscr{V}$ satisfying $|c_\alpha|_p \to 0$ as $|\alpha| \to \infty$. Here $|\cdot|_p$ is the p-adic absolute value, $\alpha = (a_1, .., a_n)$ is a multi-index of nonnegative integers, $x^\alpha = x_1^{a_1} \cdots x_n^{a_n}$, and $|\alpha| = a_1 + \cdots + a_n$.

DEFINITION 2.3. Let A be a finitely generated $\mathscr{V}$-algebra, and let $\{x_1, ..., x_n\}$ be a set of generators. We define the *dagger algebra of* A, denoted $A^\dagger$, to be the subalgebra of $\hat{A}$ consisting of the *overconvergent power series*. That is, $A^\dagger$ consists of elements representable as

$$\sum c_\alpha x^\alpha$$

with $c_\alpha \in \mathscr{V}$, such that there exist real numbers C, r with $C > 0$, $0 < r < 1$, satisfying $|c_\alpha|_p \le Cr^{|\alpha|}$ for all α.

DEFINITION 2.4. Let X be a smooth affine variety over k with coordinate ring $\overline{A}$. By a theorem of Elkik [**Elk74**], there exists a $\mathscr{V}$-algebra A such that $A/pA \cong \overline{A}$ and $\mathfrak{X} = \operatorname{Spec}(A)$ is a smooth affine $\mathscr{V}$-scheme. The i-th cohomology group of the complex $\Omega^\bullet_{A^\dagger/\mathscr{V}}$ is denoted $\mathrm{H}^i_{\mathrm{MW}}(X/\mathscr{V})$. We define the *$i$-th Monsky-Washnitzer cohomology group of* X, denoted by either $\mathrm{H}^i_{\mathrm{MW}}(X/K)$ or $\mathrm{H}^i_{\mathrm{MW}}(\overline{A}/K)$, to be $\mathrm{H}^i_{\mathrm{MW}}(X/\mathscr{V}) \otimes_{\mathscr{V}} K$.

REMARK 2.5. Monsky and Washnitzer prove [**MW68**, Theorem 5.6] that the map sending X to $\mathrm{H}^i_{\mathrm{MW}}(X/K)$ is a contravariant functor. In particular, it is independent of the choices that were made (namely, the choice of the $\mathscr{V}$-algebra A and its generators). Additionally, σ can be extended to a σ-linear map on $A^\dagger$, which induces a σ-linear map F_p on $\mathrm{H}^i_{\mathrm{MW}}(X/K)$. More precisely, there is a p-power morphism

$$X \to X^\sigma \tag{2.1}$$

where X^σ is the variety obtained from X by applying the p-power map to the coefficients of its defining equation. That is, X^σ is equal to the fibred product of X with $\operatorname{Spec}(k)$, where $\operatorname{Spec}(k) \to \operatorname{Spec}(k)$ is induced by the p-power map $\overline{\sigma} : k \to k$. If $X = \operatorname{Spec}(\overline{A})$, then (2.1) is induced by a semilinear map $\overline{A} \otimes_k^{\overline{\sigma}^{-1}} k \to \overline{A}$ (here we use the superscript $\overline{\sigma}^{-1}$ to indicate that scalar multiplication is carried out as $c \cdot (x \otimes b) = x \otimes cb = \overline{\sigma}^{-1}(c)x \otimes b$ for $x \in \overline{A}$, $b, c \in k$). If A is a smooth lift of $\overline{A}$, then the Monsky-Washnitzer cohomology of X^σ is the de Rham cohomology of $A^\dagger \otimes_K^{\sigma^{-1}} K$, and there is a natural identification

$$\mathrm{H}^i_{\mathrm{MW}}(X^\sigma/K) \cong \mathrm{H}^i_{\mathrm{MW}}(X/K) \otimes_K^{\sigma^{-1}} K.$$

Therefore the map induced on cohomology by the Frobenius morphism in Equation (2.1) is a semilinear map

$$\mathrm{H}^i_{\mathrm{MW}}(X/K) \otimes_K^{\sigma^{-1}} K \xrightarrow{F_p} \mathrm{H}^i_{\mathrm{MW}}(X/K).$$

In the original theory of Monsky and Washnitzer, it was not assumed that the spaces $\mathrm{H}^i_{\mathrm{MW}}(X/K)$ are finite-dimensional, however it was later proven by Berthelot [**Ber97**] that this is indeed the case. Suppose then that M is the matrix of the map F_p with respect to some basis. The q-power Frobenius on $\mathrm{H}^i_{\mathrm{MW}}(X/K)$ is the composition

$$\mathrm{H}^i_{\mathrm{MW}}(X/K) \xrightarrow{F_p} \mathrm{H}^i_{\mathrm{MW}}(X/K) \otimes_K^{\sigma} K \xrightarrow{F_p} \mathrm{H}^i_{\mathrm{MW}}(X/K) \otimes_K^{\sigma^2} K \to \cdots \to \mathrm{H}^i_{\mathrm{MW}}(X/K)$$

whose matrix representation is the "norm matrix" $M^{\sigma^{a-1}} \cdots M^\sigma M$.

We now discuss the notions and relevant theorems of rigid cohomology. We do not give the technical background, as it is lengthy and unnecessary for putting the theory to practical use. The explicit constructions may be found in the work of Berthelot [**Ber96**].

DEFINITION 2.6. Let X be a smooth k-scheme. One can define a finite-dimensional K-vector space $\mathrm{H}^i_{\mathrm{rig}}(X/K)$, contravariantly functorial in X, called the *i-th rigid cohomology group* of X, with a σ-linear map induced by the Frobenius on k.

From Berthelot [**Ber97**, Proposition 1.10], we have the following comparison isomorphism.

PROPOSITION 2.7. Let X be a smooth, affine k-scheme. There is a functorial, Frobenius-equivariant isomorphism

$$\mathrm{H}^i_{\mathrm{rig}}(X/K) \cong \mathrm{H}^i_{\mathrm{MW}}(X/K)$$

Fundamental to all p-adic point counting algorithms is the ability to compute a p-adic cohomology group as the de Rham cohomology of some scheme, hence the comparison isomorphism below is essential.

THEOREM 2.8. [**BC94**, Cor 2.6] *Let $\mathfrak{X}$ be a smooth proper $\mathscr{V}$-scheme of finite type with geometrically connected fibres, and let $\mathfrak{Z}$ be a normal crossings divisor on $\mathfrak{X}$ relative to $\mathscr{V}$. Put $\mathfrak{U} = \mathfrak{X} \setminus \mathfrak{Z}$, and let U and $\mathfrak{U}_K$ denote the special and generic fibres of $\mathfrak{U}$. Then there is an isomorphism*

$$H^i_{rig}(U/K) \cong H^i_{dR}(\mathfrak{U}_K/K)$$

In particular, when $\mathfrak{Z}$ is empty the theorem reduces to an isomorphism

$$\mathrm{H}^i_{\mathrm{rig}}(X/K) \cong \mathrm{H}^i_{\mathrm{dR}}(\mathfrak{X}_K/K).$$

3. The cohomology of a superelliptic curve

3.1. Preliminaries on superelliptic curves. We define a *superelliptic curve* $C \to S$ of genus g to be a smooth projective morphism of schemes such that the fibre over each point $\eta \to S$ is connected, has dimension 1, genus g, and is birationally equivalent to an affine plane curve C^0 given by an equation

$$y^r = x^d + a_{d-1}x^{d-1} + \cdots + a_0 = \prod_{i=1}^{n}(x-\alpha_i)^{e_i} \in k(\eta)[x]$$

where r is a fixed prime and d is a positive integer such that $(r,d)=1$. We require the singular points of the planar equation to be rational over $k(\eta)$ ($\alpha_i \in k(\eta)$ if $e_i > 1$) and if $\mathrm{char}(k(\eta)) = p > 0$, we require $(p,r)=1$ and $(p,e_i)=1$ for each i.

If C is a superelliptic curve over a field K, then there is an automorphism ρ of order r on the geometric points of C, given by

$$\rho : (x,y) \mapsto (x, \zeta y)$$

where $\zeta \in \overline{K}$ is a primitive r^{th} root of unity. This leads to corresponding map in cohomology. In general, let A be a $\mathscr{V}$-module, suppose $\mathbb{Z}/r\mathbb{Z}$ acts on

$$A' := A \otimes_{\mathscr{V}} \mathscr{V}[\zeta],$$

and let ρ be a generator for this action. Set $\alpha = 1-\rho$ and $\beta = 1+\rho+\cdots+\rho^{r-1}$.

DEFINITION 3.1. We will denote by A^+ the kernel of $A \xrightarrow{\mathrm{id}\otimes 1} A' \xrightarrow{\alpha} A'$, and by A^- the kernel of $A \xrightarrow{\mathrm{id}\otimes 1} A' \xrightarrow{\beta} A'$.

REMARK 3.2. Suppose that A is generated over $\mathscr{V}$ by eigenvectors of ρ. These are also eigenvectors for β with corresponding eigenvalues 0 or r, so β acts on A. For any element $\omega \in A_\zeta$ write $r\omega = \beta\omega + (r-\beta)\omega$. One can then verify easily that $\beta\omega \in A^+$ and $(r-\beta)\omega \in A^-$. Thus there exists a decomposition $A = A^+ \oplus A^-$, since r is a unit in $\mathscr{V}$.

Suppose K is a perfect field and C is a superelliptic curve defined over K with equation $y^r = \prod_{i=1}^n (x-\alpha_i)^{e_i}$. If one writes $e_i = r\delta_i + \lambda_i$, then C is birationally equivalent to a superelliptic curve over K defined by the equation

$$y^r = \prod_{i=1}^n (x-\alpha_i)^{\lambda_i}$$

via the map

$$(a,b) \mapsto (a, b\prod_{i=1}^n (a-\alpha_i)^{\delta_i}).$$

We therefore assume hereafter that $1 \le e_i \le r-1$ for all i.

Suppose C_k is a superelliptic curve over k defined by an equation

$$y^r = \overline{f}(x) = \prod_{i=1}^n (x-\overline{\alpha}_i)^{e_i}.$$

Suppose $\overline{\alpha}_1, ..., \overline{\alpha}_m$ are the multiple roots of $\overline{f}$, and put

$$\overline{\tau}(x) = \prod_{i=m+1}^n (x-\alpha_i).$$

Let $\alpha_1, ..., \alpha_m$ be lifts of the multiple roots to $\mathscr{V}$ and let $\tau(x)$ be a polynomial obtained from $\overline{\tau}(x)$ by lifting its coefficients to $\mathscr{V}$.

PROPOSITION 3.3. The equation $y^r = \tau(x)\prod_{i=1}^m (x-\alpha_i)^{e_i}$ induces a superelliptic curve C over $\mathscr{V}$ of genus $g = \dfrac{(r-1)(n-1)}{2}$ and special fibre C_k.

PROOF. This can be seen by embedding the scheme defined by

$$y^r = \tau(x)\prod_{i=1}^m (x-\alpha_i)^{e_i}$$

in $\mathbb{P}^2_{\mathscr{V}}$ and blowing up the closed point at each α_i as well as the closed point at infinity. One finds that a finite succession of blowups resolves each of the singular points, and the coprimality conditions $(d,r)=1$ and $(e_i,r)=1$ imply that there is a unique point lying above each singularity. The genus is computed by noting that C minus $n+1$ points (the points above $y=0$ as well as the point above infinity) is an r-fold cover of the projective line minus $n+1$ points. One can then compute

$$\begin{aligned} \chi(C) - n - 1 &= r\chi(\mathbb{P}^1 - \{n+1 \text{ points}\}) \\ &= r(2-n-1) \end{aligned}$$

which gives $\chi(C) = 2 - (r-1)(n-1)$. Then use the Hurwitz genus formula

$$2 - 2g = \chi(C)$$

to obtain the expression for g. □

3.2. Reduction method. Let C'_k denote the open subvariety of C_k minus the points along $y = 0$, and let $\overline{\mathcal{A}}$ denote its coordinate ring. Then the ring

$$\mathcal{A} = \frac{\mathscr{V}[x, y, 1/y]}{(y^r - f(x))}$$

is a smooth lift of $\overline{\mathcal{A}}$, and can therefore be used to calculate Monsky-Washnitzer cohomology. By Proposition 2.7 and Theorem 2.8, to calculate the Monsky-Washnitzer cohomology of C'_k it is enough to calculate the de Rham cohomology of $\mathcal{A}_K := \mathcal{A} \otimes_{\mathscr{V}} K$. Let

$$0 \to \mathcal{A}_K \xrightarrow{d} \Omega^1_{\mathcal{A}_K} \to 0$$

denote the de Rham complex of $\mathcal{A}_K$. Using the relation $dy = \dfrac{f'(x)dx}{ry^{r-1}}$ one can write any element of $\Omega^1_{\mathcal{A}_K}$ as a finite sum of the form

$$\sum_{\substack{0 \leq i < d \\ -\infty < j < \infty}} a_{ij} x^i y^j dx.$$

Let $S_k = \{i : e_i \geq k\}$. We will assume the set $\{\alpha_i\}$ is ordered so that $S_2 = \{1, 2, ..., m\}$. Define the polynomial $h(x) \in \mathscr{V}[x]$ to be the greatest common divisor of $f(x)$ and $f'(x)$. Then

$$\begin{aligned} h(x) &= \prod_{i=1}^{n} (x - \alpha_i)^{e_i - 1} \\ &= \prod_{k=2}^{e} T_k(x) \end{aligned}$$

where we have defined $T_k(x) = \prod_{i \in S_k} (x - \alpha_i)$ and $e = \max\{e_i\}$.

REMARK 3.4. In the sections that follow we will regularly use the notation $K[x]_{<d}$ to denote the K-vector space of polynomials with degree strictly less than d.

Define polynomials $u(x) = \dfrac{f(x)}{h(x)} = T_1(x)$ and $v(x) = \dfrac{f'(x)}{h(x)}$. For $i \geq 0$, we can use division with remainder to find polynomials $a_i(x), b_i(x)$ such that

$$x^i v(x) = a_i(x) u(x) + b_i(x)$$

with $\deg(b_i(x)) < n$. Note that $a_0 = 0, b_0 = v(x)$, and the leading term of $a_i(x)$ is $d \cdot x^{i-1}$ for $i \geq 1$. Thus for any positive k, $\{a_i(x)\}_{i=1}^k$ is a basis for $K[x]_{<k}$, the K-module of polynomials of degree less than k. Define

$$\begin{aligned} a_{i,j}(x) &= \frac{j}{r} a_i(x) + i x^{i-1} \\ b_{i,j}(x) &= \frac{j}{r} b_i(x). \end{aligned}$$

Then the leading term of $a_{i,j}(x)$ is $\left(\frac{jd+ri}{r}\right)x^{i-1}$ which is nonzero as $(d,r)=1$ and $d \nmid i$. Therefore $\{a_{i,j}(x)\}_{i=1}^{k}$ is also a basis for $K[x]_{<k}$ for all integers j.

Consider now the K-vector space homomorphism

$$\hat{v} : K[x] \to \frac{K[x]}{(u(x))}$$

defined by multiplication by $v(x)$. Since $u(x)$ and $v(x)$ have no common roots, $v(x)$ is not a zero divisor in $K[x]/(u(x))$. It follows that a polynomial is in the kernel of $\hat{v}$ if and only if it is divisible by $u(x)$. Therefore $\hat{v}$ induces a K-automorphism of $K[x]/(u(x)) \cong K[x]_{<n}$. Since the image of x^i is $b_i(x)$, it follows that $\{b_i(x)\}_{i=0}^{n-1}$ is a K-basis for $K[x]_{<n}$.

Using this notation, we can write

$$\begin{aligned} d(x^i y^j) &= ix^{i-1}y^j\,dx + \frac{j}{r}f'(x)x^i y^{j-r}\,dx \\ &= ix^{i-1}y^j\,dx + \frac{j}{r}a_i(x)y^j\,dx + \frac{j}{r}b_i(x)h(x)y^{j-r}\,dx \\ &= a_{i,j}(x)y^j\,dx + b_{i,j}(x)h(x)y^{j-r}\,dx. \end{aligned} \tag{3.1}$$

By repeatedly taking appropriate linear combinations of the polynomials $a_{i,j}(x)$, for any differential $A(x)y^j dx$ with $j \geq 0$ and $\deg(A(x)) < d$ one can use Equation (3.1) to write

$$A(x)y^j\,dx = \frac{B(x)h(x)dx}{y^l} + d\nu \tag{3.2}$$

for some $\nu \in \mathcal{A}_K$, $B(x) \in K[x]$ with $\deg(B(x)) < n$, and $0 < l \leq r$. By the following proposition, we can also use (3.1) to increase the order in y of a differential.

PROPOSITION 3.5. *Let $A(x) \in K[x]$ be any polynomial of degree less than d and let $j \neq r$ be an integer. Then there exists a polynomial $B(x)$ of degree less than n, and an element $\nu \in \mathcal{A}_K$ such that*

$$\frac{A(x)dx}{y^j} = \frac{B(x)h(x)dx}{y^j} + d\nu$$

PROOF. Since $h(x) = T_2(x)T_3(x)\cdots T_e(x)$, it is sufficient to prove the following claim: For an integer k, $0 \leq k \leq e-2$, and any polynomial $A^{[k]}(x)$ such that

$$\deg(T_2(x)\cdots T_{k+1}(x)A^{[k]}(x)) < d,$$

there exists a polynomial $A^{[k+1]}(x)$ and an exact differential $d\nu$ such that

$$\deg(T_2(x)\cdots T_{k+2}(x)A^{[k+1]}(x)) < d$$

and

$$T_2(x)\cdots T_{k+1}(x)A(x)\frac{dx}{y^j} = T_2(x)\cdots T_{k+2}(x)B(x)\frac{dx}{y^j} + d\nu.$$

For $s = 1, ..., e-2$ define $f^{[t]}(x) = f(x)/(T_1\cdots T_{t+1})$, and for $i = 1, ..., m$ define $f_i(x) = f(x)/(x-\alpha_i)^{e_i}$. Using a partial fraction decomposition one can find constants A_i for each $i \in S_{k+2}$ and a polynomial $Q^{[k+1]}(x)$ of degree less than $\deg(f^{[k+1]}(x))$ such that

$$\frac{A^{[k]}(x)}{f^{[k]}(x)} = \frac{Q^{[k+1]}(x)}{f^{[k+1]}(x)} + \sum_{i \in S_{k+2}} \frac{A_i}{(x-\alpha_i)^{(e_i-k)}}. \tag{3.3}$$

Multiplying by $f(x)dx/y^j$ one can then write

$$\begin{aligned} T_2(x)\cdots T_{k+1}(x)A^{[k]}(x)\frac{dx}{y^j} &= T_2(x)\cdots T_{k+2}(x)Q^{[k+1]}(x)\frac{dx}{y^j} \\ &\quad + \sum_{i\in S_{k+2}} A_i(x-\alpha_i)^k f_i(x)\frac{dx}{y^j}. \end{aligned}$$

A straightforward calculation shows that

$$(x-\alpha_i)^k f_i(x)\frac{dx}{y^j} = c_i(x-\alpha_i)^{k+1} f_i'(x)\frac{dx}{y^j} + d\nu_i \tag{3.4}$$

where $c_i = \dfrac{(j-r)}{(r(k+1)-je_i)}$ and $\nu_i = \dfrac{c_i f(x)}{(1-\frac{j}{r})(x-\alpha_i)^{e_i-k-1}y^j}$.

Writing

$$(x-\alpha_i)^{k+1} f_i'(x) = T_2(x)\cdots T_{k+2}(x)R_i(x),$$

the result follows using $A^{[k+1]}(x) = Q^{[k+1]}(x) + \sum_{i\in S_{k+2}} A_i c_i R_i(x)$. □

3.3. A choice of basis. Denote by $\mathcal{B}^-$ the elements $\{x^i h(x)dx/y^l\}$ for $0 \le i \le n-2$, $0 < l < r$. Let $\mathcal{B}^+$ denote the elements $\{x^i dx/u(x)\}$ for $0 \le i \le n-1$.

COROLLARY 3.6. *The cohomology group $\mathrm{H}^1_{\mathrm{MW}}(\overline{\mathcal{A}}/K)$ is generated over K by the differentials $\mathcal{B}^- \cup \mathcal{B}^+$.*

PROOF. Use Proposition 3.5 and (3.1) along with (3.2) to write an element $\omega \in \Omega^1_{\mathcal{A}_K}$ as a linear combination of the elements $\{x^i h(x)dx/y^l\}$, $0 \le i \le n-1$, $0 < l \le r$. If $l \ne r$, make a further reduction using the relation

$$d(y^{r-l}) = \left(\frac{r-l}{l}\right)\frac{v(x)h(x)dx}{y^l}$$

so that $i \le n-2$. The result then follows either from Berthelot's comparison between de Rham and Monsky-Washnitzer cohomology, or Lemmas . □

Denote by Z_k the union of the point of C_k at infinity as well as the points along $y = 0$.

PROPOSITION 3.7. *There is a Frobenius-equivariant isomorphism of K-vector spaces*

$$\mathrm{H}^1_{\mathrm{rig}}(C_k/K) \cong \mathrm{H}^1_{\mathrm{MW}}(C_k'/K)^- \tag{3.5}$$

with basis represented by the elements of $\mathcal{B}^-$, as well as a short exact sequence

$$0 \to \mathrm{H}^1_{\mathrm{MW}}(C_k'/K)^+ \to \mathrm{H}^0_{MW}(Z_k)(-1) \to \mathrm{H}^2_{\mathrm{rig}}(C_k/K) \to 0 \tag{3.6}$$

with basis of the first space represented by the elements of $\mathcal{B}^+$.

PROOF. The Gysin sequence in rigid cohomology (see [**AKR09**, Proposition 2.4.1 b]) for the inclusion $C' \hookrightarrow C$ gives a long exact sequence which, after using p-adic comparison isomorphisms, includes

$$0 \to \mathrm{H}^1_{\mathrm{rig}}(C_k/K) \to \mathrm{H}^1_{\mathrm{MW}}(C_k'/K) \to \mathrm{H}^0_{MW}(Z_k)(-1) \to \mathrm{H}^2_{\mathrm{rig}}(C_k/K) \to 0.$$

Since ρ acts as the identity on Z_k, it follows that $\mathrm{H}^0_{MW}(Z_k/K) = 0$. Since formation of cohomology commutes with ρ we also have

$$\mathrm{H}^1_{\mathrm{rig}}(C_k/K)^+ \cong \mathrm{H}^1_{\mathrm{rig}}(C_k/K)/\langle\rho\rangle \cong \mathrm{H}^1_{\mathrm{rig}}(\mathbb{P}^1_k/K) = 0,$$

yielding an injection

$$\mathrm{H}^1_{\mathrm{rig}}(C_k/K) \hookrightarrow \mathrm{H}^1_{\mathrm{MW}}(C'_k/K)^-.$$

By comparison to de Rham cohomology one has that $\dim_K(\mathrm{H}^1_{\mathrm{rig}}(C_k/K)) = 2g$. Since $\mathcal{B}^-$ has $2g$ elements and generates $\mathrm{H}^1_{\mathrm{MW}}(C'_k/K)^-$, the first part of the proposition follows.

Sequence (3.6) is the part of the Gysin sequence fixed under ρ. Further dimensional comparison gives $\dim_K(\mathrm{H}^2_{\mathrm{rig}}(C_k/K)) = 1$, and $\dim(\mathrm{H}^0_{MW}(Z_k/K)) = n+1$, from which it follows that $\dim_K(\mathrm{H}^1_{\mathrm{MW}}(C'_k/K))^+$ is equal to n. Since $\mathcal{B}^+$ has n elements and generates the cohomology fixed by ρ, it must be a basis. □

The following lemmas are adaptations of Lemmas 2 and 3 from [**Ked01**], and give precision loss estimates from the reduction procedure. The proofs are similar but lengthy, so we do not include them here. For details see Lemmas 3.2.3 and 3.2.4 in [**Bur14**].

LEMMA 3.8. *Suppose* $\omega = A(x)\frac{dx}{y^j} \in \Omega^1_{\mathcal{A}_K}$, *where* $A(x)$ *is a polynomial of degree less than* d *with coefficients in* $\mathcal{V}$, *and* j *is a positive integer with* $j \equiv l \mod r$, $0 < l < r$. *Set* $N = p^{\lfloor \log_p |ej-r| \rfloor}$. *There exist* $\nu \in \mathcal{A}$ *and* $\widetilde{A}(x) \in K[x]_{<n-1}$ *such that*

$$\omega = \widetilde{A}(x)h(x)\frac{dx}{y^l} + d\nu \tag{3.7}$$

and $N\widetilde{A}(x)$ *has coefficients in* $\mathcal{V}$.

LEMMA 3.9. *Let* $\omega = A(x)y^j dx \in \Omega^1_{\mathcal{A}_K}$, *where* $A(x)$ *is a polynomial of degree* $d' < d$ *with coefficients in* $\mathcal{V}$, *and* j *is a positive integer with* $j \equiv -l \mod r$, $0 < l < r$. *Set* $N = p^{\lfloor \log_p (r(d'+1)+jd) \rfloor}$. *There exist* $\nu \in \mathcal{A}_l$ *and* $\widetilde{A}(x) \in K[x]_{<n-1}$ *such that*

$$\omega = \widetilde{A}(x)h(x)\frac{dx}{y^l} + d\nu \tag{3.8}$$

and $N\widetilde{A}(x)$ *has coefficients in* $\mathcal{V}$.

3.4. A Frobenius-stable lattice. From the theory of crystalline cohomology, $\mathrm{H}^1_{\mathrm{rig}}(C_k/K)$ has a Frobenius-stable $\mathcal{V}$-lattice $\mathrm{H}^1_{\mathrm{crys}}(C_k/\mathcal{V})$ isomorphic to $\mathrm{H}^1_{\mathrm{dR}}(C/\mathcal{V})$ such that the restriction map

$$\mathrm{H}^1_{\mathrm{dR}}(C/\mathcal{V}) \to \mathrm{H}^1_{\mathrm{dR}}(C'_K/K) \cong \mathrm{H}^1_{\mathrm{MW}}(C'_k/K)$$

is Frobenius-equivariant.

Let $\mathcal{D}$ denote a divisor of C corresponding to a $\mathcal{V}$-valued point. We have the following theorem due to Edixhoven:

THEOREM 3.10. *For any integer* $s \geq 2g$, *there exists a* $\mathcal{V}$*-module* Υ_s *killed by* $p^{\lfloor \log_p (s-1) \rfloor}$ *and a homomorphism*

$$\varphi_s : \Gamma(C, \Omega^1_{C/\mathcal{V}}(s\mathcal{D})) \to \Upsilon_s$$

such that

$$H^1_{dR}(C/\mathcal{V}) \cong \ker \varphi_s / (\operatorname{im} d_s \cap \ker \varphi_s).$$

Here d_s *represents the derivation*

$$d_s : \mathcal{O}_C((s-1)\mathcal{D}) \to \Omega^1_{C/V}(s\mathcal{D}).$$

PROOF. The details of the proof can be found in van den Bogaart [**vdB08**], Lemma 3.10 and the discussion preceding the lemma, replacing $2g$ by s. □

It follows that for $s \geq 2g$, there is a surjective map $\ker \varphi_s \to \mathrm{H}^1_{\mathrm{dR}}(C/\mathscr{V})$. In fact, $(\ker \varphi_s)^- \to \mathrm{H}^1_{\mathrm{dR}}(C/\mathscr{V})$ is a surjection by Remark 3.2 paired with the fact that r is a unit in $\mathscr{V}$. As a result, we have the following.

COROLLARY 3.11. *For $s \geq 2g$, the images of $(\ker \varphi_s)^-$ and $\mathrm{H}^1_{\mathrm{dR}}(C/\mathscr{V})$ in $\mathrm{H}^1_{\mathrm{dR}}(C'_K/K)$ are equal.*

The following proposition compares the crystalline lattice with our chosen basis. We use the notation $\Gamma(C, \Omega^1_{C/\mathscr{V}}(s\mathscr{D}))^l$ and $\Gamma(C', \Omega^1_{C/\mathscr{V}})^l$ to denote the submodules of $\Gamma(C, \Omega^1_{C/\mathscr{V}}(s\mathscr{D}))$ and $\Gamma(C', \Omega^1_{C/\mathscr{V}})$, respectively, consisting of differentials which can be written $\sum_{i \in -l + r\mathbb{Z}} A_i(x) y^i$.

PROPOSITION 3.12. *Fix an integer l with $0 < l < r$, let $s = (d-1)(r-1)$. Then*

i) *The $\mathscr{V}$-module $\Gamma(C, \Omega^1_{C/\mathscr{V}}(s\mathscr{D}))^l$ contains the $\mathscr{V}$-span of the set*

$$\left\{ \frac{x^i h(x) dx}{y^l} \right\}_{i=0}^{n-2},$$

and

ii) *the image of the restriction map*

$$\Gamma(C, \Omega^1_{C/\mathscr{V}}(s\mathscr{D}))^l \to \Gamma(C', \Omega^1_{C/\mathscr{V}})^l$$

is contained in the $\mathscr{V}$-span of the set

$$\left\{ \frac{x^i dx}{y^l} \right\}_{i=0}^{m_l}$$

where $m_l = \lfloor \frac{d(l+r-1)}{r} - 2 \rfloor$.

PROOF. We consider when an element $\omega = \dfrac{x^i h(x) dx}{y^l} \in \Gamma(C', \Omega^1_{C/\mathscr{V}})^l$ can be extended to all of C, with a pole of order at most s along $\mathrm{Supp}\mathscr{D}$.

Let P_i denote the generic point in C lying above $(\alpha_i, 0)$ and let P_∞ denote the generic point above infinity. Order calculations give $\mathrm{ord}_{P_i}(y) = e_i$, $\mathrm{ord}_{P_\infty}(y) = -d$, $\mathrm{ord}_{P_i}(x - \alpha_i) = r$, and $\mathrm{ord}_{P_\infty}(x) = -r$. An element $\omega = \dfrac{x^j h(x) dx}{y^l}$ can be extended to $\Omega^1_{C/\mathscr{V}}(s\mathscr{D})$ if $\mathrm{ord}_{P_i}\omega \geq 0$ for each i, and $\mathrm{ord}_{P_\infty}\omega \geq -s$.

For $0 \leq j \leq n-2$ we have

$$\begin{aligned}
\mathrm{ord}_{P_\infty}(\omega) &= (j + d - n) \cdot \mathrm{ord}_{P_\infty}(x) + \mathrm{ord}_{P_\infty}(dx) - l \mathrm{ord}_{P_\infty}(y) \\
&= -r(j + d - n) - r - 1 + ld \\
&\geq -r(n - 2 + d - n) - r - 1 + d \\
&= -r(d-2) - r + (d-1) \\
&= -(r-1)(d-1) = -s
\end{aligned}$$

and

$$\begin{aligned}\mathrm{ord}_{P_i}(\omega) &= \mathrm{ord}_{P_i}(x^j) + \mathrm{ord}_{P_i}(h(x)) + \mathrm{ord}_{P_i}(dx) - l \cdot \mathrm{ord}_{P_i}(y) \\ &\geq \mathrm{ord}_{P_i}(h(x)) + r - 1 - le_i \\ &= (e_i - 1)r + r - 1 - le_i \\ &= e_i(r - l) - 1 \geq 0\end{aligned}$$

which proves part (i). For part (ii), note that an element of $\Gamma(C', \Omega^1_{C/\mathscr{V}})$ can be written

$$\omega = \sum_{L_1 \leq j \leq L_2} A_j(x) y^j dx$$

where $A_j(x) \in \mathscr{V}[x]$ has degree less than $d - 1$. An order calculation gives

$$\mathrm{ord}_{P_i}(\omega) \leq \mathrm{ord}_{P_i}(A_{L_1}(x) y^{L_1} dx) = \mathrm{ord}_{P_i}(A_j(x))r + L_1 e_i + r - 1.$$

Since $\sum_{i=1}^n e_i = d$ and $\sum_{i=1}^n \mathrm{ord}_{P_i}(A_{L_1}(x)) < d$, there must be some i such that $e_i > \mathrm{ord}_{P_i}(A_j(x))$. In order for ω to extend to this P_i we must have

$$\begin{aligned}0 &\leq \mathrm{ord}_{P_i}(\omega) \\ &\leq (e_i - 1)r + L_1 e_i + r - 1 \\ &= e_i r + L_1 e_i - 1,\end{aligned}$$

which gives

$$L_1 \geq \frac{1 - e_i r}{e_i} > -r.$$

Now if ω extends to $\Gamma(C, \Omega^1_{C/\mathscr{V}}(s\mathscr{D}))$ we must also have $-s \leq \mathrm{ord}_{P_\infty}\omega$, so that

$$-(d-1)(r-1) \leq -L_2 d - \deg(A_{L_2}(x))r - r - 1.$$

In particular $L_2 < r$. For l satisfying $0 < l < r$, let ω_l denote the term in ω whose y-exponent is equal to $-l$ mod r. Then we have

$$\begin{aligned}\omega_{r-l} &= \frac{A_{-l}(x)dx}{y^l} + A_{r-l}(x) y^{r-l} dx \\ &= \frac{B_l(x)dx}{y^l}\end{aligned}$$

For some polynomial $B_l(x) \in \mathscr{V}[x]$ whose degree satisfies

$$-(d-1)(r-1) \leq ld - \deg(B_l(x))r - r - 1,$$

that is,

$$\deg(B_l(x)) \leq \frac{(d-1)(r-1) + ld - r - 1}{r} = \frac{d(r+l-1)}{r} - 2.$$

□

COROLLARY 3.13. Let V denote the $\mathscr{V}$-span of of the differentials

$$\left\{\frac{x^i h(x) dx}{y^l}\right\}_{\substack{0 \leq i \leq n-2 \\ 0 < l < r}}$$

in $\mathrm{H}^1_{\mathrm{dR}}(C'_K/K)$ and let W denote the image of $\mathrm{H}^1_{\mathrm{dR}}(C/\mathscr{V})$ in $\mathrm{H}^1_{\mathrm{dR}}(C'_K/K)$. Set $e = \max\{e_i\}$, $N = \lfloor \log_p((d-1)(r-1) - 1) \rfloor$. Then

$$p^N W \subset V \subset p^{-N} W.$$

PROOF. By Theorem 3.10 and Proposition 3.12(i), the differentials

$$p^N \frac{x^i h(x)dx}{y^l}$$

for $0 \le i \le n-2$, $0 < l < r$, belong to $\ker \varphi_s$, where $s = (d-1)(r-1)$. Therefore, by Corollary 3.11, $V \subset p^{-\lfloor \log_p((d-1)(r-1)-1) \rfloor} W$. By Proposition 3.12($ii$), $\ker \varphi_s$ is contained in the $\mathscr{V}$-span of $\dfrac{x^i dx}{y^l}$, for $0 < l < r$, $0 \le i \le \lfloor \frac{d(l+r-1)}{r} - 2 \rfloor$. Fix some $0 < l < r$, and suppose $A(x)$ is a polynomial of degree less than $\lfloor \frac{d(l+r-1)}{r} - 2 \rfloor$ with coefficients in $\mathscr{V}$. Using division with remainder we can write

$$\frac{A(x)dx}{y^l} = \frac{B_1(x)dx}{y^l} + B_2(x)y^{r-l}dx$$

with $B_1(x), B_2(x) \in \mathscr{V}[x]$ where $B_1(x)$ is a polynomial of degree less than d. If $\dfrac{d(l-1)}{r} - 2 \ge 0$, then $B_2(x)$ is a polynomial with degree less than or equal to $\dfrac{d(l-1)}{r} - 2$. Otherwise $B_2(x) = 0$.

By Lemma 3.8 (ii) and Lemma 3.9 (ii), and using the computation

$$r\left(\frac{d(l-1)}{r} - 2 + 1\right) + (r-l)d = (d-1)(r-1) - 1,$$

it follows that there exist exact differentials $d\nu_1, d\nu_2$, and polynomials $\widetilde{B}_1(x), \widetilde{B}_2(x) \in \mathscr{V}[x]$ of degree less than $n-1$ such that

$$\begin{aligned} p^{\lfloor \log_p((d-1)(r-1)-1) \rfloor} \frac{B_2(x)dx}{y^l} &= \frac{\widetilde{B}_2(x)h(x)dx}{y^l} + d\nu_1, \text{ and} \\ p^{\lfloor \log_p |el-r| \rfloor} \frac{B_1(x)dx}{y^l} &= \frac{\widetilde{B}_1(x)h(x)dx}{y^l} + d\nu_2. \end{aligned}$$

Since $e \le d$ and $l \le r-1$, it follows that $|el - r| \le (d-1)(r-1) - 1$, hence the image of $p^N \dfrac{A(x)dx}{y^l}$ lies in V. □

4. The Matrix of Frobenius

Following Gaudry and Gürel [**GG01**], we can compute a representative for the action of the semi-linear p-power Frobenius automorphism

$$F : H^1_{MW}(C'_k/K)^- \to H^1_{MW}(C'_k/K)^-$$

on our selected basis $\mathcal{B}^-$ from the previous section. Let $\sigma : \mathscr{V} \to \mathscr{V}$ denote the Witt vector Frobenius automorphism. We can extend it to polynomials by taking $x^\sigma = x^p$, and to differentials by taking $(dx)^\sigma = px^{p-1}dx$. Setting

$$E(x) = \frac{1}{p}(f(x)^\sigma - f(x)^p)$$

we can now extend σ to $\mathcal{A}^\dagger$ by taking

$$\left(\frac{1}{y^l}\right)^\sigma = y^{-lp}\left(1 + \frac{pE(x)}{y^{rp}}\right)^{-l/r} = \sum_{k=0}^{\infty} \binom{-l/r}{k} \frac{p^k E(x)^k}{y^{pl+prk}}$$

where

$$\binom{-l/r}{k} = \frac{(-l/r)(-l/r-1)\cdots(-l/r-k+1)}{k!}$$

which is an element of $\mathbb{Z}_p$ since $(r,p)=1$. Finally we have

$$\left(\frac{x^i h(x)dx}{y^l}\right)^\sigma = \sum_{k=0}^{\infty}\binom{-l/r}{k}\frac{p^{k+1}x^{pi+p-1}h(x)^\sigma E^k(x)}{y^{pl+prk}}dx. \tag{4.1}$$

For each i and l such that $0 \le i \le n-2$ and $0 < l < r$, the reduction algorithm is applied to a truncation of the above sum, which yields a $(r-1)(n-1) \times (r-1)(n-1)$ matrix M_p, an approximation of F with respect to the chosen basis. By semilinearity, the matrix M_q for the q-power Frobenius F^a is calculated as $M_p^{\sigma^{a-1}} M_p^{\sigma^{a-2}} \cdots M_p$.

THEOREM 4.1. *Let $N = \lfloor \log_p((d-1)(r-1)-1)\rfloor$. Then*

i) *$p^{2N} M_p$ has entries in $\mathscr{V}$, and*

ii) *if $\widetilde{M}$ is an approximation of the matrix M_p, correct to precision $p^{N'}$ for some integer $N' \ge 4N$, then the matrix $\widetilde{M}^{\sigma^{a-1}}\widetilde{M}^{\sigma^{a-2}}\cdots\widetilde{M}$ is an approximation of M_q, correct to precision $p^{N'-4N}$.*

PROOF. Let H denote a $\mathscr{V}$-basis for $\mathrm{H}^1_{\mathrm{dR}}(C/\mathscr{V})$, viewed as a $\mathscr{V}$-lattice in $\mathrm{H}^1_{\mathrm{MW}}(C_k/K)^-$, and let Q denote the change of basis matrix from H to $\mathcal{B}$. Then $M_p = QAQ^{-1}$ and $M_q = QA^{\sigma^{a-1}}\cdots A^\sigma A Q^{-1}$ for some matrix A with entries in $\mathscr{V}$. By Corollary 3.13, Q and Q^{-1} have entries in $p^{-N}\mathscr{V}$, so part (i) follows immediately.

For part (ii) note that the matrix $\widetilde{A} := Q^{-1}\widetilde{M}Q$ is an approximation of A, with

$$\widetilde{A} \equiv A \mod p^{N'-2N}.$$

The entries of $\widetilde{A}$ are therefore in $\mathscr{V}$, hence

$$Q\widetilde{A}^{\sigma^{a-1}}\cdots\widetilde{A}^\sigma\widetilde{A}Q^{-1} \equiv QA^{\sigma^{a-1}}\cdots A^\sigma AQ^{-1} \mod p^{N'-4N}.$$

The right side of this equation is the matrix M_q, therefore one can write

$$\begin{aligned}\widetilde{M}^{\sigma^{a-1}}\cdots\widetilde{M} &= (Q\widetilde{A}^{\sigma^{a-1}}Q^{-1})(Q\widetilde{A}^{\sigma^{a-2}}Q^{-1})\cdots(Q\widetilde{A}Q^{-1})\\ &= Q\widetilde{A}^{\sigma^{a-1}}\widetilde{A}^{\sigma^{a-2}}\cdots\widetilde{A}Q^{-1}\\ &\equiv M_q \mod p^{N'-4N},\end{aligned}$$

which completes the theorem. □

4.1. Precision loss estimates. By the Weil conjectures [**Har77**, Appx C], the numerator of the zeta function of C_k can be written

$$P(T) = \prod_{i=1}^{2g}(1-\mu_i T) = \sum_{i=0}^{2g} a_i T^i \in \mathbb{Z}[T]$$

where $|\mu_i| = q^{1/2}$ for all i and g is the genus of C_k. Furthermore, we can order the μ_i's such that $\mu_{g+i} = q/\mu_i$ for $i = 1,..,g$.

Let $\mathcal{I}_k$ denote the subsets of $\{1, ..., 2g\}$ of length k. For $I = (j_1, ..., j_k) \in \mathcal{I}_k$, write $\mu_I = \mu_{j_1} \cdots \mu_{j_k}$. Then for each $1 \leq i \leq 2g$, we can write

$$\begin{aligned} a_i &= (-1)^i \sum_{I \in \mathcal{I}_i} \mu_I \\ &= (-1)^i \sum_{I \in \mathcal{I}_i} \frac{q^g}{\mu_{\{1,...,2g\} \setminus I}} \\ &= (-1)^i q^g \sum_{I \in \mathcal{I}_{2g-i}} \frac{1}{\mu_I} \\ &= (-1)^i q^g \sum_{I \in \mathcal{I}_{2g-i}} \frac{1}{q^{2g-i}} \mu_I \\ &= q^{i-g} a_{2g-i}. \end{aligned} \tag{4.2}$$

Therefore the polynomial $P(T)$ can be obtained by computing each a_i for $1 \leq i \leq g$. For each a_i, we have

$$|a_i| \leq \binom{2g}{i} q^{i/2} \leq \binom{2g}{g} q^{i/2},$$

which gives $|a_i| \leq \binom{2g}{g} q^{g/2}$ for $1 \leq i \leq g$.

We also have that

$$P(T) = \det(1 - qF^{-1}T | \mathrm{H}^1_{\mathrm{MW}}(C'_k/K)^-)$$

so if we define $Q(T) = T^{2g}P(1/T)$, then $Q(T)$ is the characteristic polynomial of qF^{-1} with roots $\mu_1, ..., \mu_{2g}$. In fact, $Q(T)$ is the characteristic polynomial of F, since the sets $\{\mu_i\}$ and $\{q/\mu_i\}$ are equal.

It follows that our calculations for the matrix of F should be correct modulo p^{N_1}, where $N_1 := \lceil \log_p \left(2\binom{2g}{g} q^{g/2} \right) \rceil$. In order to ensure that no p-adic denominators arise in our calculation, we can appeal to Theorem 4.1, which gives that the denominators of the p-power Frobenius with respect to the basis $\mathcal{B}^-$ are cleared upon multiplication by p^{N_2}, where $N_2 := 2\lfloor \log_p((d-1)(r-1)-1) \rfloor$. To ensure the correct precision of the q-power Frobenius, by the same theorem, we must increase the precision by p^{2N_2}. Therefore we must compute the matrix of $p^{N_2}F$ with accuracy up to $p^{N_1+3N_2-1}$.

By Equation (4.1), the k-th term in the expansion of $F(p^{N_2} \dfrac{x^i h(x) dx}{y^l})$ is divisible by p^{k+N_2}, and by Lemma 3.8 the coefficients of its reduction are divisible by $p^{k+N_2-\lfloor \log_p(ep(l+r(k-1))-r) \rfloor}$ where $e = \max\{e_i\}$. Therefore it is sufficient to calculate the first N_3 terms with

$$N_3 + N_2 - \lfloor \log_p(ep(l + r(N_3 - 1)) - r) \rfloor \geq N_1 + 3N_2 - 1$$

and all calculations should be performed in $\mathcal{V}/p^{N_3+N_2}\mathcal{V}$.

5. The Algorithm

For p an odd prime, set $q = p^a$, $k = \mathbb{F}_q$, and $\mathscr{V} = W(k)$. Let $\overline{f}(x) \in k[x]$ be a polynomial of degree d that has n distinct roots, m of which are k-rational multiple roots, such that the multiplicity of any root is not divisible by p. Denote by e the greatest multiplicity of any root of $\overline{f}$. Let $r > e$ be a prime different from p which does not divide d, and set $g = (n-1)(r-1)/2$. The following algorithm outlines the procedure for computing the zeta function of the k-curve defined by the equation $y^r = \overline{f}(x)$.

Step 1. Using the procedure of section 3.1, lift $\overline{f}$ to a polynomial $f \in \mathscr{V}[x]$ such that $y^r = f(x)$ defines a superelliptic curve over $\mathscr{V}$.

Step 2. Letting e denote the greatest multiplicity of any linear factor of f, compute integers $N_1 = \lceil \log_p(2\binom{2g}{g}q^{g/2})\rceil$, $N_2 = 2\lfloor \log_p((d-1)(r-1)-1)\rfloor$, and N_3, where N_3 is the smallest integer such that $N_3 - \lfloor \log_p(ep(l + r(N_3-1)) - r)\rfloor \geq N_1 + 2N_2 - 1$.

Step 3. Performing computations in $\mathscr{V}/p^{N_3+N_2}\mathscr{V}$, compute the first N_3 terms of $p^{N_2}F\left(\frac{x^i h(x)}{y^l}\right)$ for $0 < l < r$, $0 \leq i \leq n-2$. One effective way to do this, following [**GG01**] is to set $\tau = 1/y^r$, $E = \frac{1}{p}(f(x)^\sigma - f(x)^p)$, and $S = 1 + pE\tau^p$, so that

$$F\left(\frac{x^i h(x)}{y^l}\right) = \tau^{(pl \text{ div } r)} p x^{pi+p-1} h(x)^\sigma S^{-l/r} \frac{dx}{y^{pl}}.$$

One can use Newton's method to quickly compute $X = S^{-1/r}$ as a root of the function $F(X) = 1 - \frac{1}{SX^r}$: set $X_0 = 1$ for the initial value, and let

$$X_{i+1} = X_i - F(X_i)/F'(X_i) = \frac{1}{r}((r+1)X_i - SX_i^{r+1}).$$

After each iteration one should use the relation $f = \tau^{-1}$ to rewrite the resulting expression so that the degree in x is less than d, and reduce modulo τ^{pN_3}.

Step 4. For $i \in S_2, k = 0, 1, ..., e_i - 2$, precompute the polynomials $f^{[k]}(x)/(x - \alpha_i)^{e_i - k}$ and $R_{i,k+1}(x)$. For $i, k = 0, ..., d-n-1$ compute the polynomials $b_i(x), a_{i+1}(x)$ from section 3.2 and constants $\lambda_{i,k}$ such that $\sum_{i=0}^{d-n-1} \lambda_{i,k} b_i(x) = x^k$. Use the stored data and the processes of section 3.2 to reduce the degree in τ of the polynomials found in Step 3, until one can write it as a linear combination of differentials of the form $\frac{x^i h(x)}{y^l}$. Let M' denote the resulting matrix, and let $M = \frac{1}{p^{N_2}}M'$ (with entries viewed in $\frac{1}{p^{N_2}}\mathscr{V}$ up to precision $p^{N_1+2N_2}$).

Step 5. Compute the norm matrix $|M| = M^{\sigma^{a-1}} M^{\sigma^{a-2}} \cdots M$, and $\det(|M| - T)$. Compute integers in $[-\frac{1}{2}p^{N_1}, \frac{1}{2}p^{N_1}]$ congruent to the integers of the resulting polynomial modulo p^{N_1}.

5.1. Complexity analysis. Assume p is fixed. Assume also that multiplication of two objects of bit size N can be performed in time $O(N^{1+\varepsilon})$. Let $R = \mathscr{V}/p^{N_3+N_2}\mathscr{V}$. Then any element of R can be stored using $a(N_3 + N_2) = O(a^{2+\varepsilon}g^{1+\varepsilon})$ bits. Note that $N_3 = O(a^{1+\varepsilon}g^{1+\varepsilon})$, and $d \leq nr = O(g)$.

Step 1 and 2 can clearly be performed in $O(1)$ time and space. For Step 3, one can compute S in $O(a)$ ring operations. The expression we need to represent $S^{-j/r}$ is a polynomial over R of degree less than d in x and of degree less than pN_3 in τ, and thus requires $O(dpN_3a^{2+\varepsilon}g^{1+\varepsilon}) = O(a^{3+\varepsilon}g^{3+\varepsilon})$ bits of storage. Only $O(1)$ operations are required to compute $S^{-1/r}$, therefore the computation of $S^{-j/r}$ for $j = 1, ..., r-1$ is $O(ra^{3+\varepsilon}g^{3+\varepsilon})$ in runtime and space.

The precomputations for Step 4 can be done in comparable time to the rest of the step, and requires storing $O(d-n)$ polynomials of degree less than d over R, as well as another $(d-n)^2$ elements of R for a total of $O(d(d-n)a^{2+\varepsilon}g^{1+\varepsilon})$ bits. One uses the reduction process of section 3.2 for each of the $2g$ polynomials found in the previous step. For $j = pN_3, pN_3 - 1, ..., 1$, we repeat the following process on the polynomial associated with τ^j, which is of degree less than d over the ring R. From the proof of Proposition 3.5, one computes $A^{[k+1]}(x)$ from $A^{[k]}(x)$ using $O(|S_{k+2}|d)$ ring operations to find the values $A_{i,k+1}$ for $i \in S_{k+2}$ as well finding the quotient $Q^{[k+1]}(x)$ using polynomial long division. Repeating this for $k = 0, ..., e-2$ gives a total of $O((d-n)d)$ ring operations until one obtains a polynomial $B(x)$ of degree less than $d-n$. Another $O((d-n)^2)$ ring operations are required to compute $B(x)$ as linear combination of the polynomials $b_{i,r-j}(x), i = 0, ..., d-n-1$. In total, this step is $O(a^{3+\varepsilon}g^{5+\varepsilon})$ in complexity.

The final step in computing the norm matrix $|M|$ and its characteristic polynomial adds a time cost of $O(a^{3+\varepsilon}g^{4+\varepsilon})$ (see the resource analysis of Kedlaya [**Ked01**]). Overall, the complexity of this algorithm is $O(a^{3+\varepsilon}g^{5+\varepsilon})$.

5.2. Examples. We begin with an example of a genus 5 curve defined over $\mathbb{F}_{5^{10}}$ by the equation

$$y^3 = x^2(x^5 + x^4 + x + t)$$

where t generates $\mathbb{F}_{5^{10}}$ over $\mathbb{F}_5$ and has minimal polynomial $z^{10} + 3z^5 + 3z^4 + 2z^3 + 4z^2 + z + 2$. In this case we set $N_1 = 29$, $N_2 = 2$, and $N_3 = 36$. The highest working precision we will need is 5^{38}, and the zeta function should be calculated modulo 5^{29}. After 3 minutes and 18 seconds, and using 80MB of space, the algorithm revealed the first six coefficients of the numerator of the zeta function of the normalized curve to be $1 - 1253T + 10171416T^2 + 10359663716T^3 + 177276031807004T^4 - 154385140679896875T^5$, and one can then compute the size of its Jacobian as $N = 888064554752583506264372336514315O9$.

As a second example, we compute the zeta function of a genus 6 curve over $\mathbb{F}_{3^{20}}$ with planar equation

$$y^5 = x^4(x-1)^4(x-t)^4(x-t^2)^4$$

where $\mathbb{F}_{3^{20}}$ generated over $\mathbb{F}_3$ by t with minimal polynomial $z^{20} + 2z^{13} + z^{11} + z^{10} + z^9 + z^8 + 2z^5 + 2z^4 + 2z^3 + z + 2$. Here $N_1 = 67$, $N_2 = 6$ and $N_3 = 85$. The program outputted

$$\begin{aligned} &1 - 222102T + 26390120031T^2 - 2538440313760890T^3 \\ &+ 216995406474678950790T^4 - 15512966168826218109491232T^5 \\ &+ 960365390614818936440153417604T^6 \end{aligned}$$

giving a total of

$$N = 1796895837436906734331716960381629479650658407297220522000$$

points on the Jacobian. The running time was 22 minutes and the memory usage for this example was approximately 1 GB.

Consider now a curve defined over $\mathbb{F}_7$ with affine equation given by

$$y^3 = x^2(x^4 - x - 1).$$

This does not define a superelliptic curve by the definition of section 3.1, since $(d, r) = 3$. We apply the same algorithm with $N_1 = 5$, $N_2 = 2$, $N_3 = 11$, and compute with precision 7^{13} the reduction of the first 11 terms of the Frobenius multiplied by 7^2. At the final step we divide by 7^2, and take the result modulo 7^5. The processing time was 1.82 seconds with 33 MB of memory usage, and the program outputted the polynomial

$$(1 - T)^2(1 + T + 7T^2)(1 + 2T - 3T^2 + 14T^3 + 49T^4).$$

This result can be interpreted as follows. The normal model is a curve C of genus 3, with three points at infinity which are permuted by the automorphism

$$\rho : (x, y) \mapsto (x, \zeta y),$$

where ζ is a primitive cube root of unity. If we let Σ denote the points at infinity, then the Gysin sequence in rigid cohomology gives

$$0 \to \mathrm{H}^1_{\mathrm{rig}}(C/K) \to \mathrm{H}^1_{\mathrm{rig}}(C'/K)^- \to \mathrm{H}^0_{\mathrm{rig}}(\Sigma/K)^- \to \cdots$$

where C' denotes the affine curve minus the points along $y = 0$. In this case $\mathrm{H}^0_{\mathrm{rig}}(\Sigma/K) = K[w]/(w^3 - 1)$ is non-trivial (cf. Proposition 3.5), with ρ given as the K-linear map $\rho(w) = \zeta w$. This decomposes into eigenspaces $K \oplus Kw \oplus Kw^2$, with the Frobenius acting trivially on each, and $\mathrm{H}^0_{\mathrm{rig}}(\Sigma/K)^- = Kw \oplus Kw^2$. We conclude that $(1 - T)^2$ is the characteristic polynomial of the Frobenius acting on $\mathrm{H}^0_{\mathrm{rig}}(\Sigma/K)^-$, and that the zeta function of C is

$$\frac{(1 + T + 7T^2)(1 + 2T - 3T^2 + 14T^3 + 49T^4)}{(1 - T)(1 - 7T)}.$$

References

[AKR09] Timothy G. Abbott, Kiran S. Kedlaya, and David Roe, *Bounding Picard numbers of surfaces using p-adic cohomology* (English, with English and French summaries), Arithmetics, geometry, and coding theory (AGCT 2005), Sémin. Congr., vol. 21, Soc. Math. France, Paris, 2010, pp. 125–159. MR2856564

[BC94] Francesco Baldassarri and Bruno Chiarellotto, *Algebraic versus rigid cohomology with logarithmic coefficients*, Barsotti Symposium in Algebraic Geometry (Abano Terme, 1991), Perspect. Math., vol. 15, Academic Press, San Diego, CA, 1994, pp. 11–50. MR1307391

[Ber96] Pierre Berthelot, *Cohomologie rigide et cohomologie rigide à supports propres, Première partie*, Prépublication, Université de Rennes (1996).

[Ber97] Pierre Berthelot, *Finitude et pureté cohomologique en cohomologie rigide* (French), Invent. Math. **128** (1997), no. 2, 329–377, DOI 10.1007/s002220050143. With an appendix in English by Aise Johan de Jong. MR1440308

[Bur14] Robert M. Burko, *Computing the zeta functions of two classes of singular curves*, ProQuest LLC, Ann Arbor, MI, 2014. Thesis (Ph.D.)–University of Toronto (Canada). MR3260071

[DV06a] Jan Denef and Frederik Vercauteren, *An extension of Kedlaya's algorithm to hyperelliptic curves in characteristic 2*, J. Cryptology **19** (2006), no. 1, 1–25, DOI 10.1007/s00145-004-0231-y. MR2210897

[DV06b] Jan Denef and Frederik Vercauteren, *Counting points on C_{ab} curves using Monsky-Washnitzer cohomology*, Finite Fields Appl. **12** (2006), no. 1, 78–102, DOI 10.1016/j.ffa.2005.01.003. MR2190188

[Elk74] Renée Elkik, *Solutions d'équations à coefficients dans un anneau hensélien* (French), Ann. Sci. École Norm. Sup. (4) **6** (1973), 553–603 (1974). MR0345966

[GG01] Pierrick Gaudry and Nicolas Gürel, *An extension of Kedlaya's point-counting algorithm to superelliptic curves*, Advances in cryptology—ASIACRYPT 2001 (Gold Coast), Lecture Notes in Comput. Sci., vol. 2248, Springer, Berlin, 2001, pp. 480–494, DOI 10.1007/3-540-45682-1_28. MR1934859

[GG03] Pierrick Gaudry and Nicolas Gürel, *Counting points in medium characteristic using Kedlaya's algorithm*, Experiment. Math. **12** (2003), no. 4, 395–402. MR2043990

[Har77] Robin Hartshorne, *Algebraic geometry*, Springer-Verlag, New York-Heidelberg, 1977. Graduate Texts in Mathematics, No. 52. MR0463157

[Har07] David Harvey, *Kedlaya's algorithm in larger characteristic*, Int. Math. Res. Not. IMRN **22** (2007), Art. ID rnm095, 29, DOI 10.1093/imrn/rnm095. MR2376210

[Ked01] Kiran S. Kedlaya, *Counting points on hyperelliptic curves using Monsky-Washnitzer cohomology*, J. Ramanujan Math. Soc. **16** (2001), no. 4, 323–338. MR1877805

[Lau06] Alan G. B. Lauder, *A recursive method for computing zeta functions of varieties*, LMS J. Comput. Math. **9** (2006), 222–269, DOI 10.1112/S1461157000001261. MR2261044

[MW68] P. Monsky and G. Washnitzer, *Formal cohomology. I*, Ann. of Math. (2) **88** (1968), 181–217, DOI 10.2307/1970571. MR0248141

[Tui] Jan Tuitman, *Counting points on curves using a map to $\mathbf{P}^1$*, Math. Comp. **85** (2016), no. 298, 961–981, DOI 10.1090/mcom/2996. MR3434890

[vdB08] Theo van den Bogaart, *About the choice of basis in Kedlaya's algorithm*, ArXiv e-prints (2008).

Department of Mathematics, University of Toronto, Toronto, Ontario

Contemporary Mathematics
Volume **701**, 2018
http://dx.doi.org/10.1090/conm/701/14145

A refinement of a conjecture of Gross, Kohnen, and Zagier

Carlos Castaño-Bernard

To the memory of Fumiyuki Momose

ABSTRACT. Consider a Heegner point $P \in E(\mathbb{Q})$ of minimal height on an elliptic curve $E/\mathbb{Q}$ of analytic rank one. The purpose of this note is to describe and motivate a conjecture on the 2-part $\mathcal{I}_2$ of the index $\mathcal{I}$ of the subgroup generated by P on $E(\mathbb{Q})$, under the assumption that the conductor N of E is prime. It predicts that if $|\text{Ш}(E)_2| = 1$ then $\mathcal{I}_2$ may be completely determined by local, effectively computable data at $p = \infty$.

1. Introduction

Let E be an elliptic curve defined over $\mathbb{Q}$ such that its L-function $L(E/\mathbb{Q}, s)$ has a simple zero at $s = 1$. By the work of Gross-Zagier [**11**] we know that the Heegner point construction yields $\mathbb{Q}$-rational points P on E with canonical Néron-Tate height $\hat{h}(P) > 0$. From Kolyvagin [**13**] we know that the index

$$\mathcal{I} = [E(\mathbb{Q}) : \mathbb{Z}P]$$

is finite; from now on assume P is such that $\hat{h}(P)$ is minimal so that $\mathcal{I}$ measures the failure of the Heegner point construction to generate $E(\mathbb{Q})$. The purpose of this note is to describe and motivate a conjecture on the effective determination of the 2-part $\mathcal{I}_2 = ord_2(\mathcal{I})$ of the index $\mathcal{I}$ for E of prime conductor N. The conjecture may be stated in terms of local information at $p = 2$ and ∞. So let us consider the Tate-Šafarevič group $\text{Ш}(E)$ of E, which is a finite group by Kolyvagin [**12**], given our assumption on $L(E/\mathbb{Q}, s)$. Also consider the canonical map

$$\kappa_{E,\infty} : E(\mathbb{Q}) \longrightarrow \Phi_{E,\infty},$$

where $\Phi_{E,\infty} = E(\mathbb{R})/E(\mathbb{R})^0$, i.e. the component group of E at ∞.

CONJECTURE 1.1. *Let E be as above. If we further assume that N is prime, then*

$$\mathcal{I}_2 = \begin{cases} |coker\, \kappa_{E,\infty}|, & \text{if } |\text{Ш}(E)_2| = 1, \\ 4, & \text{if } |\text{Ш}(E)_2| = 4. \end{cases}$$

Moreover, there is a homomorphism $\xi_E : Cl_{4N} \longrightarrow \widetilde{\Phi_{E,\infty}}$ such that

$$|coker(\kappa_{E,\infty})| > 1 \Leftrightarrow |coker(\xi_E)| > 1.$$

2010 *Mathematics Subject Classification.* Primary 11G05, 11G18; Secondary 11G40.
Key words and phrases. Elliptic curves, Heegner divisors, Heegner cycles.

Here Cl_{4N} is the class group attached to the real quadratic order $\mathcal{O}_{4N}$ and ξ_E is defined near the end of Section 3 using only basic topological properties of the real-analytic map $\varphi_\infty : X_0^+(N)(\mathbb{R}) \longrightarrow E(\mathbb{R})$ associated to a modular parametrization

$$\varphi : X_0^+(N) \longrightarrow E$$

and a description of the connected components of the real locus $X_0^+(N)(\mathbb{R})$ of the Atkin-Lehner quotient $X_0^+(N) = X_0(N)/w_N$ in terms of Heegner objects of discriminants dividing $4N$. We will see that ξ_E is effectively computable. In particular, if $\text{Ш}(E)_2$ is trivial a prediction of Conjecture 1.1 is the complete determination of $\mathcal{I}_2$ by effectively computable data at $p = \infty$ only.

As we assumed that $L(E, s)$ has a simple zero at $s = 1$, the Birch and Swinnerton-Dyer formula predicts

$$L'(E,1) = \hat{h}(g_E) \cdot |\text{Ш}(E)| \cdot \left(\prod_p c_p \right) \cdot \Omega_E \cdot |E(\mathbb{Q})_{tors}|^{-2}, \tag{1.1}$$

where $g_E \in E(\mathbb{Q})$ is a generator of $E(\mathbb{Q})/E(\mathbb{Q})_{tors}$, and $\Omega_E = \int_{E(\mathbb{R})^0} |\omega_E|$, where ω_E is the Néron differential of E. From the work of Gross, Kohnen and Zagier [**10**, p. 560] we know that Equation 1.1 may be restated as

$$\mathcal{I}^2 = c_E \cdot n_E \cdot |\text{Ш}(E)| \cdot \left(\prod_{p<\infty} c_p \right), \tag{1.2}$$

where $c_E \in \mathbb{Z}$ is Manin's constant, and n_E is expected to be the index of a certain subgroup of the -1-eigenspace $H_1(E, \mathbb{Z})^-$ of $H_1(E, \mathbb{Z})$ they constructed using Heegner geodesic cycles that live in $H_1(X_0^+(N), \mathbb{Z})^-$. Note that Equation 1.1 includes the c_∞ factor while Equation 1.2 does not. This may suggest that some local information at $p = \infty$ is encoded in n_E. Hoping to pin down the nature of this local information, let us assume that the conductor N of E is prime so that Equation 1.2 becomes $\mathcal{I}^2 = n_E \cdot |\text{Ш}(E)|$.[1] If we further assume that $\text{Ш}(E)$ is trivial, then one may expect that the natural global-to-local map $\kappa_{E,\infty}$ is related to the 2-part $n_{E,2}$ of the index n_E and, more precisely, that

$$\mathcal{I}_2 = |\mathit{coker}(\kappa_{E,\infty})|.$$

This together with some evidence discussed in Section 4 suggested the second part of Conjecture 1.1. (See Section 4 for further motivation.)

The problem of determining the 2-part of $\mathcal{I}$ is an interesting one, as computational data suggests that the index $\mathcal{I}$ usually consists of only small prime factors and by far the most frequent factor is 2. To illustrate this let us consider the set $\mathcal{C}$ of 2127 curves E of rank one and prime conductor in the current range of Cremona's online tables [**7**]. Out of the 268 curves E in $\mathcal{C}$ such that $\mathcal{I} > 1$, precisely one of them is such that $\mathcal{I}$ is odd, namely curve 322249a1 with $\mathcal{I} = 3$ and predicted $|\text{Ш}(E)| = 9$. The distribution of the elliptic curves E in $\mathcal{C}$ with $\mathcal{I} \neq 3$ is described in Table 1.

Conjecture 1.1 may be regarded as a refinement of the conjecture of Gross, Kohnen, and Zagier, albeit only for the 2-part and under the assumption that N is prime. However, as Brumer and McGuinness [**4**, p. 376] point out, the set of elliptic curves $E/\mathbb{Q}$ of prime conductor appears to be a typical sample of the set of

[1] Indeed, the work of Mestre and Oesterlè [**15**] implies that all the Tamagawa numbers $c_p = 1$, except possibly for $p = \infty$. Also, by the work of Abbes and Ullmo [**1**] we may conclude that $c_E = 1$.

TABLE 1. Distribution of curves E with $\mathcal{I} \neq 3$

$\|Ш(E)\|$ = \ n_E =	1	4
1	1859	246
4	0	21

all elliptic curves over $\mathbb{Q}$. It is hoped that Conjecture 1.1 is just a special case of a product formula for $\mathcal{I}$ expressed in terms of $Ш(E)$ and local information coming mostly from the canonical maps

$$\kappa_{E,p} : E(\mathbb{Q}) \longrightarrow \Phi_{E,p}$$

for all primes p, including $p = \infty$, where N is not necessarily prime. Here as usual $\Phi_{E,p}$ denotes the group of components of the Néron model of $E/\mathbb{Q}$, if $p \neq \infty$ and $\Phi_{E,\infty} = E(\mathbb{R})/E(\mathbb{R})^0$, as above. This is consistent with the fact that for N prime a result of Mestre and Oesterlè [**15**] implies that the Tamagawa numbers $c_p = |\Phi_p(\mathbb{F}_p)|$ are all trivial, for each prime $p \neq \infty$.

The rest of the note is organised as follows. In Section 2 we introduce notation, recall some basic properties of Heegner objects, and prove two lemmas. Section 3 includes a description of the real locus of the Atkin-Lehner quotient $X_0^+(N)$. The map ξ_E is introduced just after Theorem 3.3, under some technical assumptions. In Section 4 we further discuss the evidence for Conjecture 1.1.

2. Heegner objects

2.1. Heegner divisors. Let N be a positive integer and as usual let $Y_0(N)$ be the (coarse) moduli space of pairs (A, A') of elliptic curves equipped with an isogeny $A \longrightarrow A'$ of degree N with cyclic kernel. We have a natural map

$$\mathcal{H} \xrightarrow{\quad\psi\quad} Y_0(N)(\mathbb{C})$$

$$\tau \longmapsto \left(\mathbb{C}/(\mathbb{Z}\tau + \mathbb{Z}), \mathbb{C}/(\mathbb{Z}\tau + \tfrac{1}{N}\mathbb{Z})\right)$$

Let $X_0(N)$ be the compactification of the affine curve $Y_0(N)$ and recall that the natural map ψ induces a bijection $\mathcal{H}^*/\Gamma_0(N) \cong X_0(N)(\mathbb{C})$, where $\mathcal{H}^* = \mathcal{H} \cup \mathbb{P}^1(\mathbb{Q})$, $\mathcal{H} = \{\tau \in \mathbb{C} : \Im(\tau) > 0\}$ and

$$\Gamma_0(N) = \{\left(\begin{smallmatrix} \alpha & \beta \\ \gamma & \delta \end{smallmatrix}\right) \in \mathrm{SL}_2(\mathbb{Z}) : \gamma \equiv 0 \pmod{N}\}.$$

A point $x = (A, A')$ of $Y_0(N)$ is a Heegner point if $\mathrm{End}(A) = \mathrm{End}(A') = \mathcal{O}_D$ is an imaginary quadratic order of discriminant $D = f^2 d$ such that $(f, N) = 1$, where f is the conductor of $\mathcal{O}_D$. Heegner points attached to $\mathcal{O}_D$ exist if and only if

$$D \equiv r^2 \pmod{4N},$$

for some $r \in \mathbb{Z}$ uniquely defined $\pmod{2N}$, as $\mathfrak{n}_r = \mathbb{Z}N + \mathbb{Z}\frac{-r+\sqrt{D}}{2} \subset K = \mathbb{Q}(\sqrt{d})$ is a rank one projective $\mathcal{O}_D$-module such that $\mathcal{O}_D/\mathfrak{n}_r \cong \mathbb{Z}/N\mathbb{Z}$ and $x = (\mathcal{O}_D, \mathfrak{n}_r, [\mathfrak{a}]) = (\mathbb{C}/\mathfrak{a}, \mathbb{C}/(\mathfrak{n}_r^{-1}\mathfrak{a}))$ where $[\mathfrak{a}]$ lies in the ring class group Cl_D of the order $\mathcal{O}_D$. Moreover, the map

$$P \mapsto z(P) = \tfrac{-B+\sqrt{D}}{2A}$$

induces a bijection between the set of orbits $\mathcal{Q}^0_{D,r}/\Gamma_0(N)$ and the set of Heegner points associated to the pair (D, r), where $\mathcal{Q}^0_{D,r}$ is the set of primitive binary quadratic forms $P = [A, B, C]$ such that $D = B^2 - 4AC$, $N|A > 0$, and $B \equiv r \pmod{2N}$. The action of complex conjugation on Heegner points corresponds with the map $P \mapsto P'$ defined on positive definite binary quadratic forms $P = [A, B, C]$ by

$$P' = [A, -B, C]. \tag{2.1}$$

From the theory of Complex Multiplication we know that the Heegner points $x = (j(\tau), j(N\tau))$ attached to (D, r) are defined over the ring class field $H = K(j(\tau))$ of $\mathcal{O}_D$ and constitute a single Galois orbit over K. For simplicity let us assume that D is fundamental and $D < -4$. We have the Heegner divisor

$$P_{D,r} = \sum_{[\mathfrak{a}]\in\mathrm{Cl}_D} (\mathcal{O}_D, \mathfrak{n}_r, [\mathfrak{a}]).$$

Consider the canonical embedding of $X_0(N) \longrightarrow J_0(N)$, where $J_0(N)$ is the Jacobian of $X_0(N)$ over $\mathbb{Q}$ and $i\infty$ is mapped to O. Let $h(D) = |\mathrm{Cl}_D|$. The divisor of degree zero $y_{D,r} = P_{D,r} - h(D)i\infty$ defines a point in $J_0(N)(K)$ via the canonical isomorphism between the set of points of the Jacobian and the group of divisor of degree zero on the curve $X_0(N)$ modulo the principal ones. Let $y^+_{D,r}$ be the image of $y_{D,r}$ in the Jacobian $J^+_0(N)$ of the Atkin-Lehner quotient $X^+_0(N) = X_0(N)/w_N$, where w_N is the Fricke involution, i.e. the morphism induced by $(A, A') \mapsto (A', A)$ on the modular curve $X_0(N)$. It is well-known that the actions induced by the involutions w_N and complex conjugation on $X_0(N)$ have the same effect on $y_{D,r}$, so $y^+_{D,r} \in J^+_0(N)(\mathbb{Q})$.

Let $\mathbb{T} \subset \mathrm{End}(J^+_0(N))$ be the subring generated by the Hecke operators T_p, for all primes p not dividing N and the operators U_q, for all primes q dividing N. As we assumed that the L-function $L(E/\mathbb{Q}, s)$ has a simple zero at $s = 1$, the idempotent $\mathbf{t}_f \in \mathbb{T} \otimes \mathbb{Q}$ attached to the rational newform $f(\tau)$ of $E/\mathbb{Q}$ is non-trivial. Following Bertolini and Darmon [**2**, p. 201], we normalise the idempotent so that it lies in $\mathbb{T}$ and it is not divisible by an integer in $\mathbb{T}$. Let $E_f/\mathbb{Q}$ be the elliptic curve given by a classical theorem of Eichler and Shimura and define

$$(y^+_{D,r})_f = \mathbf{t}_f\, y^+_{D,r} \in E_f(\mathbb{Q})$$

But it is known that there is an isogeny $E_f \longrightarrow E$ defined over $\mathbb{Q}$. So the above construction yields a $\mathbb{Q}$-rational point on E.

As we assumed that N is prime, the curves E_f and E are isomorphic (and the computation of Heegner points on E is simpler). Indeed, by the work of Kolyvagin we know that $E/\mathbb{Q}$ has rank one, so E may not be a Neumann-Setzer curve.

2.2. Heegner geodesic cycles. Suppose (Δ, ρ) satisfies the Heegner condition $\Delta \equiv \rho^2 \pmod{4N}$, but now the discriminant $\Delta > 0$. We will assume for simplicity that Δ is not a square. Let $\mathcal{Q}^0_{N,\Delta,\rho}$ be the set of primitive binary quadratic forms $Q = [A, B, C]$ such that $\Delta = B^2 - 4AC$, $N|A$, and $B \equiv \rho \pmod{2N}$.

Following Darmon [**8**, p. 48], for each solution $(u, v) \in \mathbb{Z} \times \mathbb{Z}$ of Pell's equation

$$u^2 - \Delta v^2 = 1 \tag{2.2}$$

the automorph

$$M_Q = \begin{pmatrix} u - Bv & -2Cv \\ 2Av & u + Bv \end{pmatrix} \in \Gamma_0(N)$$

of $Q \in \mathcal{Q}^0_{N,\Delta,\rho}$ acts on the geodesic $\{z^-(Q) \to z^+(Q)\} \subset \mathcal{H}$ defined by the endpoints

$$z^-(Q) = \tfrac{-B-\sqrt{\Delta}}{2A}, \qquad z^+(Q) = \tfrac{-B+\sqrt{\Delta}}{2A},$$

so the restriction

$$\gamma_{Q,\tau} = \psi|\{\tau \to M_Q\tau\}$$

is a closed curve on $X_0(N)(\mathbb{C})$, where $\tau \in \{z^-(Q) \to z^+(Q)\}$ is arbitrary. We shall restrict our attention to automorphs M_Q coming from solutions (u, v) of Equation 2.2 such that the eigenvalue $\lambda_{u,v} = u + v\sqrt{\Delta}$ of M_Q is greater than 1. This means that $\{z^-(Q) \to z^+(Q)\}$ and $\{\tau \to M_Q\tau\}$ share the same orientation. We further assume that $\lambda_{u,v}$ is the smallest one. Then M_Q is uniquely defined. By a slight abuse of notation we shall write γ_Q instead of $\gamma_{Q,\tau}$. Let us also introduce the closed path on $X_0^+(N)$ defined by the restriction

$$\gamma_Q^+ = \pi|\gamma_Q,$$

where $\pi : X_0(N) \longrightarrow X_0^+(N) = X_0(N)/w_N$ is the canonical map.

LEMMA 2.1. *Suppose N is a prime number such that $N \equiv 3 \pmod 4$ and let $Q \in \mathcal{Q}^0_{N,\Delta,\rho}$, where $\Delta = 4N$ and $\rho = 0$. Then γ_Q traces a closed path on $X_0(N)$ twice.*

PROOF. Let $\epsilon = \alpha + \beta\sqrt{\Delta} \in \mathcal{O}_\Delta^\times$ be a fundamental unit, where $\Delta = 4N$. As we assumed that $N \equiv 3 \pmod 4$, the norm $\mathbf{N}(\epsilon) = \alpha^2 - \beta^2\Delta = 1$, where $\alpha \in 2\mathbb{Z}$ and β is a half-integer, i.e. $\beta = \frac{\beta_1}{2}$, where $\beta_1 \in 2\mathbb{Z} + 1$. But $\epsilon^2 = u + v\sqrt{\Delta}$, where

$$\begin{aligned} u &= \alpha^2 + \beta^2\Delta \in \mathbb{Z}, \\ v &= 2\alpha\beta \in \mathbb{Z}, \end{aligned}$$

is a fundamental solution to Equation 2.2. So for each $Q = [A, B, C] \in \mathcal{Q}^0_{N,\Delta,\rho}$ the matrix

$$\begin{pmatrix} u - Bv & -2Cv \\ 2Av & u + Bv \end{pmatrix} \in \Gamma_0(N)$$

is the automorph M_Q we have attached to Q, provided $\epsilon > 1$. Moreover, the relation $B^2 - 4AC = 4N$ implies that $2|B$ and

$$M_Q^{\frac{1}{2}} = \begin{pmatrix} \alpha - B\beta & -2C\beta \\ 2A\beta & \alpha + B\beta \end{pmatrix} \in \Gamma_0(N),$$

as a short calculation shows that its square is M_Q. Therefore

$$\psi| \left\{\tau \to M_Q^{\frac{1}{2}}\tau\right\}$$

is a closed path on $X_0(N)$ and thus γ_Q traces that same path twice. □

LEMMA 2.2. *Suppose N is a prime number such that $N \equiv 1 \pmod 4$ and let $Q \in \mathcal{Q}^0_{N,\Delta,\rho}$, for $(\Delta, \rho) = (4N, 0)$ and (N, N). Then γ_Q^+ traces a closed path on $X_0^+(N)$ twice.*

PROOF. Let $Q = [A, B, C] \in \mathcal{Q}^0_{N,\Delta,\rho}$, where $\Delta = 4N$ and $\rho = 0$. Note that $\tau_0 \in \{z^-(Q) \to z^+(Q)\}$, where $\tau_0 = z(P)$ is the point in the upper-half plane $\mathcal{H}$ attached to the positive definite binary quadratic form

$$P = \left[A^2, AB, \tfrac{1}{2}\left(B^2 - 2AC\right)\right] \in \mathcal{Q}^0_{-A^2\Delta, A\rho}$$

We will show that $\pi \circ \psi$ restricted to the geodesic path

$$\left\{\tau_0 \to M_Q^{\frac{1}{2}}\tau_0\right\}$$

is a closed path on the Atkin-Lehner quotient $X_0^+(N)$. Again let $\epsilon = \alpha + \beta\sqrt{\Delta} \in \mathcal{O}_\Delta^\times$ be a fundamental unit. But as we assumed that $N \equiv 1 \pmod 4$ and N is prime, a result of Legendre [**14**, p. 64] implies that the norm $\mathbf{N}(\epsilon) = \alpha^2 - \beta^2\Delta = -1$, where again $\alpha \in 2\mathbb{Z}$ and β is a half-integer. Straight forward computations shows that

$$M_Q^{\frac{1}{2}} = \frac{1}{\sqrt{\Delta}}\begin{pmatrix} \Delta\beta - B\alpha & -2C\alpha \\ 2A\alpha & \Delta\beta + B\alpha \end{pmatrix} \in \mathrm{SL}_2(F)$$

and also that $P_1 = M_Q^{\frac{1}{2}} P \in \mathcal{Q}^0_{-A^2\Delta, A\rho}$, where $P_1 = (A_1, B_1, C_1)$,

$$\begin{aligned} A_1 &= A^2 u \\ B_1 &= A(Bu - \Delta v) \\ C_1 &= \tfrac{1}{4}(B^2 u + \Delta(u - 2Bv)) \\ u &= \alpha^2 + \beta^2\Delta, \\ v &= 2\alpha\beta, \end{aligned}$$

and $F = \mathbb{Q}(\sqrt{N})$. But it turns out that $R_Q P_1 = P'$, where

$$R_Q = \begin{pmatrix} \alpha + B\beta & 2C\beta \\ 2A\beta & -\alpha + B\beta \end{pmatrix} \in \Gamma_0(N)$$

and P' is defined using Equation 2.1. In other words, the point $M_Q^{\frac{1}{2}}\tau_0$ is $\Gamma_0(N)$-equivalent to the point $-\overline{\tau_0}$. But clearly the images of τ_0 and $-\overline{\tau_0}$ under $\pi \circ \psi$ coincide. Therefore the restriction of $\pi \circ \psi$ to the above geodesic path is a closed path on $X_0^+(N)$. The case $\Delta = N$ with $\rho = N$ is entirely similar an it is left as an exercise for the reader. □

By a further abuse of notation, we write γ_Q to express its homology class. Given pairs (D_0, r_0) and (D_1, r_1) that satisfy the Heegner condition $D_i \equiv r_i^2 \pmod{4N}$, where each D_i is a fundamental discriminant for $i = 0, 1$, and such that $(D_0, D_1) = 1$ and $D_0 D_1 > 0$, Gross, Kohnen, and Zagier defined the Heegner geodesic cycle

$$\gamma_{D_0, D_1; r_0 r_1} = \sum_{[Q]} \chi_{D_0}(Q)\gamma_Q,$$

where the sum is over all the $\Gamma_0(N)$-orbits $[Q]$ of forms $Q \in \mathcal{Q}^0_{N, D_0 D_1, r_0 r_1}$ and χ_{D_0} is the generalised genus character of Gross, Kohnen, and Zagier [**10**, p. 508]. The cycle $\gamma_{D_0, D_1; r_0 r_1}$ is invariant with respect to the action induced by the Fricke involution w_N. So $\gamma_{D_0, D_1; r_0 r_1}$ lies in $H_1(X_0^+(N), \mathbb{Z})$. If we assume further that $D_0 < 0$ and $D_1 < 0$, then $\gamma_{D_0, D_1; r_0 r_1}$ is anti-invariant under the action of complex conjugation on $X_0^+(N)$, so it lies in the -1-eigenspace $H_1(X_0^+(N), \mathbb{Z})^-$ with respect to the action of complex conjugation.

3. On the real locus of $X_0^+(N)$

3.1. Key idea and basic facts. Our description of the real locus of $X_0^+(N)$ relies on the work of Ogg [**16**] and also on dictionary between classes of projective $\mathcal{O}_D$-modules of rank one and certain equivalence classes of embeddings $\mathcal{E}$ of the real quadratic field $F = \mathbb{Q}(\sqrt{N})$ into the central simple algebra $M_2(\mathbb{Q})$ satisfying some technical conditions, for real quadratic orders $\mathcal{O}_D$ containing $\mathcal{O}_{4N}$, due to Bertolini and Darmon [**2**]. This will lead us to the construction of the map ξ_E of Conjecture 1.1.

By the work of Shimura [**18**] we know that the real structure of $X_0(N)$, i.e. the action of complex conjugation on $X_0(N)$, satisfies

$$\overline{\psi(\tau)} = \psi\left(M \cdot \overline{\tau}\right)$$

where $\psi : \mathcal{H} \longrightarrow X_0(N)(\mathbb{C})$ is the natural map and $M \in M_0(N)^\times$ is such that $det(M) = -1$ and $\tau \in \mathcal{H}$. Here $M_0(N)$ is the Eichler order of $M_2(\mathbb{Q})$ consisting of 2 by 2 matrices with entries in $\mathbb{Z}$ and lower left entry divisible by N, and as usual $M_0(N)^\times$ denotes the group of invertible elements of $M_0(N)$. Consider the natural map

$$\pi : X_0(N) \longrightarrow X_0^+(N)$$

A real point of $X_0^+(N)$ is either the image of a point $x \in X_0(N)(\mathbb{R})$ under π or the image of a point $x \in X_0(N)(\mathbb{C})$ such that

$$\overline{x} = w_N x.$$

Following Ogg, we call the former type of real point of $X_0^+(N)$ *old* and call the latter type of real point *new*. Let Φ_N be the set of connected components of $X_0(N)(\mathbb{R})$ and let Φ_N^+ be the set of connected components of $X_0^+(N)(\mathbb{R})$. The work of Ogg [**16**, pp. 290–294] yields

$$|\Phi_N^+| = \begin{cases} \frac{h(4N)+h(N)}{2}, & \text{if } N \equiv 1 \pmod 4 \\ \frac{h(4N)+1}{2}, & \text{otherwise,} \end{cases}$$

where $h(\Delta)$ is the class number attached to the quadratic order $\mathcal{O}_D$. In particular, the Cohen-Lenstra heuristics [**6**] predict that the set of primes N such that $|\Phi_N^+| > 1$ has positive density.

We will find convenient to deal with the connected component C_0^+ of $X_0^+(N)(\mathbb{R})$ that contains the cusp $i\infty$ separately from the rest of the components.

3.2. The connected component of $X_0^+(N)(\mathbb{R})$ that contains $i\infty$. First we shall explicitly determine the old points of the connected component C_0^+. Then we will use the fixed points of the Fricke involution w_N of $X_0(N)$ that are defined over $\mathbb{R}$ to connect the old part with the new part of C_0^+. The component C_0^+ is the only one that contains both new and old points, which explains why we discussed it separately.

LEMMA 3.1 (Ogg). *If N is an odd prime then $\Phi_N = \{C_0\}$. Moreover, the restriction of ψ to the concatenation*

$$\{i\infty \to 0\} \cdot \{\tfrac{1}{2} \to i\infty\} \subset \mathcal{H}^*,$$

is a simple close path along C_0.

PROOF. (Cf. Ogg [**16**, pp. 290–291].) Note that the natural image on $X_0(N)(\mathbb{C})$ of the path $\{i\infty \to 0\}$ is pointwise fixed by complex conjugation. The same is true for the path $\{\frac{1}{2} \to i\infty\}$, as the action of complex conjugation on it has the same effect as the action of the matrix $\left(\begin{smallmatrix}1&1\\0&1\end{smallmatrix}\right) \in \Gamma_0(N)$. But the images on $X_0(N)(\mathbb{C})$ of the two paths intersect precisely at their respective extremities, thus defining a simple closed curve. Indeed, it suffices to note that the action of the matrix

$$\begin{pmatrix} -\frac{N-1}{2} & 1 \\ \\ -N & 2 \end{pmatrix} \in \Gamma_0(N)$$

on $\mathbb{P}^1(\mathbb{Q})$ has the effect $0 \mapsto \frac{1}{2}$. But the curve $X_0(N)$ is non-singular. So the connected component of $X_0(N)(\mathbb{R})$ is homeomorphic to the circle S^1, hence it must be described by the above simple closed path and the lemma follows. $\square$

Gross [**9**, p. 347] has shown that the ramification points of the natural map π are the Heegner points whose discriminant D is divisible by N and divides $-4N$. Precisely two of these (2-ramification) points are real, namely the ones represented by the points

$$\tau_0 = \tfrac{i}{\sqrt{N}} \in \{i\infty \to 0\} \quad \text{and} \quad \tau_1 = \tfrac{1}{2} + \tfrac{i}{2\sqrt{N}} \in \{\tfrac{1}{2} \to i\infty\}.$$

Therefore the old part of C_0^+ is the image of the union

$$\{i\infty \to \tau_0\} \cup \{\tau_1 \to i\infty\}$$

under the composition $\pi \circ \psi$. But the component C_0^+ also contains new points. Indeed, the Fricke involution w_N is induced by an involution on $\mathcal{H}$ with fixed point τ_0, namely $\tau \mapsto -\frac{1}{N\tau}$, so the image of

$$\gamma_0 = \left\{\tau_0 \to \tfrac{1}{\sqrt{N}}\right\}$$

under $\pi \circ \psi$ is contained in C_0^+. But w_N is also induced by an involution on $\mathcal{H}$ with fixed point τ_1, thus also the image of

$$\gamma_1 = \left\{\tau_1 \to \tfrac{1}{2} + \tfrac{1}{2\sqrt{N}}\right\}$$

under $\pi \circ \psi$ is contained in C_0^+. As we shall recall below, by making γ_0 and γ_0 suitably "shorter" we may obtain a nice description of the new part of C_0^+. More precisely, there are $\tau_0' \in \gamma_0$ and $\tau_1' \in \gamma_1$ such that both $\{\tau_0 \to \tau_0'\}$ and $\{\tau_1 \to \tau_1'\}$ map injectively into C_0^+, the union

$$(\pi \circ \psi)\{\tau_0 \to \tau_0'\} \cup (\pi \circ \psi)\{\tau_1 \to \tau_1'\}$$

is the old part of C_0^+, and the intersection

$$(\pi \circ \psi)\{\tau_0 \to \tau_0'\} \cap (\pi \circ \psi)\{\tau_1 \to \tau_1'\}$$

is a singlet, as shown in the proof of the following proposition.

PROPOSITION 3.2. *Let N be an odd prime number. The restriction of $\pi \circ \psi$ to either*

$$\{i\infty \to \tau_0\} \cdot \{\tau_0 \to \tau_0'\} \cdot \{\tau_1' \to \tau_1\} \cdot \{\tau_1 \to i\infty\},$$

if $N \equiv 1 \pmod 4$ or

$$\{i\infty \to \tau_0\} \cdot \{\tau_0 \to \tau_0'\} \cdot \{\tau_1 \to i\infty\},$$

otherwise, is a simple closed curve along C_0^+. *Here*

$$\begin{aligned} Q_0 &= [N,0,-1], \\ Q_1 &= \left[N,-N,\tfrac{1}{4}(N-1)\right], \quad \textit{if } N\equiv 1 \pmod 4 \textit{ and} \\ \tau_0' &= M_{Q_0}^{\frac14}\tau_0, \\ \tau_1' &= M_{Q_1}^{\frac14}\tau_1, \qquad\qquad \textit{if } N\equiv 1 \pmod 4 \end{aligned}$$

and $M_Q \in \Gamma_0(N)$ *is the automorph of* Q *of Subsection* 2.2.

PROOF. Suppose $N\equiv 1 \pmod 4$. From the proof of Lemma 2.2 we may see that a square root $M_{Q_0}^{\frac12}$ of the automorph M_{Q_0} attached to the primitive binary quadratic form $Q_0=[N,0,-1]$ of discriminant $\Delta = 4N$ is given by

$$M_{Q_0}^{\frac12} = \frac{1}{\sqrt{\Delta}}\begin{pmatrix}\Delta\beta & 2\alpha \\ 2N\alpha & \Delta\beta\end{pmatrix} \in \mathrm{SL}_2(F)$$

where $\alpha\in 2\mathbb{Z}$ and β is a half-integer coming from the fundamental unit $\epsilon=\alpha+\beta\sqrt{\Delta}\in\mathcal{O}_\Delta^\times$, which we know (from Lemma2.2) that has norm $\mathbf{N}(\epsilon)=\alpha^2-\beta^2\Delta=-1$ and again $F=\mathbb{Q}(\sqrt{N})$. So a 4-th root of the automorph M_Q is given by

$$M_{Q_0}^{\frac14} = \tfrac{1}{t}\left(1+M_Q^{\frac12}\right)$$

where $t^2 = 2+2\sqrt{\Delta}\beta$. But $P=[N,0,1]\in\mathcal{Q}^0_{-4N,0}$ is such that $\tau_0 = z(P)$ and a straight forward calculation yields

$$M_{Q_0}^{\frac14}P = \sqrt{\Delta}\cdot J_0,$$

where

$$J_0 = [N\beta,-\alpha,\beta].$$

Clearly $2\cdot J_0\in\mathcal{Q}^0_{-4,r}$, where $r\in\mathbb{Z}$ such that $r^2\equiv -4 \pmod{4N}$. Thus the image of

$$\tau_0' = M_{Q_0}^{\frac14}\tau_0$$

under ψ is one of the two Heegner points of discriminant $D=-4$. Using an entirely similar argument we may see that also $\psi(\tau_1')$ is a Heegner point of discriminant $D=-4$. But the images under $\pi\circ\psi$ of these two Heegner points coincide on $X_0^+(N)$. Thus $\pi\circ\psi$ restricted to

$$\{\tau_0\to\tau_0'\}\cdot\{\tau_1'\to\tau_1\}$$

is a path on $X_0^+(N)$ that parametrizes the new part of the component C_0^+ and the assertion of the proposition for the case $N\equiv 1 \pmod 4$ follows. Now suppose $N\equiv 3 \pmod 4$. From the proof of Lemma 2.1 we may see that a square root of the automorph M_{Q_0} attached to the primitive binary quadratic form Q_0 of discriminant $\Delta = 4N$ is given by

$$M_{Q_0}^{\frac12} = \begin{pmatrix}\alpha & 2\beta \\ 2N\beta & \alpha\end{pmatrix} \in \Gamma_0(N)$$

where again $\alpha \in 2\mathbb{Z}$ and β is a half-integer coming from the fundamental unit $\epsilon = \alpha + \beta\sqrt{\Delta} \in \mathcal{O}_\Delta^\times$ of norm $\mathbf{N}(\epsilon) = \alpha^2 - \beta^2\Delta = 1$. So a 4-th root of the automorph M_Q is given by

$$M_{Q_0}^{\frac{1}{4}} = \tfrac{1}{t}\left(1 + M_Q^{\frac{1}{2}}\right)$$

where $t^2 = 2\alpha + 2$. As before, we consider $P = [N, 0, 1] \in \mathcal{Q}^0_{-4N,0}$ is such that $\tau_0 = z(P)$ and a straight forward calculation yields

$$M_{Q_0}^{\frac{1}{4}} P = J_1,$$

where

$$J_1 = [N\alpha, -4N\beta, \alpha]$$

and $\frac{1}{2}J_1 \in \mathcal{Q}^0_{-N,N}$. So in this case

$$\tau_0' = M_{Q_0}^{\frac{1}{4}} \tau_0$$

represents a Heegner point $(\mathcal{O}_{-N}, \mathfrak{n}, [\mathfrak{a}])$ such that $\mathbf{N}(\mathfrak{a}) = N$. The same assertion holds for τ_1 defined just before the proposition. But all ideals of norm N in the maximal order $\mathcal{O}_{-N}$ are principal, hence $(\pi \circ \psi)(\tau_0') = (\pi \circ \psi)(\tau_1)$. Therefore the restriction of $\pi \circ \psi$ to the geodesic path

$$\{\tau_0 \to \tau_0'\}$$

parametrizes the new part of the connected component C_0^+. □

3.3. The remaining components of the real locus of $X_0^+(N)$. Define the set $\widetilde{\Phi_N^+}$ whose elements are C_0^+ and the ordered pairs (C, s), where $C \in \Phi_N^+$ is different from C_0^+, and $s \in \{-1, 1\}$. Fix an orientation for the real 1-dimensional manifold $X_0^+(N)(\mathbb{R})$.

THEOREM 3.3. *Let N be an odd prime number. For each real quadratic order $\mathcal{O}_\Delta$ that contains $\mathcal{O}_{4N}$ we have an injective map*

$$\iota_\Delta : Cl_\Delta \longrightarrow \widetilde{\Phi_N^+}$$

where

$$\iota_\Delta[\mathfrak{a}] = \begin{cases} C_0^+, & \text{if } [\mathfrak{a}] = 1 \\ (C, s) & \text{otherwise,} \end{cases}$$

and $C \in \Phi_N^+$ is the image under $\pi \circ \psi$ of the geodesic path

$$\left\{\tau \to M_Q^{\frac{1}{2}}\tau\right\}.$$

Here the point $\tau \in \gamma_Q$ is arbitrary, the binary quadratic form $Q \in \mathcal{Q}_{\Delta,\rho}$ corresponds to the rank-one projective $\mathcal{O}_\Delta$-module $\mathfrak{a}$, and M_Q is the automorph attached to Q; the sign of s is the winding direction of the simple closed path

$$(\pi \circ \psi)|\{\tau \to M_Q^{\frac{1}{2}}\tau\}.$$

Moreover, the sets

$$\iota_\Delta\left(Cl_\Delta - \{1\}\right)$$

where Δ is such that $\mathcal{O}_\Delta$ contains $\mathcal{O}_{4N}$, constitute a partition of the set of signed components $\widetilde{\Phi_N^+} - \{C_0^+\}$.

PROOF. We begin by recalling a natural bijective map of Bertolini and Darmon [**2**, pp. 185–186]. Given an order $\mathcal{O}_\Delta$ of a real quadratic field F, an embedding $\mathcal{E}$ of F into the central simple algebra $M_2(\mathbb{Q})$ is optimal if $\mathcal{E}(F) \cap M_0(N) = \mathcal{E}(\mathcal{O}_\Delta)$, where $M_0(N)$ is the algebra of 2 by 2 matrices with entries in $\mathbb{Z}$ and lower left entry divisible by N, i.e. the Eichler order of level N in $M_2(\mathbb{Q})$. Fix a pair (ρ, Δ) that satisfies the Heegner condition $\rho^2 \equiv \Delta \pmod{4N}$ so that

$$\mathfrak{n}_\rho = \mathbb{Z}N + \mathbb{Z}\tfrac{-\rho+\sqrt{\Delta}}{2} \subset F$$

is a rank-one projective $\mathcal{O}_\Delta$-module of norm N. Hence the canonical map

$$\mathcal{O}_\Delta \longrightarrow \mathcal{O}_\Delta/\mathfrak{n}_\rho$$

induces an isomorphism $\mathcal{O}_\Delta/\mathfrak{n}_\rho \cong \mathbb{Z}/N\mathbb{Z}$. We say that an optimal embedding $\mathcal{E}$ is oriented with respect to the choice of $\rho \pmod{2N}$ if the diagram

$$\begin{array}{ccccc} \mathcal{O}_\Delta & \xrightarrow{\ \mathcal{E}\ } & M_0(N) & \qquad & \left(\begin{smallmatrix} a & b \\ c & d \end{smallmatrix}\right) \\ \Big\downarrow & & \Big\downarrow & & \Big\downarrow \\ \mathcal{O}_\Delta/\mathfrak{n}_\rho & \xrightarrow{\ \sim\ } & \mathbb{Z}/N\mathbb{Z} & & d \pmod{N} \end{array}$$

commutes. Let $\mathrm{Emb}_\rho(\mathcal{O}_\Delta, M_0(N))$ denote the set of $M_0(N)^\times$-conjugacy classes of oriented optimal embeddings attached to $\rho \pmod{2N}$. The bijective map mentioned at the beginning of the proof is

$$\begin{array}{ccc} \mathrm{Cl}_\Delta & \xrightarrow{\ \sim\ } & \mathrm{Emb}_\rho(\mathcal{O}_\Delta, M_0(N)) \\ & & \\ [\mathfrak{a}] & \longmapsto & \mathcal{E}_\mathfrak{a} \end{array}$$

where $\mathcal{E}_\mathfrak{a}(\alpha) = \left(\begin{smallmatrix} a & b \\ c & d \end{smallmatrix}\right)$, for each $\alpha \in \mathcal{O}_\Delta$, is the matrix whose coefficients are defined by

$$\begin{aligned} \alpha e_1 &= a e_1 + c e_2 \\ \alpha e_2 &= b e_1 + d e_2 \end{aligned}$$

where $\{e_1, e_2\}$ is a $\mathbb{Z}$-basis of $\mathcal{O}_\Delta$ picked in such a way that $e_1 \in \mathfrak{n}_\rho$, and $\mathfrak{a}$ is picked so that its norm is prime to N. But if we put $F = \mathbb{Q}(\sqrt{N})$, then results of Ogg [**16**, pp. 290–294], together with Lemma 2.1 and Lemma 2.2, say that we have an injective map

$$\begin{array}{ccc} \mathrm{Emb}_\rho(\mathcal{O}_\Delta, M_0(N)) - \{1\} & \longrightarrow & \widetilde{\Phi_N^+} - \{C_0^+\} \\ & & \\ \mathcal{E}_Q & \longmapsto & \left((\pi \circ \psi)\left\{\tau \to M_Q^{\frac{1}{2}}\tau\right\}, s\right) \end{array}$$

where the sign of s is the winding direction of the path

$$(\pi \circ \psi)|\{\tau \to M_Q^{\frac{1}{2}}\tau\},$$

on the quotient $X_0^+(N)$, and as before the point $\tau \in \{z_-(Q) \to z_+(Q)\}$ is arbitrary, and $Q \in \mathcal{Q}_{\Delta,\rho}$ corresponds with the rank-one, projective $\mathcal{O}_\Delta$, for $(\Delta, \rho) = (4N, 0)$ and also that $(\Delta, \rho) = (N, N)$, if $N \equiv 1 \pmod 4$. Composing the former (bijective) map with the latter (injective) map yields the map ι_Δ stated in the theorem, provided we let $\iota_\Delta(1) = C_0^+$. The rest of the claims in the theorem are an immediate consequence of Ogg's work [**16**]. □

Let E be an elliptic defined over $\mathbb{Q}$ and define $\widetilde{\Phi_{E,\infty}}$ as the set whose elements are the trivial component $E(\mathbb{R})^0$ and also the two ordered pairs

$$(E(\mathbb{R}) - E(\mathbb{R})^0, \pm 1),$$

provided $E(\mathbb{R}) - E(\mathbb{R})^0 \neq \emptyset$, and endow each of the two sets that may arise with the obvious group structure (i.e. either the trivial group or μ_3, with $E(\mathbb{R})^0$ the neutral element). Define for Δ as in the above theorem

$$\xi_E : \mathrm{Cl}_\Delta \longrightarrow \widetilde{\Phi_{E,\infty}}$$

by the commutativity of the diagram

$$\begin{array}{ccc} \mathrm{Cl}_\Delta & \xrightarrow{\ \iota_\Delta\ } & \widetilde{\Phi_N^+} \\ & \searrow^{\xi_E} & \downarrow{\scriptstyle \widetilde{\varphi}} \\ & & \widetilde{\Phi_{E,\infty}} \end{array}$$

where $\widetilde{\varphi}$ is defined by letting $\widetilde{\varphi}(C_0^+) = E(\mathbb{R})^0$, and for each signed component $(C, s) \in \widetilde{\Phi_N^+}$ we let $\widetilde{\varphi}(C)$ be the restriction of φ to

$$(\pi \circ \psi)| \left\{ \tau \to M_Q^{\frac{1}{2}} \tau \right\},$$

provided it contains no ramification points of *even* index, so that it has a well-defined winding direction, and thus defines an element of $\widetilde{\Phi_{E,\infty}}$. Here Q is a binary quadratic form in $\mathcal{Q}_{\Delta,\rho}^0$ that corresponds to C via the above theorem, with $\tau \in \{z_-(Q) \to z_+(Q)\}$, and as before $\varphi : X_0^+(N) \longrightarrow E$ is a modular parametrization taking the cusp $i\infty$ to O_E, where E is assumed such that its L-function has a simple zero at $s = 1$.

As discussed below, there is substantial numerical evidence that suggests that ξ_E is well-defined and also, more significantly, that it is a group homomorphism.

4. Motivation and evidence

4.1. On the index n_E. As before, let $E/\mathbb{Q}$ be an elliptic curve such that its Hasse-Weil L-function $L(E/\mathbb{Q}, s)$ has a simple zero at $s = 1$. So the Heegner construction yields points $P \in E(\mathbb{Q})$ such that $\hat{h}(P) > 0$. (See Subsection 2.1.) As before, we pick one such P with $\hat{h}(P)$ minimal. Let f be the rational newform attached to E. The analytic rank one condition on E implies that the corresponding modular parametrization of E taking the cusp $i\infty$ to O_E factors through the natural map $\pi : X_0(N) \longrightarrow X_0^+(N)$ and thus induces a non-constant morphism

$$\varphi : X_0^+(N) \longrightarrow E$$

defined over $\mathbb{Q}$. There is $\gamma \in H_1(E(\mathbb{C}), \mathbb{Z})^-$ such that

$$c_f(D_0, r_0) c_f(D_1, r_1) \gamma = \varphi_*[\gamma_{D_0, D_1, r_0, r_1}]$$

for all (D_i, r_i) that satisfy the Heegner condition $r_i^2 \equiv D_i \pmod{4N}$ with $D_i < 0$ $(i = 0, 1)$,

$$(y_{D,r}^+)_f = c_f(D, r)P$$

and, after suitably normalising a Jacobi $\phi_f \in J_{2,N}^{new}$ that corresponds to the new form f,

$$\phi_f = \sum_{\substack{n,r \\ D=r^2-4Nn<0}} c_f(D, r)q^n\zeta^r.$$

This is a consequence of the "ideal statement" of theorem C conjectured by Gross, Kohnen, and Zagier [**10**, p. 503] and proved by Borcherds [**3**, p. 229] using a special case of his main result together with the Gross-Zagier theorem [**11**]. The index n_E may be defined as

$$n_E = [H_1(E(\mathbb{C}), \mathbb{Z})^- : \mathbb{Z}\gamma].$$

Note that the Heegner geodesic cycles $\gamma_{D_0,D_1;r_0r_1}$ all lie in $H_1(X_0^+(N), \mathbb{Z})^-$, since $D_i < 0$ (for $i = 0, 1$). But the Heegner pairs (D_i', r_i') with $D_i' > 0$ (for $i = 0, 1$)

$$\gamma_{D_0',D_1';r_0'r_1'} \in H_1(X_0^+(N), \mathbb{Z})^+,$$

just as the classes of the Heegner geodesic paths in our description of the connected components of $X_0^+(N)(\mathbb{R})$. There is a close relation between the Heegner geodesic cycles belonging to the -1-eigenspace with the ones that belong to the $+1$-eigenspace, which we discuss as follows. Let

$$H_1(X_0^+(N), \mathbb{Z}) \times H_1(X_0^+(N), \mathbb{Z}) \longrightarrow \mathbb{Z}$$

be the antisymmetric bilinear form defined by the intersection number $\alpha \cdot \beta$ of homology classes α and β. The following generalises a result of Skoruppa [**19**] and is proved in [**5**].

THEOREM 4.1. *With the notation and assumptions just introduced, if we fix (D_0, r_0) and also fix the two pairs (D_i', r_i') (for $i = 0, 1$), but let the pair (D_1, r_1) vary, then the intersection product*

$$c(D_1, r_1) = \gamma_{D_0,D_1;r_0r_1} \cdot \gamma_{D_0',D_1';r_0'r_1'}$$

gives the Fourier coefficients of a Jacobi form

$$\phi = \sum_{\substack{n,r \\ D=r^2-4Nn<0}} c(D, r)q^n\zeta^r.$$

of weight 2 and index N. Moreover, the set of Jacobi forms ϕ thus obtained generate the $\mathbb{C}$-vector space $J_{2,N}$.

This theorem suggests relations between Heegner geodesic cycles $\gamma_{D_0,D_1;r_0r_1}$ that live in $H_1(X_0^+(N), \mathbb{Z})^-$ and the set $\widetilde{\Phi_N^+}$, which was described in terms of certain Heegner geodesic paths that live in the complementary eigenspace $H_1(X_0^+(N), \mathbb{Z})^+$. This led us to guess that there should be some nice relations between the corresponding objects attached to the elliptic curve E, as sketched by the following diagram.

$$\begin{array}{ccc} & X_0^+(N) & E \\ & & \\ -1 & \{\gamma_{D_0,D_1;r_0r_1}\} & \longmapsto n \\ & \vdots & \vdots \\ +1 & \widetilde{\Phi_N^+} & \longmapsto |coker(\xi_E)| \end{array}$$

The above, together with numerical evidence, led us to conjecture

$$n_{E,2} > 1 \Leftrightarrow |coker(\xi_E)| > 1,$$

where $n_{E,2}$ is the 2-part of n_E, under the assumption that $\text{Ш}(E)_2 = 1$. The condition on the 2-part of $\text{Ш}(E)$ comes from the Gross, Kohnen, and Zagier conjecture.

4.2. Further numerical evidence. We have strong numerical evidence to support the first statement of Conjecture 1.1. More precisely, using SAGE [**20**] we obtained numerical evidence that supports that statement for each of the 2127 relevant curves E in Cremona's online tables [**7**], and also for each of the 10412 curves E of rank one, prime conductor, and predicted $\text{Ш}(E) = 1$ in the Stein-Watkins database up to the curve E with minimal Weierstraß equation

$$y^2 + xy + y = x^3 + x^2 - 68x - 282$$

and conductor $N = 7614833$. We have also verified that

$$\xi_E : \text{Cl}_\Delta \longrightarrow \widetilde{\Phi_{E,\infty}}$$

is a homomorphism for all elliptic curves in Cremona's online tables [**7**] such that $N \leq 10333$, up to sign. More precisely, for each of these curves E we have computed the component of $E(\mathbb{R})$ determined by $\xi_E(\mathfrak{a})$, for each $[\mathfrak{a}] \in \text{Cl}_{4N}$, but not the associated winding directions s. For example, for the two elliptic curves of rank one and conductor $N = 10333$ we obtained

ξ_{10333A}	a	a^2	a^3
b	1	1	0
b^2	1	1	0
b^3	1	1	0

ξ_{10333B}	a	a^2	a^3
b	1	0	1
b^2	0	1	1
b^3	1	1	0

For both curves $\Phi_{E,\infty}$ is non-trivial. Here 0 denotes the component $E(\mathbb{R})^0$ and 1 denotes the component $E(\mathbb{R}) - E(\mathbb{R})^0$. The class group Cl_{4N} is the elementary abelian 3-group of rank 2 generated by

$$\begin{aligned} a &= [3, 200, -111] \\ b &= [7, 198, -76] \end{aligned}$$

The kernel of ξ_{10333A} is the subgroup generated by b while the kernel of ξ_{10333B} is the one generated by $a^2 \cdot b$. Moreover, as predicted by the Conjecture 1.1, we have $\mathcal{I}_2 = 1$ for both curves.

Also, for some of the curves E in the range $N \leq 10333$ we have been able to obtain good approximations of the paths $\xi_E(\mathfrak{a})$, for each $[\mathfrak{a}] \in \text{Cl}_{4N}$, which show that the associated winding directions s are precisely those required for ξ_E to be a group homomorphism. There is work in progress to extend this computations much further.

Acknowledgements

The author would like to heartily thank the organisers of the Barcelona-Boston-Tokyo Seminar for the invitation and also for their hospitality and excellent accommodation facilities. Special thanks go to the Centro Universitario de Investigación en Ciencias Básicas, University of Colima, for all their support during the preparation of this manuscript.

We used in our computations elliptic curves E from Cremona's on-line tables [**7**], whose current range is $N \leq 400,000$. We also used part of the exceedingly large Stein-Watkins Tables [**20**]. All our computer calculations were performed by GNU/Linux computers using PARI [**17**] and SAGE [**20**].

References

[1] Ahmed Abbes and Emmanuel Ullmo, *À propos de la conjecture de Manin pour les courbes elliptiques modulaires* (French, with French summary), Compositio Math. **103** (1996), no. 3, 269–286. MR1414591

[2] Massimo Bertolini and Henri Darmon, *A Birch and Swinnerton-Dyer conjecture for the Mazur-Tate circle pairing*, Duke Math. J. **122** (2004), no. 1, 181–204, DOI 10.1215/S0012-7094-04-12216-X. MR2046811

[3] Richard E. Borcherds, *The Gross-Kohnen-Zagier theorem in higher dimensions*, Duke Math. J. **97** (1999), no. 2, 219–233, DOI 10.1215/S0012-7094-99-09710-7. MR1682249

[4] Armand Brumer and Oisín McGuinness, *The behavior of the Mordell-Weil group of elliptic curves*, Bull. Amer. Math. Soc. (N.S.) **23** (1990), no. 2, 375–382, DOI 10.1090/S0273-0979-1990-15937-3. MR1044170

[5] C. Castaño-Bernard, *Heegner geodesic cycles and intersection numbers*, To be submitted for publication.

[6] H. Cohen and H. W. Lenstra Jr., *Heuristics on class groups of number fields*, Number theory, Noordwijkerhout 1983 (Noordwijkerhout, 1983), Lecture Notes in Math., vol. 1068, Springer, Berlin, 1984, pp. 33–62, DOI 10.1007/BFb0099440. MR756082

[7] J. E. Cremona, *Elliptic curves of conductor* $\leq 400,000$, http://www.maths.nott.ac.uk/personal/jec/ftp/data/.

[8] Henri Darmon, *Heegner points, Heegner cycles, and congruences*, Elliptic curves and related topics, CRM Proc. Lecture Notes, vol. 4, Amer. Math. Soc., Providence, RI, 1994, pp. 45–59. MR1260954

[9] Benedict H. Gross, *Heegner points and the modular curve of prime level*, J. Math. Soc. Japan **39** (1987), no. 2, 345–362, DOI 10.2969/jmsj/03920345. MR879933

[10] B. Gross, W. Kohnen, and D. Zagier, *Heegner points and derivatives of L-series. II*, Math. Ann. **278** (1987), no. 1-4, 497–562, DOI 10.1007/BF01458081. MR909238

[11] Benedict H. Gross and Don B. Zagier, *Heegner points and derivatives of L-series*, Invent. Math. **84** (1986), no. 2, 225–320, DOI 10.1007/BF01388809. MR833192

[12] V. A. Kolyvagin, *Finiteness of* $E(\mathbf{Q})$ *and* $SH(E, \mathbf{Q})$ *for a subclass of Weil curves* (Russian), Izv. Akad. Nauk SSSR Ser. Mat. **52** (1988), no. 3, 522–540, 670–671; English transl., Math. USSR-Izv. **32** (1989), no. 3, 523–541. MR954295

[13] V. A. Kolyvagin and D. Yu. Logachëv, *Finiteness of SH over totally real fields* (Russian), Izv. Akad. Nauk SSSR Ser. Mat. **55** (1991), no. 4, 851–876; English transl., Math. USSR-Izv. **39** (1992), no. 1, 829–853. MR1137589

[14] A. M. Legendre, *Théorie des Nombres*, third ed., vol. I, Chez Firmin Didot Frères, Libraires, Rue Jacob, N° 24, Paris, 1830.

[15] J.-F. Mestre and J. Oesterlé, *Courbes de Weil semi-stables de discriminant une puissance m-ième* (French), J. Reine Angew. Math. **400** (1989), 173–184, DOI 10.1515/crll.1989.400.173. MR1013729

[16] A. P. Ogg, *Real points on Shimura curves*, Arithmetic and geometry, Vol. I, Progr. Math., vol. 35, Birkhäuser Boston, Boston, MA, 1983, pp. 277–307. MR717598

[17] The PARI Group, Bordeaux, *PARI/GP, Version 2.3.5*, http://pari.math.u-bordeaux.fr/.

[18] Goro Shimura, *On the real points of an arithmetic quotient of a bounded symmetric domain*, Math. Ann. **215** (1975), 135–164, DOI 10.1007/BF01432692. MR0572971

[19] Nils-Peter Skoruppa, *Heegner cycles, modular forms and Jacobi forms*, Sém. Théor. Nombres Bordeaux (2) **3** (1991), no. 1, 93–116. MR1116103

[20] W. A. Stein et al., *Sage Mathematics Software (Version 7.2.0)*, The Sage Development Team, 2016,
http://www.sagemath.org.

CUICBAS, Universidad de Colima, México

Email address: ccastanobernard@gmail.com

Contemporary Mathematics
Volume **701**, 2018
http://dx.doi.org/10.1090/conm/701/14141

A vanishing criterion for Dirichlet series with periodic coefficients

Tapas Chatterjee, M. Ram Murty, and Siddhi Pathak

Dedicated to F. Momose, with respect and admiration.

ABSTRACT. We address the question of non-vanishing of $L(1, f)$ where f is an algebraic-valued, periodic arithmetical function. We do this by characterizing algebraic-valued, periodic functions f for which $L(1, f) = 0$. The case of odd periodic functions was resolved by Baker, Birch and Wirsing in 1973. We apply a result of Bass to obtain a characterization for the even periodic functions. We also describe a theorem of the first two authors which says that it is enough to consider only the even and the odd periodic functions in order to obtain a complete characterization.

1. Introduction

To unravel the mysteries surrounding Dirichlet's theorem about non-vanishing of $L(1, \chi)$ for a non-principal Dirichlet character χ, Sarvadaman Chowla [**5**] made the following conjecture in the early 1960s:

CONJECTURE. Let f be a rational-valued arithmetical function, periodic with prime period p. Further assume that $f(p) = 0$ and

$$\sum_{a=1}^{p} f(a) = 0.$$

Then

$$\sum_{n=1}^{\infty} \frac{f(n)}{n} \neq 0,$$

unless f is identically zero.

Chowla [**5**] proved this conjecture in the case of odd functions i.e., $f(p - n) = -f(n)$ based on an outline of the proof by Siegel. The complete resolution of Chowla's question in a wider setting was given by Baker, Birch and Wirsing in 1973 [**2**]. They proved the following general theorem:

2010 *Mathematics Subject Classification.* Primary 11M06; Secondary 11M20.
Key words and phrases. Bass's theorem, Linear forms in logarithms.
The first author's research was supported by ISIRD grant at the Indian Institute of Technology Ropar.
The second author's research was supported by an NSERC Discovery grant.

THEOREM. If f is a non-vanishing function defined on the integers with algebraic values and period q such that (i) $f(n) = 0$ whenever $1 < (n, q) < q$ and (ii) the qth cyclotomic polynomial Φ_q is irreducible over $\mathbb{Q}(f(1), f(2), \cdots, f(q))$, then

$$\sum_{n=1}^{\infty} \frac{f(n)}{n} \neq 0.$$

Let us observe that in the case of Chowla's conjecture, condition (i) is vacuous as q is prime and $f(q) = 0$. The condition (ii) is also satisfied as f is rational-valued and the q-th cyclotomic polynomial is irreducible over $\mathbb{Q}$. Thus, the Baker-Birch-Wirsing theorem implies Chowla's conjecture.

Chowla's question can be asked in the most general setting as follows. Fix a positive integer q. Let f be an algebraic-valued arithmetical function, periodic with period q. It is useful to define an L-function associated to the function f, namely,

$$L(s, f) = \sum_{n=1}^{\infty} \frac{f(n)}{n^s}.$$

Using the analytic continuation of the Hurwitz zeta function, we can deduce that $L(1, f)$ exists if and only if $\sum_{a=1}^{q} f(a) = 0$. Thus, we can ask the following question: if f is not identically zero, then is it true that

$$\sum_{n=1}^{\infty} \frac{f(n)}{n} \neq 0\ ?$$

The answer in this case turns out to be negative. As an example, consider the function f defined such that

$$\sum_{n=1}^{\infty} \frac{f(n)}{n^s} = \left(1 - p^{(1-s)}\right)^2 \zeta(s), \tag{1.1}$$

for $\Re(s) > 1$. In particular,

$$f(n) = \begin{cases} 1 & \text{if } (n, p) = 1, \\ 1 - 2p & \text{if } p|n, p^2 \nmid n, \\ (p-1)^2 & \text{otherwise.} \end{cases}$$

Note that the function f is periodic with period p^2. Taking limit of the right hand side of (1.1) as $s \to 1^+$, we get

$$L(1, f) = 0,$$

because $\zeta(s)$ has a simple pole at $s = 1$.

In their paper, Baker, Birch and Wirsing [**2**] also give a characterization of all odd algebraic-valued periodic arithmetical functions f that satisfy $L(1, f) = 0$. Since their argument is short and elegant, we describe it in the third section. Their approach suggests a change in perspective. Instead of trying to prove the non-vanishing of an expression, we will try to characterize the functions f for which $L(1, f) = 0$.

Let f be any function. Then it can be written as the sum of an even function and an odd function as follows. Define $f_o(a) := [f(a) - f(-a)]/2$ and $f_e(a) := [f(a) + f(-a)]/2$. Then clearly $f = f_o + f_e$. In 2014, Tapas Chatterjee and Ram Murty [**4**] made an important observation. They proved,

THEOREM 1.1. *For a periodic, algebraic-valued arithmetical function f,*

$$L(1,f)=0 \iff L(1,f_o)=0 \text{ and } L(1,f_e)=0.$$

In the view of this theorem, to understand the vanishing of $L(1,f)$ in our case, it is enough to consider even algebraic-valued periodic arithmetical functions. In the fourth section, we apply a beautiful result of Bass [**3**] to obtain a set of functions that act as building blocks for even algebraic-valued periodic arithmetical functions f with $L(1,f)=0$. This completes, to an extent, the characterization we were after.

2. Preliminaries

The aim of this section is to introduce notation and some fundamental results that will be used in the later part of the paper.

Let q be a fixed positive integer. Consider $f:\mathbb{Z}\to\overline{\mathbb{Q}}$, periodic with period q. Define

$$L(s,f)=\sum_{n=1}^{\infty}\frac{f(n)}{n^s}.$$

Let us observe that $L(s,f)$ converges absolutely for $\Re(s)>1$. Since f is periodic,

$$\begin{aligned}L(s,f)&=\sum_{a=1}^{q}f(a)\sum_{k=0}^{\infty}\frac{1}{(a+kq)^s}\\&=\frac{1}{q^s}\sum_{a=1}^{q}f(a)\zeta(s,a/q),\end{aligned}$$

where $\zeta(s,x)$ is the Hurwitz zeta function. For $\Re(s)>1$ and $0<x\le 1$, the Hurwitz zeta function is defined as

$$\zeta(s,x)=\sum_{n=0}^{\infty}\frac{1}{(n+x)^s}.$$

In 1882, Hurwitz [**7**] proved that $\zeta(s,x)$ has an analytic continuation to the entire complex plane except for a simple pole at $s=1$ with residue 1. This implies that $L(s,f)$ has an analytic continuation to the entire complex plane except for a simple pole at $s=1$ with residue $\frac{1}{q}\sum_{a=1}^{q}f(a)$. Thus, $\sum_{n=1}^{\infty}\frac{f(n)}{n}$ converges whenever $\sum_{a=1}^{q}f(a)=0$, which we will assume henceforth. Thus, $L(s,f)$ is an entire function.

Given a function f which is periodic mod q, we define the Fourier transform of f as

$$\hat{f}(x):=\frac{1}{q}\sum_{a=1}^{q}f(a)\zeta_q^{-ax},$$

where $\zeta_q=e^{2\pi i/q}$. This can be inverted using the identity

$$f(n)=\sum_{x=1}^{q}\hat{f}(x)\zeta_q^{xn}. \tag{2.1}$$

Thus, the condition for convergence of $L(1, f)$, i.e., $\sum_{a=1}^{q} f(a) = 0$ can be interpreted as $\hat{f}(q) = 0$. Substituting (2.1) in the expression for $L(s, f)$ we have,

$$L(s,f) = \sum_{n=1}^{\infty} \frac{1}{n^s} \sum_{x=1}^{q} \hat{f}(x) \zeta_q^{xn}$$
$$= \sum_{x=1}^{q} \hat{f}(x) \sum_{n=1}^{\infty} \frac{\zeta_q^{xn}}{n^s}.$$

Assuming that $\hat{f}(q) = 0$, specializing at $s = 1$ and using the Taylor series expansion of log we conclude that

$$L(1,f) = -\sum_{x=1}^{q-1} \hat{f}(x) \log(1 - \zeta_q^x), \tag{2.2}$$

where log is the principal branch.

We will also apply the following theorem of Baker [**1**] concerning linear forms in logarithms.

THEOREM 2.1. *If $\alpha_1, \alpha_2, \cdots, \alpha_n$ are non-zero algebraic numbers, such that $\log \alpha_1$, $\log \alpha_2$, $\cdots$, $\log \alpha_n$ are $\mathbb{Q}$-linearly independent, then* 1, $\log \alpha_1$, $\cdots$, $\log \alpha_n$ *are $\bar{\mathbb{Q}}$-linearly independent.*

A useful corollary of this statement is

COROLLARY 2.2. If $\alpha_1, \alpha_2, \cdots, \alpha_n$ are algebraic numbers different from 0 and 1, and $\beta_1, \beta_2, \cdots, \beta_n$ are $\mathbb{Q}$-linearly independent algebraic numbers then,

$$\beta_1 \log \alpha_1 + \beta_2 \log \alpha_2 + \cdots + \beta_n \log \alpha_n$$

is non-zero and hence, transcendental.

Another helpful corollary of Baker's theorem is the following statement about linear forms in logarithms of positive algebraic numbers. For a proof of the corollary, we refer the reader to [**8**].

COROLLARY 2.3. Let $\alpha_1, \cdots, \alpha_n$ be positive algebraic numbers. If $c_0, \cdots, c_n$ are algebraic numbers and $c_0 \neq 0$, then

$$c_0 \pi + \sum_{j=1}^{n} c_j \log \alpha_j$$

is non-zero and hence, a transcendental number.

REMARK. The proof for the above corollary given in [**8**] also goes through when the branch of logarithm chosen is the principal branch since $i\pi$ can be replaced by $2 \log i$.

In an unpublished paper, Milnor predicted the complete set of multiplicative relations among cyclotomic numbers in the set $\{1 - \zeta_q^x : 1 \leq x \leq q-1\}$ for a fixed positive integer q. Here ζ_q denotes a primitive q^{th} root of unity. This conjecture was proved by Hyman Bass [**3**] in 1965. In 1972, Veikko Ennola [**6**, Theorem 3, pg.238] realized that the conjecture was true up to a factor of 2 and not in general. He gave a different proof of the conjecture in his paper. Since his formulation of the theorem is easier to apply in our setting, we will state it here. Let $a_x := \log(|1 - \zeta_q^x|)$. The following theorem characterizes all additive relations among the numbers $\{a_x | 1 \leq x \leq q-1\}$.

THEOREM 2.4. *Consider the following two relations: For* $1 \le x \le \left\lfloor \frac{(q-1)}{2} \right\rfloor$,

$$\mathfrak{R}_1 : a_x - a_{q-x} = 0, \tag{2.3}$$

and for any divisor d *of* q *and* $1 < d < q$ *and* $1 \le c \le d-1$,

$$\mathfrak{R}_2 : a_{\frac{q}{d}c} - \sum_{j=0}^{\frac{q}{d}-1} a_{c+dj} = 0. \tag{2.4}$$

Let R *be an additive relation among the* a_x*'s over the integers. Then* $2R$ *is a* $\mathbb{Z}$*-linear combination of relations of the form* $\mathfrak{R}_1$ *and* $\mathfrak{R}_2$.

To distinguish between the cases when q is odd and even in sections four and five, let $\mathbb{I}_e$ denote the indicator function of even integers, i.e.,

$$\mathbb{I}_e(n) = \begin{cases} 1 & \text{if } n \text{ is even.} \\ 0 & \text{otherwise.} \end{cases}$$

3. Odd functions

In this section, we reproduce a simple argument of Baker, Birch and Wirsing [**2**] that gives us a necessary and sufficient condition on an odd periodic function f such that $L(1, f) = 0$.

THEOREM 3.1. *Let* f *be an odd algebraic-valued arithmetical function, periodic with period* q*. Then* $L(1, f) = 0$ *if and only if* $\sum_{x=1}^{q-1} x\hat{f}(x) = 0$.

PROOF. Let us note that

$$1 - \zeta_q^x = -(\zeta_q^{x/2} - \zeta_q^{-x/2})\zeta_q^{x/2} = -2i\Big(\sin\left(\frac{x\pi}{q}\right)\Big)e^{x\pi i/q}, \tag{3.1}$$

and so the principal value of the logarithm is

$$\log(1 - \zeta_q^x) = \log\left(2\sin\frac{x\pi}{q}\right) + \left(\frac{x}{q} - \frac{1}{2}\right)\pi i \tag{3.2}$$

for $1 \le x < q$. Substituting (3.2) in the expression for $L(1, f)$ as a linear form in logarithms (2.2), we get

$$\begin{aligned} L(1,f) &= -\sum_{x=1}^{q-1} \hat{f}(x)\left[\log\left(2\sin\frac{x\pi}{q}\right) + \left(\frac{x}{q} - \frac{1}{2}\right)\pi i\right] \\ &= -\sum_{x=1}^{q-1} \hat{f}(x)\log\left(2\sin\frac{x\pi}{q}\right) - \frac{i\pi}{q}\sum_{x=1}^{q-1} x\hat{f}(x) + \frac{i\pi}{2}\sum_{x=1}^{q-1}\hat{f}(x). \end{aligned} \tag{3.3}$$

Since f is an odd function, $\hat{f}$ is also an odd function. Hence,

$$2\sum_{x=1}^{q-1} \hat{f}(x) = \sum_{x=1}^{q-1}[\hat{f}(x) + \hat{f}(q-x)] = 0.$$

Therefore, the last term of (3.3) is zero. Now, note that $\sin(\pi-\theta)=\sin(\theta)$. Thus, $\sin(x\pi/q)$ is an even function. Hence, $\log(2\sin\frac{x\pi}{q})$ is even which implies that $\hat{f}(x)\log(2\sin\frac{x\pi}{q})$ is an odd function.
Therefore, the first term of (3.3),

$$\sum_{x=1}^{q-1}\hat{f}(x)\log\left(2\sin\frac{x\pi}{q}\right)=0.$$

The result is immediate from here. □

REMARK. The condition obtained above is on the Fourier transform of the function and not the function itself. Using the Fourier inversion formula, we can deduce a condition on the function. The condition obtained in Theorem 3.1 can be simplified to

$$\begin{aligned}\sum_{x=1}^{q-1}x\hat{f}(x) &=\frac{1}{q}\sum_{x=1}^{q-1}x\sum_{n=1}^{q}f(n)\zeta_q^{-nx}\\ &=\frac{1}{q}\sum_{n=1}^{q}f(n)\sum_{x=1}^{q-1}x\zeta_q^{-nx}\\ &=\frac{1}{q}\sum_{n=1}^{q-1}f(n)\sum_{x=1}^{q-1}x\zeta_q^{-nx},\end{aligned}\tag{3.4}$$

as $f(q)=f(0)=f(-q)=-f(q)=0$ since f is odd. The innermost sum can be evaluated as follows. Let T be an indeterminate. Observe that

$$\sum_{x=0}^{q-1}T^x=\frac{T^q-1}{T-1}.$$

Differentiating the above identity with respect to T, we have

$$\sum_{x=1}^{q-1}xT^{x-1}=\frac{qT^{q-1}}{T-1}-\frac{T^q-1}{(T-1)^2}.$$

Multiplying the above equation by T and substituting $T=\zeta_q^{-n}$, we get

$$\sum_{x=1}^{q-1}x\zeta_q^{-nx}=\frac{q}{\zeta_q^{-n}-1}=\frac{q\zeta_q^n}{1-\zeta_q^n}.$$

Substituting this in (3.4), we obtain

$$\frac{1}{q}\sum_{n=1}^{q-1}f(n)\sum_{x=1}^{q-1}x\zeta_q^{-nx}=\sum_{n=1}^{q-1}\frac{f(n)\zeta_q^n}{1-\zeta_q^n}=\sum_{n=1}^{q-1}\left(\frac{f(n)}{1-\zeta_q^n}\right)-\sum_{n=1}^{q-1}f(n)=\sum_{n=1}^{q-1}\frac{f(n)}{1-\zeta_q^n},$$

as the condition for convergence of the series $\sum_{n=1}^{\infty}f(n)/n$ is $\sum_{a=1}^{q}f(a)=0$ and $f(q)=0$ since f is odd. This observation along with Theorem 3.1 gives: for f odd and periodic with period q,

$$L(1,f)=0\iff\sum_{n=1}^{q-1}\frac{f(n)}{1-\zeta_q^n}=0.\tag{3.5}$$

Let us note that

$$\frac{1}{1-\zeta_q^n} = \frac{i}{2}\cot\left(\frac{n\pi}{q}\right) - \frac{1}{2}$$

and that since f is odd,

$$2\sum_{n=1}^{q-1} f(n) = \sum_{n=1}^{q-1}[f(n) + f(q-n)] = 0.$$

Hence, condition (3.5) can be translated as: For odd algebraic-valued functions f that are periodic with period q,

$$L(1,f) = 0 \iff \sum_{n=1}^{q-1} f(n)\cot\left(\frac{n\pi}{q}\right) = 0.$$

We would like to mention that Baker, Birch and Wirsing also obtained a basis for the $\bar{\mathbb{Q}}$-vector space of odd algebraic-valued arithmetical functions f, periodic with period q and $L(1,f) = 0$ in their paper [**2**]. Thus, the characterization of odd functions is complete.

4. Even functions

In this section, we will use a theorem of Bass that characterizes all the multiplicative relations among cyclotomic numbers modulo torsion ([**3**], [**6**]). We will give a necessary condition for even periodic functions f to satisfy $L(1,f) = 0$.

We define the following functions that serve as building blocks for even algebraic-valued functions, periodic with period q. Fix a divisor $d > 1$ of q. For $c \in \{1, 2, \cdots, d-1\}$, we define $F_{d,c} := F_{d,c}^{(1)} - F_{d,c}^{(2)}$ where the two functions, $F_{d,c}^{(1)}$ and $F_{d,c}^{(2)}$ are arithmetical functions, periodic with period q, defined as follows:

$$F_{d,c}^{(1)}(x) = \begin{cases} 1/2 & \text{if } x \equiv c \bmod d, \\ 0 & \text{otherwise.} \end{cases}$$

$$F_{d,c}^{(2)}(x) = \begin{cases} 1/2 & \text{if } x \equiv (qc/d) \bmod q, \\ 0 & \text{otherwise.} \end{cases}$$

We will prove the following result in this section:

THEOREM 4.1. *Let f be an even algebraic-valued arithmetical function, periodic with period q. If $L(1,f) = 0$, then f is an algebraic linear combination of the functions*

$$\left\{\widehat{F_{d,c}} : d\,|\,q,\ 1 < d < q,\ 1 \le c \le d-1\right\}.$$

Here, $\widehat{F_{d,c}}$ denotes the Fourier transform of $F_{d,c}$ which can be computed as follows: For $1 \leq y \leq q$,

$$\begin{aligned}\widehat{F^{(1)}_{d,c}}(y) &= \frac{1}{q}\sum_{a=1}^{q} F^{(1)}_{d,c}(a)\zeta_q^{-ay} \\ &= \frac{1}{2q}\sum_{j=0}^{\frac{q}{d}-1} \zeta_q^{-(c+dj)y} \\ &= \frac{\zeta_q^{-yc}}{2q}\sum_{j=0}^{\frac{q}{d}-1} \zeta_q^{-djy} \\ &= \frac{\zeta_q^{-yc}}{2q}\sum_{j=0}^{\frac{q}{d}-1} \zeta_{\frac{q}{d}}^{-jy}.\end{aligned}$$

Note that the sum

$$\sum_{j=0}^{\frac{q}{d}-1} \zeta_{\frac{q}{d}}^{-jy} = \begin{cases} \frac{q}{d} & \text{if } y \equiv 0 \bmod \frac{q}{d}, \\ 0 & \text{otherwise.}\end{cases}$$

Similarly, for $1 \leq y \leq q$,

$$\begin{aligned}\widehat{F^{(2)}_{d,c}}(y) &= \frac{1}{q}\sum_{a=1}^{q} F^{(2)}_{d,c}(a)\zeta_q^{-ay} \\ &= \frac{1}{2q}\zeta_q^{-\frac{q}{d}cy} \\ &= \frac{1}{2q}\zeta_d^{-cy}.\end{aligned}$$

More precisely,

$$\widehat{F_{d,c}}(y) = \begin{cases} \frac{\zeta_q^{-cy}}{2d} - \frac{\zeta_d^{-cy}}{2q} & \text{if } y \equiv 0 \bmod \frac{q}{d}, \\ -\frac{1}{2q}\zeta_d^{-cy} & \text{otherwise.}\end{cases}$$

Thus, we note that $\widehat{F_{d,c}}$'s are in fact, simple functions.

PROOF. Let f be an even algebraic-valued, periodic function with period q, not identically zero. Let $M := \mathbb{Q}(f(1), \cdots, f(q), \zeta_q)$. Let $\{\omega_1, \omega_2, \cdots, \omega_r\}$ be a basis for M over $\mathbb{Q}$. There exists $d_j(x) \in \mathbb{Q}$ such that,

$$\hat{f}(x) = \sum_{j=1}^{r} d_j(x)\omega_j.$$

We can choose an integer N such that $\forall\ 1 \leq x \leq q$, $\forall\ 1 \leq j \leq r$, $c_j(x) := Nd_j(x) \in \mathbb{Z}$. Hence, $N\hat{f}(x) = \sum_{j=1}^{r} c_j(x)\omega_j$. Thus,

$$\begin{aligned}NL(1,f) &= -\sum_{x=1}^{q-1}\sum_{j=1}^{r} \omega_j c_j(x) \log(1-\zeta_q^x) \\ &= -\sum_{j=1}^{r}\omega_j \sum_{x=1}^{q-1} c_j(x)\log(1-\zeta_q^x).\end{aligned}$$

Let

$$R_j := \sum_{x=1}^{q-1} c_j(x) \log(1 - \zeta_q^x).$$

Therefore,

$$-NL(1, f) = \sum_{j=1}^{r} \omega_j R_j.$$

As $\hat{f}$ is even and $\omega_1, \cdots, \omega_r$ is a basis,

$$c_j(x) = c_j(q - x).$$

Therefore,

$$R_j = \sum_{x=1}^{\lfloor \frac{(q-1)}{2} \rfloor} \left(c_j(x)[\log(1 - \zeta_q^x) + \log(1 - \zeta_q^{-x})] \right) + \mathbb{I}_e(q) c_j \left(\frac{q}{2} \right) \log 2.$$

Note that log denotes the principal branch of logarithm. Thus, $\arg(1 - \zeta_q^x) = -\arg(1 - \zeta_q^{-x})$. For $1 \leq x \leq q - 1$, define $a_x := \log(|1 - \zeta_q^x|)$. Thus,

$$R_j = \sum_{x=1}^{\lfloor \frac{(q-1)}{2} \rfloor} \left(2c_j(x) a_x \right) + \mathbb{I}_e(q) c_j \left(\frac{q}{2} \right) a_{q/2}.$$

Since $c_j(x) = c_j(q - x)$, R_j can be written as

$$R_j = \sum_{x=1}^{q-1} c_j(x) a_x = \log \left(\prod_{x=1}^{q-1} (|1 - \zeta_q^x|)^{c_j(x)} \right).$$

Let

$$\alpha_j := \prod_{x=1}^{q-1} (|1 - \zeta_q^x|)^{c_j(x)}.$$

Therefore,

$$(-N)L(1, f) = \sum_{j=1}^{r} \omega_j \log \alpha_j. \tag{4.1}$$

Let us note that α_j is non-zero, algebraic. Thus, by Corollary 2.2, if $\alpha_j \neq 1$ for some $1 \leq j \leq r$, then $L(1, f)$ will be transcendental and hence non-zero. But $L(1, f) = 0$ by assumption. Hence, $\alpha_j = 1$ and in turn $R_j = 0$, $\forall\ 1 \leq j \leq r$. Thus, we are led to consider relations among logarithms of the cyclotomic numbers, $1 - \zeta_q^x$. Let $\mathscr{R}$ denote the relation $2R_j = 0$. By Theorem 2.4, $\mathscr{R}$ belongs to the $\mathbb{Z}$-module generated by relations of the form (2.3) and (2.4). Since $c_j(x) = c_j(q-x)$, $\mathscr{R}$ belongs to the $\mathbb{Z}$-module generated by (2.4). Indeed all relations $R := \sum_{x=1}^{q-1} C_x a_x = 0$ in the $\mathbb{Z}$-module generated by (2.3) satisfy $C_x = -C_{q-x}$, which along with the fact that $c_j(x) = c_j(q - x)$ (which stem from the evenness of f) imply that $c_j(x) = 0$ $\forall\ 1 \leq x \leq q - 1$. Thus, $\mathscr{R}$ is a $\mathbb{Z}$-linear combination of (2.4). Observe that the functions $F_{d,c}$ are precisely those that represent the relation (2.4). This implies that the functions c_j are integer linear combinations of $F_{d,c}$'s, say

$$c_j(x) = \sum_{\substack{d|q,\\ 1<d<q}} \sum_{c=1}^{d-1} m_{j,d,c}\, F_{d,c}(x),$$

where $m_{j,d,c} \in \mathbb{Z}$. Then,

$$\begin{aligned} N\hat{f}(x) &= \sum_{j=1}^{r} c_j(x)\omega_j \\ &= \sum_{j=1}^{r} \omega_j \sum_{\substack{d|q\\ 1<d<q}} \sum_{c=1}^{d-1} m_{j,d,c}\, F_{d,c}(x) \\ &= \sum_{\substack{d|q\\ 1<d<q}} \sum_{c=1}^{d-1} F_{d,c}(x) \sum_{j=1}^{r} m_{j,d,c}\omega_j. \end{aligned}$$

Taking the Fourier transform of both sides and using the Fourier inversion formula, the result follows. □

REMARK. On examining the above proof, we observe that we also obtain a sufficiency condition for $L(1, f) = 0$. More precisely, suppose f is a given algebraic linear combination of the functions

$$\left\{\widehat{F_{d,c}} : d \,|\, q,\ 1 < d < q,\ 1 \le c \le d-1\right\},$$

such that f is an even function, then $L(1, f) = 0$. But it does not seem easy to determine when a given function is an algebraic linear combination of these particular functions.

As a corollary, we have:

COROLLARY 4.2. Let q be a prime and f be an even algebraic-valued arithmetical function, periodic with period q. Then $L(1, f) \neq 0$.

PROOF. If q is prime, then the set

$$\left\{\widehat{F_{d,c}} : d \,|\, q,\ 1 < d < q,\ 1 \le c \le d-1\right\}$$

is an empty set. The conclusion follows from Theorem 4.1. □

5. Conclusion

The previous two sections give a characterization of algebraic-valued periodic arithmetic functions, odd and even respectively whose L-series vanish at $s = 1$. In this section, we mention a theorem of Tapas Chatterjee and Ram Murty [**4**] which nicely ties the non-vanishing of the odd and the even functions to give us the characterization required. The proof in their paper is incorrect as care was not taken regarding the branch of logarithm. The proof given below follows a different argument. As seen earlier, given any function f, we write it as $f = f_o + f_e$, where f_o is the odd part of f and f_e is the even part of f.

THEOREM 5.1. *If f is an algebraic-valued arithmetical function, periodic with period q, then $L(1, f) = 0 \iff L(1, f_o) = 0$ and $L(1, f_e) = 0$.*

PROOF. Let us note that

$$\sum_{n=1}^{\infty} \frac{f(n)}{n} = \sum_{n=1}^{\infty} \frac{f_e(n)}{n} + \sum_{n=1}^{\infty} \frac{f_o(n)}{n}.$$

Thus, it is clear that if $L(1, f_o)$ and $L(1, f_e)$ are both zero, then so is $L(1, f)$.

Suppose now that $L(1, f) = 0$. The arguments in the third section imply

$$L(1, f_o) = \frac{-i\pi}{q} \sum_{x=1}^{q-1} x\hat{f}_o(x). \tag{5.1}$$

We claim that,

$$L(1, f_e) = -2\left[\sum_{x=1}^{\lfloor \frac{(q-1)}{2} \rfloor} \hat{f}_e(x) \log\left(2\sin\frac{x\pi}{q}\right)\right] + \mathbb{I}_e(q)\hat{f}_e\left(\frac{q}{2}\right)\log 2, \tag{5.2}$$

when q is even. Indeed, by (2.2), we know that

$$\sum_{n=1}^{\infty} \frac{f_e(n)}{n} = -\sum_{x=1}^{q-1} \hat{f}_e(x)\log(1 - \zeta_q^x).$$

Since f_e is even, so is $\hat{f}_e$, i.e, $\hat{f}_e(q - x) = \hat{f}_e(x)$. Thus,

$$L(1, f_e) = -2 \sum_{x=1}^{\lfloor \frac{(q-1)}{2} \rfloor} \hat{f}_e(x) \log|1 - \zeta_q^x| + \mathbb{I}_e(q)\hat{f}_e\left(\frac{q}{2}\right)\log 2.$$

By (3.1),

$$|1 - \zeta_q^x| = \left|2\sin\left(\frac{x\pi}{q}\right)\right|.$$

Hence,

$$L(1, f_e) = -2 \sum_{x=1}^{\lfloor \frac{(q-1)}{2} \rfloor} \hat{f}_e(x) \log\left|2\sin\left(\frac{x\pi}{q}\right)\right| + \mathbb{I}_e(q)\hat{f}_e\left(\frac{q}{2}\right)\log 2.$$

Note that, for $1 \le x \le \left\lfloor \frac{(q-1)}{2} \right\rfloor$, $\sin(x\pi/q) > 0$.

Thus (5.1) and (5.2) imply

(5.3)

$$L(1, f) = \frac{-i}{q}\left(\sum_{x=1}^{q-1} x\hat{f}_o(x)\right)\pi - 2\sum_{x=1}^{\lfloor \frac{(q-1)}{2} \rfloor} \hat{f}_e(x)\log\left(2\sin\frac{x\pi}{q}\right) + \mathbb{I}_e(q)\hat{f}_e\left(\frac{q}{2}\right)\log 2.$$

Suppose $L(1, f_o) \neq 0$. By Corollary 2.3, $L(1, f) \neq 0$, which contradicts our assumption. Thus, $L(1, f_o) = 0$ and hence, $L(1, f_e) = 0$. □

This completes the characterization of algebraic-valued, periodic arithmetical functions with $L(1, f) = 0$.

Acknowledgments

The authors thank Sanoli Gun and Purusottam Rath for their valuable comments. We also extend our gratitude to the referee for insightful suggestions on an earlier version of this paper.

References

[1] Alan Baker, *Transcendental number theory*, Cambridge University Press, London-New York, 1975. MR0422171

[2] A. Baker, B. J. Birch, and E. A. Wirsing, *On a problem of Chowla*, J. Number Theory **5** (1973), 224–236, DOI 10.1016/0022-314X(73)90048-6. MR0340203

[3] Hyman Bass, *Generators and relations for cyclotomic units*, Nagoya Math. J. **27** (1966), 401–407. MR0201414

[4] Tapas Chatterjee and M. Ram Murty, *Non-vanishing of Dirichlet series with periodic coefficients*, J. Number Theory **145** (2014), 1–21, DOI 10.1016/j.jnt.2014.05.010. MR3253290

[5] S. Chowla, *A special infinite series*, Norske Vid. Selsk. Forh. (Trondheim) **37** (1964), 85–87. MR0171759

[6] Veikko Ennola, *On relations between cyclotomic units*, J. Number Theory **4** (1972), 236–247, DOI 10.1016/0022-314X(72)90050-9. MR0299585

[7] A. Hurwitz, *Einige Eigenschaften der Dirichlet Funktionen* $F(s) = \sum(D/n)n^{-s}$, *die bei der Bestimmung der Klassenzahlen Binärer quadratischer Formen auftreten*, Zeitschrift f. Math. u. Physik, **27** (1882), 86–101.

[8] M. Ram Murty and N. Saradha, *Euler-Lehmer constants and a conjecture of Erdös*, J. Number Theory **130** (2010), no. 12, 2671–2682, DOI 10.1016/j.jnt.2010.07.004. MR2684489

Department of Mathematics, Indian Institute of Technology, Ropar, Rupnagar, India 140001

Email address: tapasc@iitrpr.ac.in

Department of Mathematics and Statistics, Queen's University, Kingston, ON Canada K7L 3N6.

Email address: murty@mast.queensu.ca

Department of Mathematics and Statistics, Queen's University, Kingston, ON Canada K7L 3N6.

Email address: siddhi@mast.queensu.ca

Contemporary Mathematics
Volume **701**, 2018
http://dx.doi.org/10.1090/conm/701/14142

Rational families of 17-torsion points of elliptic curves over number fields

Maarten Derickx, Barry Mazur, and Sheldon Kamienny

Fumiyuki Momose is very much missed. He was a generous warm human being, with immense energy and generosity of spirit, and an extremely gifted mathematician. One of his abiding interests was rational torsion on elliptic curves over number fields, as in [**Mom84**], [**KM88**]. *This article is written in his memory.*

1. Introduction

Rational torsion points on elliptic curves present challenges that one can come back to again and again since the topic simply continues to be a source of extremely interesting diophantine issues. If E is an elliptic curve over a number field k, its Mordell–Weil group, $E(k)$, is finitely generated. Moreover, any finite subgroup of $E(k)$ is of the form $\mathbf{Z}/N\mathbf{Z}\times\mathbf{Z}/m\mathbf{Z}$ where N, m are positive integers with m dividing N. Ogg's Conjecture, proved thirty-five years ago, might be phrased as saying that there is no rational torsion on elliptic curves over $\mathbf{Q}$ except as directly forced by the underlying algebraic geometry. More specifically: any example of an elliptic curve over $\mathbf{Q}$ with its Mordell–Weil group containing a subgroup isomorphic to $\mathbf{Z}/N\mathbf{Z} \times \mathbf{Z}/m\mathbf{Z}$ is a member of a rationally parametrized family, in the sense that the modular curve $X(N, m)$ classifying such examples is isomorphic to $\mathbf{P}^1$.

In a paper published over two decades ago, written jointly with M. Kenku, Momose inaugurated an analogous investigation of certain types of subgroups of torsion points on elliptic curves rational over quadratic fields [**KM88**]. Kenku and Momose proved the following theorem:

THEOREM 1.1. *(Kenku, Momose) For integers N that factor as a product of powers of prime numbers < 17, and for integers m dividing N the following statements are equivalent.*

(1) *There exists a quadratic field k and an elliptic curve E defined over k such that $E(k)$ contains a subgroup isomorphic to $\mathbf{Z}/N\mathbf{Z} \times \mathbf{Z}/m\mathbf{Z}$.*
(2) *The modular curve that classifies such torsion, $X(N,m)$, is rational or hyperelliptic.*

Following on this work, one of the authors of the present paper established general classification results for torsion in the Mordell–Weil group of elliptic curves

2010 *Mathematics Subject Classification.* Primary 11G05; Secondary 11G18, 14G35, 14H51.
The authors are thankful to Ken Ribet for very helpful comments regarding an early draft of this paper.

over quadratic fields ([**Kam92b**], [**KN12**]; for a slightly different problem regarding torsion in elliptic curves and quadratic fields, see [**LL85**]).

Nowadays, one considers even more general questions from theoretical and computational perspectives.

- We might fix N and m and ask for a structural and numerical understanding of the collection of elliptic curves defined over fields of some fixed degree d over $\mathbf{Q}$—or over a fixed base number field k—for which its Mordell–Weil group over those fields contains a subgroup isomorphic to $\mathbf{Z}/N\mathbf{Z} \times \mathbf{Z}/m\mathbf{Z}$.
- Or more specifically, we might ask to classify rationally parametrized families of elliptic curves defined over number fields K_t of degree d over k and which possess N-torsion points rational over K_t[1]. In particular, we might study functions of degree d on $X_1(N)$ defined over $\mathbf{Q}$.

This paper will focus on the latter type of problem[2] as related to a diophantine analysis of appropriate Brill–Noether varieties attached to the modular curves $X_1(N)$.

A substantial amount of computation has been done. Intriguing examples have been discovered ([**DvH14**], [**Hoe14**], [**Sut**], [**JKL11a**], [**JKL11b**]). In work in progress the authors of this paper will be treating a number of explicit examples related to modular Brill–Noether curves. The present expository article, focusing on 17-torsion—dedicated to the memory of Fumiyuki Momose—is a report on a piece of that work in progress.

2. Rational N-torsion over fields of degree d

Fix two positive integers (N, d) and darken the point (N, d) in the plane if there exists a *non-CM elliptic curve*[3] defined over a number field of degree $\leq d$ having an N-torsion point rational over that field; call such points (N, d) simply: **data points**. One would like to know anything that stands out in this data set: its structure and its statistics.

There are two standard ways to look for uniformity phenomena:

- *Focusing, for example, on prime torsion, fix d and let $P(d)$ be the largest prime p such that (p, d) is in the data set.*

  Specific *exact values* of $P(d)$ are known only for small d. By [**Maz77**] $P(1) = 7$. Kamienny proved that $P(2) = 13$; Parent, building on work of Kamienny, showed $P(3) = 13$. Recently, Maarten Derickx, Sheldon Kamienny, William Stein, and Michael Stoll [**Kam89**] showed that $P(4) = 17$, $P(5) = 19$ and $P(6) = 37$.

  For general values of d we have the (trivially obtained) lower bound

$$d^{1/2} \ll P(d)$$

---

[1]The word "rational," then, is used in two senses: the parameter t ranges through the k-rational points of a rational curve (over k).

[2]and even more specifically when the base field is $\mathbf{Q}$ and $N = 17$

[3]We're thankful to Andrew Sutherland who suggested that one might keep separate the study of examples of CM elliptic curves possessing rational points of order N over fields of low degree d, since they represent a very orderly collection of known examples where for each such CM-elliptic curve, d admits a linear upper bound in N—and this would simply muddle the essential data set.

and the deep upper bounds given by Merel's Theorem tellling us that $P(d) < \infty$. More specifically, Merel [**Mer96**] (and Oesterlé, Parent [**Par03**]) proved, for general d that

$$P(d) \leq (1 + 3^{d/2})^2,$$

so we have:

$$d^{1/2} \ll P(d) \ll 3^d.$$

We don't even seem yet able to come up with much more precise conjectures for the qualitative behavior and/or the volatility of $P(d)$. Is $P(d)$ bounded by a constant times d^A for any $A > 1/2$? Or for *some* finite value of A? Or does it grow more rapidly than that?

Consider the 'minimalist' attitude that any *interesting* diophantine phenomenon occurs no more often than would be predicted by general structural constraints. This viewpoint seems to lead to firmly believed conjectures, for example, for statistics regarding ranks of Mordell–Weil groups. This general viewpoint might also suggest the guess that $P(d) \ll d^A$ for any $A > 1/2$. But we don't seem to have enough experience yet to give any firm conjectures[4].

Here below is a graph computed by the first author of the present article jointly with Mark van Hoeij. It is a log-log plot where the axes are $(x, y) = (\log p, \log d)$, the data points recording examples of 'lowest' degree d for the corresponding p occurs as prime torsion in a non-CM elliptic curve (over a field of degree d). The quotation-marks around the word 'lowest' is meant to signal that the blue data points and the blue extrapolated line corresponds to the lowest d for which there is a rational family of such examples of prime torsion p over fields of degree d. The red data points correspond to the sporadic points. The green curve is the proved (exponential) lower bound relating d to p. Visibly, much more computation needs to be done if we are to be able to surmise any general behavior with some feeling that there is evidence behind our guess. We should note that this graph is the result of a computation, and is only proved to be complete for $p \leq 37$.

- *Fix N and let $D(N)$ be the smallest integer d such that (N, d) is in the data set.*

In contrast to our knowledge of the asymptotics of $P(d)$, with $D(N)$ we are in slightly better shape. There is a clear *cut-off* for $D(N)$: namely, $D(N) \leq \gamma_{\mathbf{Q}}(N)$ where $\gamma_{\mathbf{Q}}(N)$ is the $\mathbf{Q}$-*gonality*[5] of the modular curve $X_1(N)$. The basic $\mathbf{Q}$-parametrization $X_1(N) \to X(1) \simeq \mathbf{P}^1$ already gives us $\gamma_{\mathbf{Q}}(N) \leq \Phi(N)\Psi(N)/2$— where $\Phi(N)$ is the Euler phi function and

$$\Psi(N) = \Psi(\prod p_i^{e_i}) = \prod (p_i + 1) p_i^{e_i - 1}.$$

[4]Some conjectures in the literature give upper bounds for primes of torsion in elliptic curves of degree d, but since these published conjectures also include CM elliptic curves which our "$P(d)$" doesn't register, those conjectures necessarily must allow for an essentially linear lower bound. Specifically, see [**CCS13**] and [**LR13**].

[5]See The Appendix, Section 7 below.

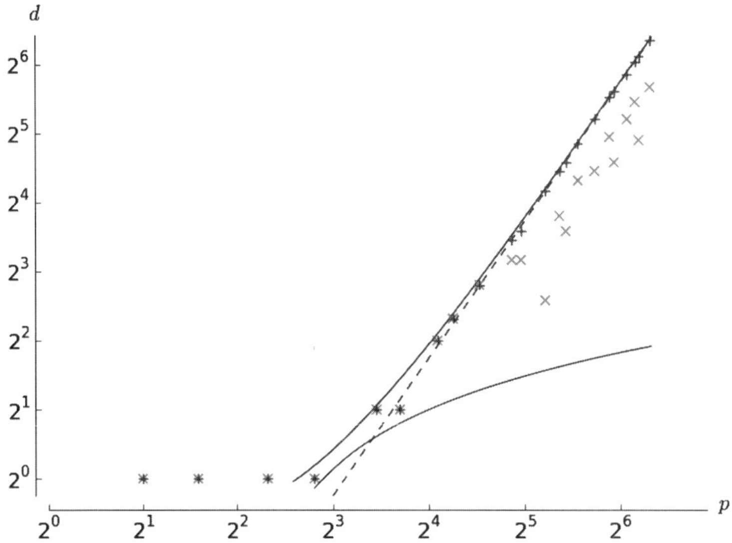

FIGURE 1. Smallest degree d such that $Y_1(p)$ has a point of that degree

In particular, we have $\gamma_{\mathbf{Q}}(N) \ll N^2$. For a discussion of the concept of gonality, see [**Abr96**], [**DvH14**] and Section 7 below.

If $d = \gamma_{\mathbf{Q}}(N)$, or more generally if there exists an $f : X_1(N) \to \mathbf{P}^1$ of degree d, then not only are there elliptic curves over fields of degree d with rational N-torsion over those fields, but there are infinitely many of them parametrized by a subset of $\mathbf{P}^1(\mathbf{Q})$. See Abramovich's basic paper [**Abr96**] where he proves the inequality

$$\frac{21}{200}(g-1) \leq \gamma_{\mathbf{C}}(N) \leq (g+3)/2,$$

where $\gamma_{\mathbf{C}}(N)$ is the **C**-*gonality*, and $g \approx N^2$ is the *genus*, of $X_1(N)$. For more elementary reasons

$$\gamma_{\mathbf{C}}(N) \leq \gamma_{\mathbf{Q}}(N) \leq g+1.$$

For the **Q**-gonalities of the modular curves $X_1(N)$ with $N \leq 40$ see [**DvH14**]. In particular

$N = p$	13	17	19	23	29	31	37
$\gamma_{\mathbf{Q}}(N)$	2	4	5	7	11	12	18

We will be considering data points (N, d) only for degrees $d \leq \gamma_{\mathbf{C}}(N)$. We will call an elliptic curve defined over a field of degree d possessing an N-torsion point rational over that field **sporadic** if $d = \gamma_{\mathbf{C}}(N)$ and it is *not* a member of a **Q**-rationally parametrized rational family of such elliptic curves defined over fields of degree d possessing N-torsion points rational over those fields. We call it **very sporadic** if $d < \gamma_{\mathbf{C}}(N)$.

Very sporadic data points exist. Here is a list of some known examples:

d	N	$\gamma_{\mathbb{C}}(N)$	*Reference*
3	21	4	[**Naj16**]
$9, 10$	29	11	[**Hoe14**]
$9, 10, 11$	31	12	[**DvH14**]

A result of Pete L. Clark, Brian Cook and James Stankewicz (which builds on the work of Dan Abramovich) [**CCS13**] implies that for a prime $p \geq 5$ there are at most finitely many points on $X_1(p)$ with degree $< \frac{7}{3200}(p^2 - 1)$. Related to this, see [**Fre94**].

3. Brill–Noether Varieties

Let X be a smooth projective curve over a characteristic zero field k. Let $\bar{k}/k$ be an algebraic closure, and $\bar{X} := X \times_{\mathrm{Spec}(k)} \mathrm{Spec}(\bar{k})$. For integers $d \geq 1$, $r \geq 0$, let

$$W_d^r(X) \subset \mathrm{Pic}^d(X)$$

denote the closed subvariety of $\mathrm{Pic}^d(X)$ (defined over k) classifying divisor classes of effective divisors D of degree d that are members of linear systems (of effective divisors of degree d) of dimension $\geq r$, or equivalently such that $h^0(X, \mathcal{O}(D)) \geq r + 1$; see [**ACGH85**], [**ACG11**], [**Dol12**].

The collection of Brill–Noether varieties $\{W_d^r(X) \mid d \geq 0, r \geq 0\}$ connect in the following ways:

(1) For $r \geq 1$ we have natural inclusions $W_d^r(X) \hookrightarrow W_d^{r-1}(X) \subset \mathrm{Pic}^d(X)$.

(2) Let α be a k-rational point of X, and let

$$f_\alpha : \mathrm{Pic}^d(X) \to \mathrm{Pic}^{d-1}(X)$$

be the morphism that sends the class of a divisor D to the class of $D - [\alpha]$. For $r \geq 1$ we have a commutative diagram of k-rational maps,

$$\begin{array}{ccc} W_d^r(X) & \xrightarrow{\subset} & \mathrm{Pic}^d(X) \\ \big\downarrow f_\alpha & & \big\downarrow f_\alpha \\ W_{d-1}^{r-1} & \xrightarrow{\subset} & \mathrm{Pic}^{d-1}(X). \end{array}$$

Statement (2) above follows from considering the global sections of the exact sequence

$$0 \to \mathcal{O}_X(D - \alpha) \to \mathcal{O}_X(D) \to \mathcal{O}_\alpha \to 0.$$

(3) We have the natural surjection

$$\eta_d : \mathrm{Sym}^d(X) \twoheadrightarrow W_d^0(X) \subset \mathrm{Pic}^d(X). \tag{3.1}$$

which is an isomorphism when restricted to

$$\mathrm{Sym}^{d,\#}(X) := \eta_d^{-1}\big(W_d^0(X) - W_d^1(X)\big) \subset \mathrm{Sym}^d(X),$$

i.e., to the inverse image of the complement of $W^1_d(X)$:

$$\begin{array}{ccccc} \mathrm{Sym}^{d,\#}(X) & \xrightarrow{\subset} & \mathrm{Sym}^d(X) & & \\ \downarrow \cong & & \downarrow & & \\ W^0_d(X) - W^1_d(X) & \xrightarrow{\subset} & W^0_d(X) & \xrightarrow{\subset} & \mathrm{Pic}^d(X). \end{array}$$

In paticular $\mathrm{Sym}^d(X)$ is a desingularization of $W^0_d(X)$. In certain cases of interest (e.g., as in our analysis of $X = X_1(17)$ below) $\mathrm{Sym}^d(X)$ is a small resolution of the singularities of $W^0_d(X)$.

By Theorem (1.1) in Chapter V of [**ACGH85**] if the genus $g > 1$ of X is in the range

$$d-1 \leq g \leq 2(d-1),$$

the Brill–Noether variety $W^1_d(X)$ is of dimension greater than or equal to $2(d-1)-g$. So, if it satisfies these conditions it can be a curve only if $2d \leq g+3$ and a surface only if $2d \leq g+4$.

We will be specifically interested in the cases $r = 0, 1$:

$$W^1_d(X) \subset W^0_d(X) \subset \mathrm{Pic}^d(X),$$

noting that a choice of k-rational point α of X will give us a (k-rational) closed immersion

$$W^1_d(X) \overset{f_\alpha}{\hookrightarrow} W^0_{d-1}(X) \subset \mathrm{Pic}^{d-1}(X).$$

Note: If X is a curve over k a number field, for any d, one has—applying a more general theorem of Faltings [**Fal94**]—that the set of k-rational points of $W^1_d(X)$ decomposes into a *finite* union,

$$W^1_d(X)(k) \quad = \quad \bigsqcup_j \mathcal{A}_j, \tag{3.2}$$

where, for each j, the Zariski closure of $\mathcal{A}_j$ is a translate of an abelian subvariety of $Pic^0(X)$. For a study of upper bounds for the dimension of such abelian subvarieties that may arise for given values of d, gonality, and genus, see [**AH91**].

3.1. The canonical involution v. An important case for us is when $d = g-1 \geq 0$ where g is the genus of X. In this situation we have the natural involution

$$\mathrm{Pic}^{g-1}(X) \overset{v}{\cong} \mathrm{Pic}^{g-1}(X)$$

defined by sending any linear equivalence class of divisors $[D]$ of degree $g-1$ to the linear equivalence class of $[K-D]$ where K is the canonical divisor of X. The involution v is 'functorially defined' and is defined over any field k over which the curve itself is defined, and commutes with any automorphism of X.

Consider the fixed locus $\mathrm{Th}(X) \subset \mathrm{Pic}^{g-1}(X)$ of the involution v. The 2^{2g} geometric points of $\mathrm{Th}(X)$ are classically referred to as theta-characteristics of X; they correspond to 'square roots' of the canonical line bundle. The finite subscheme $\mathrm{Th}(X) \subset \mathrm{Pic}^{g-1}(X)$ is a torsor over $\mathrm{Pic}^0(X)[2]$. Note that the Riemann-Roch Theorem guarantees that

$$h^0(X, \mathcal{O}(D)) = h^0(X, \mathcal{O}(K-D)), \tag{3.3}$$

so v induces an involution of $W^r_{g-1}(X)$ for any $r \geq 0$. Consider the theta divisor

$$\Theta := W^0_d(X) = W^0_{g-1}(X) \subset \mathrm{Pic}^{g-1}(X),$$

noting that choosing any theta-characteristic $\partial \in \mathrm{Th}(X) \subset \mathrm{Pic}^{g-1}(X)$ gives the commutative diagram

$$\begin{array}{ccccc} \Theta & \xrightarrow{v} & \Theta & \overset{\subset}{\longrightarrow} & \mathrm{Pic}^{g-1}(X) \\ \downarrow{\scriptstyle -\partial} & & \downarrow{\scriptstyle -\partial} & & \downarrow{\scriptstyle -\partial} \\ \Theta - \partial & \xrightarrow{-1} & \Theta - \partial & \overset{\subset}{\longrightarrow} & \mathrm{Pic}^{0}(X), \end{array}$$

the theta divisors, $\{\Theta - \partial \ \subset \ \mathrm{Pic}^0(X)\}$ for the theta-characteristics ∂ ranging through $\mathrm{Th}(X)(\bar{k})$ being—each of them—symmetric under multiplication by -1 in $\mathrm{Pic}^0(X)$ and the set of them being a torsor under the group $\mathrm{Pic}^0(X)[2]((\bar{k}))$.

Note, as well, that $W^1_{g-1}(X) \subset \Theta$ is stable under the involution v as can be seen from (3.3).

3.2. Basic Brill–Noether varieties. For X a curve defined over k, denote by $\gamma = \gamma_{\bar{k}}$, its $\bar{k}$-gonality. Call $WX := W^1_\gamma(X)$ the **Basic Brill–Noether variety** attached to X. Given a k-rational point α of X, we obtain an immersion

$$WX = W^1_\gamma(X) \ \overset{f_\alpha}{\hookrightarrow} \ W^0_{\gamma-1}(X) = \mathrm{Sym}^{\gamma-1}(X) \subset \mathrm{Pic}^{\gamma-1}(X).$$

3.3. The Basic Brill–Noether variety attached to $X_1(N)$. Consider the basic Brill–Noether variety $WX_1(N) := W(X_1(N))$. Thanks to the functorial nature of Brill–Noether varieties, the automorphism group of $X_1(N)$ viewed as finite group scheme over $\mathbf{Q}$ acts naturally on $WX_1(N)$. Thus we have the group Δ of diamond operators acting $\mathbf{Q}$-rationally on $WX_1(N)$, and the w-operators acting $\mathbf{Q}(\mu_N)^+$-rationally. When N is a prime number all these operators fit into a dihedral group that act $\mathbf{Q}(\mu_N)^+$-rationally on $WX_1(N)$.

3.4. Basic Brill–Noether curves attached to algebraic curves of genus 5 and gonality 4. *Assume from now on that X is a curve defined over $\mathbf{Q}$ of genus 5 and has $\mathbf{Q}$-gonality equal to $\mathcal{C}$-gonality $\gamma = 4$.*[6]

In this case the Basic Brill–Noether variety, $W := WX$, is a curve defined over $\mathbf{Q}$ (possibly reducible). We'll refer to it as the the Basic Brill–Noether curve attached to X. An application of Clifford's Theorem [7] guarantees that $h^0(X, \mathcal{O}_X(D)) \leq 2$ for any effective divisor D of degree 4, so $W^2_4(X)$ is empty. That is, the complete linear series that corresponds to any point in the Basic Brill–Noether curve attached to X is parametrized by a pencil.

By Lemma 7.1 of Section 7, a k-rational point of WX gives us a k-linear parametrization classof maps $X \to \mathbf{P}^1$ of degree $\gamma_{\bar{k}}(X)$, and conversely. So we have

[6] The example we treat, $X_1(17)$, is of this form, as are $X_1(21)$, and $X_1(24)$.

[7] See page 204 of **[ACGH85]** for a discussion that covers the case of interest to us: genus $=5$, gonality and degree $=4$.

PROPOSITION 3.1. *Let* $\mathrm{Aut}(X)$ *denote the group of k-rational automorphisms of X. There is a one-to-one correspondence between k-similarity classes*[8] *of maps* $X \to \mathbf{P}^1$ *(defined over k) of degree* $\gamma_{\bar{k}}(X)$ *and* $\mathrm{Aut}(X)$*-orbits of k-rational points of the Basic Brill–Noether curve WX:*

$$k\text{-similarity classes} \quad \leftrightarrow \quad WX(k)/\mathrm{Aut}(X).$$

If $d < \gamma_k(X)$ then $W^1_d(X)$ is empty (Lemma 7.1 below). In the special case where $\gamma_{\bar{k}}(X) = g-1$ we are in the situation of section 3.1 above, and we have the canonical involution v acting on $W(X)$ compatibly with its action on $\mathrm{Pic}^{g-1}(X)$ giving a commutative diagram:

$$\begin{array}{ccc} W(X) & \xrightarrow{v} & W(X) \\ \downarrow & & \downarrow \\ \mathrm{Pic}^{g-1}(X) & \xrightarrow{v} & \mathrm{Pic}^{g-1}(X) \end{array}$$

which commutes with any automorphism of X.

Consider a canonical embedding (defined over $\mathbf{Q}$)

$$\beta : X \xrightarrow{\cong} \Gamma \subset \mathbf{P}^4.$$

Since the genus of X is $g = 5$, by a theorem of Max Noether [**ACGH85**] the curve Γ lies on $3 = (g-2)(g-3)/2$ independent quadrics in $\mathbf{P}^4$.

For ease of nomenclature in this discussion (i.e., for the rest of this section) let us strictly reserve the symbols $\mathbf{P}^4$ to mean the projective 4-dimensional space which is the ambient space of the canonical embedding above, and $\mathbf{P}^2$ to mean the projective space generated by the linear space of *those* three independent quadrics just mentioned.

In the case of our interest we will be fixing bases,

$$\omega_0, \omega_1, \omega_2, \omega_3, \omega_4$$

of the 5 dimensional space $S := H^0(X, \Omega^1(X))$ such that the projectivization of $S^\vee$ is the $\mathbf{P}^4$ above. In terms of this basis we will be stipulating three independent quadratic relations

$$\begin{aligned} e_0 &:= e_0(\omega_0, \dots, \omega_4) \\ e_1 &:= e_1(\omega_0, \dots, \omega_4) \\ e_2 &:= e_2(\omega_0, \dots, \omega_4), \end{aligned}$$

generating the kernel of the natural cup product map

$$(*) \quad \mathrm{Sym}^2(S) \xrightarrow{\kappa} H^0(X, (\Omega^1(X)^{\otimes 2}),$$

and therefore representing a basis of the projective space $\mathbf{P}^2$ above.

Note that $S = H^0(X, \Omega^1(X))$ and $H^0(X, (\Omega^1(X)^{\otimes 2})$ have a natural action of $\mathrm{Aut}(X)$ with respect to which the morphism $(*)$ is equivariant.

---

[8] *k-similarity* is the natural notion of equivalence for k-parametrizations: two parametrizations are k-**similar** if one can be brought to the other by composition with appropriate k-isomorphisms of domain and range.

In the cases of our particular interest S will be the space of cuspforms of weight two and the cup product above will be given by multiplication to the space of cuspforms of weight four.

3.5. Loci of singular quadrics. For the results we are now about to quote, see page 207 of [**ACGH85**].

- Let $\mathcal{V} \subset \mathbf{P}^2$ be the sub-locus of singular quadrics[9] ‘ $\subset$ ′$\mathbf{P}^4$.
- Let $\mathcal{W} \to \mathcal{V}$ be the double cover determined by choosing one of the two systems of planes in these singular quadrics.
- Let

  $$\mathcal{G} \overset{\text{proj}}{\to} \mathcal{W}$$

  be the $\mathbf{P}^1$-bundle whose points consist of pairs $(\Pi, `)'$ where ‘ $\in$ ′$\mathcal{V}$ and $\Pi \subset$ “ is a two-plane.
- Let $v : \mathcal{W} \to \mathcal{W}$ be the involution defining the covering $\mathcal{W} \to \mathcal{V}$.
- Consider the commutative diagram

  $$\begin{array}{ccc} \mathcal{G} & \xrightarrow{\alpha} & \mathrm{Sym}^4(X) \\ \downarrow{\scriptstyle \text{proj}} & & \downarrow{\scriptstyle \text{proj}} \\ \mathcal{W} & \xrightarrow{\alpha} & \mathrm{Pic}^4(X) \end{array}$$

  where the morphism $\alpha : \mathcal{G} \to \mathrm{Sym}^4(X)$ is characterized on points by the rule that sends $(\Pi, `)'$ of $\mathcal{G}$ to the divisor (of degree four),

  $$\alpha(\Pi, `)' := \Pi \cap \Gamma \subset \Gamma,$$

  the latter being construed, via the isomorphism $\beta : X \longrightarrow \Gamma$ of subsection 3.4, as an element of $\mathrm{Sym}^4(X)$. The image of α restricted to a fiber of $\mathcal{G} \overset{\text{proj}}{\to} \mathcal{W}$ runs through a complete linear system of divisors, and therefore determines a well-defined point of $\mathrm{Pic}^4(X)$, providing a characterization (on points) of the morphism $\alpha : \mathcal{W} \to \mathrm{Pic}^4(X)$.

If ever we need to specify the curve X to which these objects are related, we indicate this in the standard manner; e.g., we write $\mathcal{W} = \mathcal{W}X, \quad \mathcal{V} = \mathcal{V}X, \quad \mathcal{G} = \mathcal{G}X$ etc.

3.6. The canonical representation of the Basic Brill–Noether curve. Let X be a curve satisfying our running hypotheses in this section, and put $W := WX$, the *Basic Brill–Noether curve* attached to X. Recall the involution $v : W \to W$ constructed in subsection 3.2. Let $\mathcal{W} = \mathcal{W}X$ be as in subsection 3.5 above, recalling the involution $v : \mathcal{W} \to \mathcal{W}$ constructed there.

The discussion of pp. 207-210 of of [**ACGH85**] gives the following identification.

[9]Recall that the ‘generic’ singular quadric threefold ‘ $\subset$ ′$\mathbf{P}^4$ has a unique singular point ϵ and is the cone ‘at ϵ’ of a (nonsingular) quadric surface given by the intersection of “ with any hyperplane not passing through ϵ. That quadric surface has two rulings by lines (possibly not rational over the base field k). Taking the cone through ϵ of each of these rulings gives us two 2-dimensional rulings, now, of the quadric threefold “ (again possibly not rational over the base field k). That is, “ is swept out by two pencils of planes (i.e., 2-dimensional projective linear subspaces).

PROPOSITION 3.2. *The image of the canonical morphism* $\alpha : \mathcal{W} \to \mathrm{Pic}^4(X)$ *is contained in* $W \subset \mathrm{Pic}^4(X)$ *and induces an isomorphism,* $\alpha : \mathcal{W} \xrightarrow{\cong} W$ *commuting with the involutions* v *on domain and image. That is, letting* $V := W/\{v\}$, *we have the commutative diagram:*

$$\begin{array}{ccc} \mathcal{W} & \xrightarrow{\cong} & W \\ \downarrow & & \downarrow \\ \mathcal{V} & \xrightarrow{\cong} & V \end{array}$$

3.7. Elliptic components and new components. Let X be a ("bi-elliptic") curve defined over $\mathbf{Q}$ satisfying our running hypotheses in this section and let $W := WX$ be—as usual—the Basic Brill–Noether curve attached to X. By an *elliptic involution* of X let us mean an involution $\iota : X \to X$ such that the quotient of X under its action,

$$\mathrm{proj}_\iota : X \to X/\{\sim \iota\} = \mathcal{E},$$

is a curve of genus one. The involution ι induces an action on W and on the constructions of Subsections 3.5 and 3.6. In particular ι commutes with the double cover mapping:

$$\begin{array}{ccc} W & \xrightarrow{\iota} & W \\ \downarrow{\scriptstyle v} & & \downarrow{\scriptstyle v} \\ V & \xrightarrow{\iota} & V \end{array}$$

For any such quotient, and any point $u \in \mathcal{E}$ let $j_u : \mathcal{E} \to \mathcal{E}$ denote the canonical (nontrivial) involution fixing the point u, and let

$$\mathrm{proj}_{j_u} : \mathcal{E} \to \mathcal{E}/\{\sim j_u\} =: \mathbf{P}^1_u$$

denote the projection to the (genus zero) quotient (which we denote $\mathbf{P}^1_u$) under the action of j_u. Denote by t_u the running parameter in the projective line $\mathbf{P}^1_u$. For any pair (u, t_u) with $u \in \mathcal{E}$ and $t_u \in \mathbf{P}^1_u$ let $D_\iota(u, t_u) \subset \mathrm{Symm}^4(X)$ be the effective divisor of degree four on X given by

$$D_\iota(u, t_u) := \mathrm{proj}_\iota^{-1} \circ \mathrm{proj}_{j_u}^{-1}(t_u).$$

For each $u \in \mathcal{E}$, then, we have a linear system of divisors of degree four on X parametrized by the variable t_u, giving us a point on W, which we denote $w_\iota(u)$.

The morphism

$$w_\iota : \mathcal{E} \longrightarrow W$$

factors through the quotient $\mathcal{E}'$ of $\mathcal{E}$ under the natural action of the 2-torsion subgroup of its jacobian, i.e., $\mathrm{Pic}^0(\mathcal{E})[2]$. The induced morphism

$$w'_\iota : \mathcal{E}' \hookrightarrow W$$

is a closed immersion, and its image is a (reduced) irreducible component of W defined over the field k. We denote this component $W_\iota \subset W$ and refer to $W_\iota \approx \mathcal{E}'$ as the **k-elliptic component of W associated to ι**. It is fixed by the action of the involution ι on W.

By a **new component** of W we will mean an irreducible component that is not elliptic in the above sense.

4. Fine Siegel units and fine Siegel points

Let $X := X_1(N)$. A **Fine Siegel unit** on X is a rational function f on X defined over $\bar{\mathbf{Q}}$ whose divisor of zeroes and poles consist only of $\mathbf{Q}$-rational cusps. Let $C(N)$ denote the set of $\mathbf{Q}$-rational cusps; so a fine Siegel unit is a unit in the ring of regular functions on $X_1(N) - C(N)$ (over $\bar{\mathbf{Q}}$). Since the action of $\mathrm{Gal}(\bar{\mathbf{Q}}/\mathbf{Q})$ preserves the divisor of zeroes-and-poles of a fine Siegel unit, an application of Hilbert's Theorem 90 guarantees that we may normalize our fine Siegel units (by multiplication by an appropriate nonzero scalar) so that they are defined over $\mathbf{Q}$. Such a normalized Siegel unit f is well-defined by its divisor of zeroes-and-poles up to a factor in $\mathbf{Q}^*$ and gives us a $\mathbf{Q}$-rational parametrization $f : X \to \mathbf{P}^1$. Let $\mathcal{Z}(N)$ denote the group of fine Siegel units modulo $\mathbf{Q}^*$. By the Manin–Drinfeld Theorem, $\mathcal{Z}(N)$ is a free abelian group of rank $|C(N)| - 1$. The action of the group Δ of diamond operators on $X_1(N)$ induces an action on $\mathcal{Z}(N)$ and there is a natural metric on $\mathcal{Z}(N)$ given by the geometric degree of the function $f : X_1(N) \to \mathbf{P}^1$. This metric satisfies a triangle inequality:

$$\deg(f \cdot g) \leq \deg(f) + \deg(g),$$

and it scales well, i.e.,

$$\deg(f^n) = |n| \cdot \deg(f)$$

for $n \in \mathbf{Z}$, and is preserved by the action of the diamond operators. See [**Yan09**] for an explicit description of the fine Siegel units in terms of their q-expansions and their expression in relation to specific modular forms.

The following two conditions on $X := X_1(N)$ hold for only a (finite) number of values of N but they do hold for the case $N = 17$.

(1) $X = X_1(N)$ contains no very sporadic points (in the terminology of Section 2) *except for the set of* $\mathbf{Q}$*-rational cusps* $C(N)$.
(2) $\gamma_C(X) < |C(N)|$.

When $N = 17$ these conditions are indeed met[10].

PROPOSITION 4.1. $X_1(17)$ *has no non-cuspidal very sporadic points.*

Proof: There are no non-cuspidal points on $X_1(17)$ of degree 1 by [**Maz77**]; none of degree 2 by [**Kam86**]; and none of degree 3 by [**Par03**].

As we shall see (Proposition 6.7) $X_1(17)$ contains no non-cuspidal sporadic points, as well.

Under hypothesis **(1)** above, every function $\phi : X_1(N) \to \mathbf{P}^1$ of degree $\gamma_C(X)$ and defined over $\mathbf{Q}$ has the property that any of its fibers above points in $\mathbf{P}^1(\mathbf{Q})$ either consists entirely of $\mathbf{Q}$-rational cusps, or contains no $\mathbf{Q}$-rational cusps at all. This is because each fiber is of degree $\gamma_C(X)$ and if it contains a $\mathbf{Q}$-rational cusp, the points of the fiber are all of degree strictly less than $\gamma_C(X)$. That is, these points are all very sporadic, so by **(1)** they are *all* rational cusps. If a fiber of such a ϕ consists entirely of rational cusps, call it a **rational-cuspidal-fiber** of ϕ. By **(2)**, ϕ has at least two rational-cuspidal-fibers. Choosing two rational-cuspidal-fibers of such a $\phi : X \to \mathbf{P}^1$ and composing ϕ with an appropriate linear fractional transformation of $\mathbf{P}^1$ that sends the image of one of those fibers to 0 and the other to ∞ we see that any such ϕ is in the $\mathbf{Q}$-similarity class of a $\mathbf{Q}$-parametrization of $X_1(N)$ by a fine Siegel unit f (of geometric degree equal to the gonality of X).

[10] As has been verified by the first author of this article and Mark van Hoeij, these two conditions are satisfied for $X_1(N)$ for $N = 32$, and for $N < 28$ but not $N = 21$.

There is a natural way of denoting such a fine Siegel unit f up to the equivalence relation defined by deeming two such Siegel units equivalent if one can be obtained from the other by composition by an appropriate $\mathbf{Q}$-automorphism $b : \mathbf{P}^1 \to \mathbf{P}^1$. Namely, one simply lists the rational-cuspidal-fibers of f giving the divisors with support on cusps that constitutes each of those fibers.

Each divisor with support on the cusps is encoded by $|C(N)|$ integers, where the i-th integer is the multiplicity of the i-th cusp. We display this data for a given f as a matrix, with exactly $|C(N)|$ columns, and as many rows as there are rational-cuspidal-fibers for f. We call it the **rational-cuspidal-fiber matrix** for the $\mathbf{Q}$-linear parametrization class of f.

We also organize these cuspidal-fiber matrices in Δ-orbits. Any such Δ-orbit determines a $\mathbf{Q}$-similarity class of $\mathbf{Q}$-parametrizations of $X_1(N)$ of geometric degree equal to the gonality of X.

Under both hypotheses above, any such f has at least two rational-cuspidal-fibers ($|C(N)| > \gamma_{\mathcal{C}}(X)$ implies that there are at least two fibers containing $\mathbf{Q}$-rational cusps). We can compose f with an appropriate $b : \mathbf{P}^1 \to \mathbf{P}^1$ sending one rational-cuspidal-fiber to 0 and another to ∞, so that $b \circ f$ is a fine Siegel unit.

Consequently

PROPOSITION 4.2. *Under the hypotheses* **(1)** *and* **(2)** *above*

(1) *Any $\mathbf{Q}$-parametrization of $X = X_1(N)$ of geometric degree $\gamma_{\mathcal{C}}(X)$ is represented by at least one fine Siegel unit in $\mathcal{Z} = \mathcal{Z}(N)$ of degree $\gamma_{\mathcal{C}}(X)$.*
(2) *There are only finitely many classes of $\mathbf{Q}$-parametrizations of $X_1(N)$ of degree $\gamma_{\mathcal{C}}(X)$.*

DEFINITION 4.3. By a **fine Siegel point** on the Basic Brill–Noether variety $WX_1(N)$ let us mean a $\mathbf{Q}$-rational point on $WX_1(N)$ represented by a linear system parametrized by a fine Siegel unit.

COROLLARY 4.4. *Under the hypotheses* **(1)** *and* **(2)** *above, the Basic Brill–Noether variety $WX_1(N)$ has only finitely many $\mathbf{Q}$-rational points. These are all fine Siegel points and are effectively obtainable.*[11].

The first statement in Corollary 4.4 follows from Proposition 4.2 simply by considering the number of pairs of possible cuspidal-fibers. Effectivity follows because there are effective methods to compute Riemann-Roch spaces of divisors on curves (cf. [**Hes02**]). The computations of cuspidal-fiber matrices can be done for some small values of N (including $N = 17$) by a combination of modular symbol computations implemented by Sage ([**Bos08**, p. 57]) and brute force computations.

When $N = 17$ we shall see that all $\mathbf{Q}$-rational points of $WX_1(N)$ are fine Siegel points. It would be interesting to understand, for more general values of N what portion of $WX_1(N)(\mathbf{Q})$ comes from Siegel (or fine) Siegel points.

A computation of Derickx and van Hoeij [**DvH14**] guarantees that for all $N \leq 40$ there is at least one modular unit of degree equal to the $\mathbf{Q}$-gonality of $X_1(N)$. It follows that if, for these values of N, the $\mathbf{Q}$-gonality were equal to the $\mathcal{C}$-gonality

[11] The same proof gives a similar finiteness and effectivity result for the set of $\mathbf{Q}$-rational points of the Basic Brill–Noether variety WX of *any* curve X defined over $\mathbf{Q}$ that has the property that all its very sporadic points are $\mathbf{Q}$- rational and $|X(\mathbf{Q})|$ is strictly greater than $\gamma_{\mathbf{Q}}(X)$.

of $X_1(N)$, the corresponding Basic Brill–Noether variety $WX_1(N)$ would contain at least one Siegel point.

5. Digression: 13-torsion

The case of $X = X_1(13)$ is simple, but still instructive, so it may be a good introduction to our discussion of $X_1(17)$. For a study of the arithmetic of $X_1(13)$ and its jacobian, see [**MT73**]. Since $X_1(13)$ is of genus two, it is hyperelliptic; $\gamma_{\mathbf{Q}}(X) = 2$. Let $\sigma : X \to X$ denote the hyperelliptic involution. We thank Andrew Sutherland who communicated to us the following hyperelliptic equation[12] for $X_1(13)$:

$$y^2 = x^6 + 4x^5 + 6x^4 + 2x^3 + x^2 + 2x + 1 := f(x).$$

So, in these coordinates, $\sigma(x, y) = (x, -y)$. The polynomial $f(x)$ is irreducible over the cyclotomic field $\mathbf{Q}(\mu_{13})$ obtained by adjoining all 13-th roots of unity to $\mathbf{Q}$. This guarantees that $J_1(13) := \mathrm{Pic}^0 X_1(13)$ has no nontrivial $\mathbf{Q}(\mu_{13})$-rational 2-torsion, and hence no $\mathbf{Q}$-rational 2-torsion. Let $\mathbf{C} \subset \mathbf{X_1(13)}$ denote the set of six $\mathbf{Q}$-rational cusps. The hyperelliptic involution $\sigma : X_1(13) \to X_1(13)$ preserves the set $\mathbf{C}$, and acts without fixed points on it (the latter statement following immediately from the irreducibility of $f(x)$ over $\mathbf{Q}$). Let

$$\mathbf{C} = \{\mathbf{u}, \sigma\mathbf{u}\} \sqcup \{\mathbf{v}, \sigma\mathbf{v}\} \sqcup \{\mathbf{w}, \sigma\mathbf{w}\}$$

be the orbit decomposition of $\mathbf{C}$ under the action of σ.

We have the natural projection,

$$\eta_2 : \mathrm{Sym}^2(X) \longrightarrow \mathrm{Pic}^2(X), \tag{5.1}$$

with the basic Brill–Noether variety $W = W(X) \subset \mathrm{Pic}^2(X)$ equal to a single point—call it $0 \in Pic^2(X)$. It can be taken to be the image of any of the three $\mathbf{Q}$-rational points

$$(u, \sigma u), (v, \sigma v), (w, \sigma w) \in \mathrm{Sym}^2(X).$$

This allows us to make the natural identification, rational over $\mathbf{Q}$, $\mathrm{Pic}^0(X) \cong \mathrm{Pic}^2(X)$, sending 0 to 0, and to view $\mathrm{Pic}^2(X)$ in a canonical way as an abelian surface over $\mathbf{Q}$. By [**MT73**] the Mordell–Weil group of $\mathrm{Pic}^2(X)$ is of order 19.

The morphism η_2 is an isomorphism on the inverse image of the complement of the origin $\{0\}$ in $\mathrm{Pic}^2(X)$ and identifies $\mathrm{Sym}^2(X)$ as the blow-up of $\mathrm{Pic}^2(X)$ at the point 0. Let

$$\mathcal{G} \subset \mathrm{Sym}^2(X)$$

denote this exceptional divisor. We may view, then, $\mathcal{G} := \eta_2^{-1}(0)$ as the locus $\{(x, \sigma x) \mid x \in X\}$, or more properly as the quotient X/σ of X under the action of the involution σ. In terms of the parameters (x, y) above, we may say that $\mathcal{G}$ is parametrized by the variable x.

We also interpret this as saying that there is a unique one-parameter family of elliptic curves over quadratic number fields containing (correspondingly rational) 13-torsion, this family being rationally parameterized by $\mathcal{G}$.

We focus, now, on the isomorphism over $\mathbf{Q}$:

$$\eta_2 : \mathrm{Sym}^{2,\#}(X) = \mathrm{Sym}^2(X) - \mathcal{G} \stackrel{\cong}{\longrightarrow} \mathrm{Pic}^2(X) - \{0\}. \tag{5.2}$$

[12]Such an equation had also been computed by J. Blass and was shown to be irreducible over Q; see the reference [1] in [**MT73**].

THEOREM 5.1. *The morphism η_2 induces a one-one correspondence between these two finite sets of cardinality* 18*:*

- $\mathrm{Sym}^2(\mathcal{C}) - \{(u,\sigma u),(v,\sigma v),(w,\sigma w)\}$
- *The set of* $\mathbf{Q}$*-rational points of* $\mathrm{Pic}^2(X) - \{0\}$.

Proof: From the above discussion, it follows that η_2 is an injection of the set $\mathrm{Sym}^2(\mathcal{C}) - \{(u,\sigma u),(v,\sigma v),(w,\sigma w)\}$ into $\mathrm{Pic}^2(X) - \{0\}$. The theorem follows by computing the cardinality of the two sets in question; i.e.,

$$\binom{6}{2} + 6 - 3 \quad = \quad 19 - 1.$$

COROLLARY 5.2. *Every sporadic point for $X_1(13)/\mathbf{Q}$ is cuspidal. Elliptic curves defined over quadratic fields possessing (correspondingly rational) 13-torsion are parametrized by the non-cuspidal $\mathbf{Q}$-rational points of $\mathcal{G} = \mathbf{P}^1$ via the natural hyperelliptic projection $X_1(13) \to \mathcal{G}$.*

6. Families of 17-torsion

The curve $X = X_1(17)$ is of genus 5 with $\mathbf{Q}$-gonality and $\mathbf{C}$-gonality both equal to 4. The basic Brill–Noether variety $WX_1(17)$ is a curve.

The curve $X_1(17)$ has no non-cuspidal very sporadic points (Proposition 4.1) and no non-cuspidal sporadic points (Proposition 6.7). Andrew Sutherland has computed elegant equations for these modular curves in [**Sut**]. The equation for $X_1(17)$ is particularly crisp[13]:

There is a birational morphism (over $\mathbf{Q}$) of $X_1(17)$ onto the bi-projective curve of bi-degree $(4,4)$ in $\mathbf{P}^1 \times \mathbf{P}^1$ cut out by the polynomial

$$(*) \quad x^4y - x^3y^3 - x^3y + x^2y^4 + x^2y - x^2 - xy^4 + xy^3 - xy^2 + xy + y^3 - 2y^2 + y = 0.$$

This morphism is an embedding when restricted to the complement of the cusps, $Y_1(17) \subset X_1(17)$ into $\mathbf{P}^1 \times \mathbf{P}^1$. Projection to the first factor is given by the modular unit[14] $x := E_5E_6/E_1E_3$ and the projection to the second factor is given by the modular unit $y := E_6E_7/E_2E_8$. Both x and y are functions of degree 4 and in fact there is another function of degree 4, namely:

$$z = \frac{y(x^2 - yx + y - 1)}{(y-1)^2x}.$$

An example of the type of result that we prove (based, of course, on the results already mentioned) is the following:

THEOREM 6.1. *Any elliptic curve defined over a field of degree ≤ 4 possessing points of order 17 defined over that field can be obtained by applying a diamond operator to a point of $X_1(17)$ for which one of the functions x, y takes on a rational value $\neq 0, 1$ or z takes a value $\neq 0$. Conversely, setting x, y to a rational value $\neq 0, 1$ or z to a value $\neq 0$ defines an elliptic curve over a field of degree four with a rational 17-torsion point.*

[13] As we understand it, this equation was originally written down by Cady and Elkies; see also a closely related description of $X_1(17)$ in [**JKL11b**].

[14] Here we are using the notation of Yang [**Yan09**], following Kubert-Lang [**KL81**].

Moreover, the rational parameters x, y, z give, up to **Q**-*similarity*[15] *all* **Q**-*rational parametrizations of* $X_1(17)$ *of degree equal to its gonality (i.e., degree = 4). The Galois group of the finite extension*

$$\mathbf{Q}(x) \subset \mathbf{Q}(X_1(17))$$

is the full symmetric group[16] S_4 *while the finite mappings*

$$y, z : X_1(17) \to \mathbf{P}^1$$

factor through the bi-elliptic representation

$$X_1(17) \longrightarrow X_1(17)/\{\text{action of } \langle 13\rangle\} = X_1(17)/\{\text{action of } \langle 3\rangle^4\}$$

and their Galois group is the dihedral group D_4.

The functions x, y, z of the theorem are in the equivalence classes of **Q**-parametrizations of type (**C**), (**A**), (**B**) described in subsection 6.2 below.

The fun here is that there are, in fact, two distinct ways of getting at the diophantine problem involved, as discussed above. They dovetail in a nice way. We can approach the problem either by considering:

- **Q**-rational points on the Basic Brill–Noether modular curve WX,

or

- rational cuspidal divisors and "fine" Siegel units.

6.1. Via the Basic Brill–Noether modular curve. We have computed the Basic Brill–Noether modular curve $W := WX_1(17)$ to be a double cover of a plane quintic (reducible) curve

$$(*) \quad V: \quad X \cdot (X^4 - 3X^2Y^2 - 3X^2Z^2 + Y^4 + 2Y^3Z + 3Y^2Z^2 - 2YZ^3 + Z^4) = 0.$$

The involution v of W that is the automorphism of the double cover $W \to V$ (the identity on V) has three descriptions. First, it is given by the diamond operator involution $\langle 13\rangle = \langle 3\rangle^4$. Secondly, it is also the involution induced on W (via the Serre duality theorem) from the transformation of divisors of degree four $D \mapsto K - D$ where K is the canonical divisor (of degree 8) on $X_1(17)$. The third description comes from what one might call the *canonical representation* of $W \to V$ as described in some generality in Subsections 3.5 and 3.6 above.

The group, Δ, of **Q**-automorphisms of X is canonically isomorphic to $(\mathbf{Z}/17\mathbf{Z})^*/\{\pm 1\}$. The operator $\langle 3\rangle \in \Delta$ is a generator.

Let $S_k := S_k(\Gamma_1(17))$ denote the **Q**-vector space of cuspforms of weight k on $\Gamma_1(17)$). Since the genus of $X_1(17)$ is 5 we have $\dim S_2 = 5$. The characteristic polynomial of $\langle 3\rangle$ acting on S_2 is $(x-1)(x^4+1)$, this means that there is a basis $\omega_0, \ldots, \omega_4 \in S_2$ such that with respect to this basis we have:

$$\langle 3\rangle = \begin{bmatrix} 1 & 0 & 0 & 0 & 0 \\ 0 & 0 & 1 & 0 & 0 \\ 0 & 0 & 0 & 1 & 0 \\ 0 & 0 & 0 & 0 & 1 \\ 0 & -1 & 0 & 0 & 0 \end{bmatrix}$$

[15]Recall the definition in Proposition 3.1: two parametrizations are **Q-similar** if one can be brought to the other by composition with appropriate **Q**-isomorphisms of domain and range.

[16]Hilbert's Irreducibility theorem would then guarantee infinitely many specializations $x \mapsto a \in \mathbf{Q}^*$ give a quartic polynomial in $\mathbf{Q}[y]$ with full symmetric Galois group.

One such basis is given by

$$\omega_0 := q - q^2 - q^4 - 2q^5 + 4q^7 + 3q^8 + O(q^9) \tag{6.1}$$

$$\omega_1 := q - q^2 - q^3 + q^6 - q^7 + q^8 + O(q^9) \tag{6.2}$$

$$\omega_2 := q^2 - q^3 - 2q^4 + q^5 + q^6 + q^7 + O(q^9) \tag{6.3}$$

$$\omega_3 := -q^2 + q^3 + q^4 + q^5 - q^6 - q^7 - q^8 + O(q^9) \tag{6.4}$$

$$\omega_4 := q^3 - 2q^4 + q^6 - q^7 + 3q^8 + O(q^9) \tag{6.5}$$

Every nonzero element in $\mathrm{Sym}^2(S_2)$ defines a quadratic form in the ω_i and and hence a quadric in $\mathbf{P}^4$. Now let $Y \subseteq \mathrm{Sym}^2(S_2)$ be the kernel of the natural map:

$$\mathrm{Sym}^2(S_2) \to S_4$$

Then Y will be a 3-dimensional space with basis e_0, e_1, e_2 given by

$$e_0 := \omega_0^2 - \omega_1^2 - \omega_2^2 - \omega_3^2 - \omega_4^2 \tag{6.6}$$

$$e_1 := 2\omega_1\omega_2 + 2\omega_1\omega_3 - 2\omega_3\omega_4 \tag{6.7}$$

$$e_2 := 2\omega_2\omega_3 + 2\omega_1\omega_4 + 2\omega_2\omega_4 \tag{6.8}$$

The matrix of $\langle 3\rangle$ acting on Y with respect to this basis is:

$$\langle 3\rangle = \begin{pmatrix} 1 & 0 & 0 \\ 0 & 0 & 1 \\ 0 & -1 & 0 \end{pmatrix}$$

Let (a_0, a_1, a_2) be coordinates of Y with respect to the basis e_0, e_1, e_2. Now consider the locus $V \subset \mathbf{P}^2 = \mathbf{P}(Y)$ corresponding to the singular quadrics in $\mathbf{P}^4$. This locus will be given by the single homogenous equation of degree 5, **(*)** above.

Each of these singular quadrics has (generally) two rulings by planes, and each of these planes intersect the canonically embedded curve X in an (effective, of course) divisor of degree 4. Each ruling, then, gives a unique linear system of effective divisors of degree 4 on X. That is, we can identify the Basic Brill–Noether curve W with the locus of rulings on these singular quadrics. The involution v simply switches rulings on the same singular quadric.

The plane quintic V breaks up into the union of a line

$$V_0 : \quad X = 0$$

and a plane quartic

$$V_1 : \quad X^4 - 3X^2Y^2 - 3X^2Z^2 + Y^4 + 2Y^3Z + 3Y^2Z^2 - 2YZ^3 + Z^4 = 0$$

and $W = W_0 \cup W_1$ is a union of two irreducible components where W_0 (a double cover of V_0) is an elliptic curve of Cremona type 17a4.

The curve of genus one, W_0 is directly related to the bi-elliptic representation of $X_1(17)$. It has four rational points, two of which yield parameterizations in the $\mathbf{Q}$-similarity class of the function y and the other two yield parameterizations in the class of z.

The more interesting component W_1 is given (birationally) as a double cover of V_1 given by extracting a "square root" of the function

$$(2Y^2Z + 2XY^2 + XZ^2 - X^3)/X^3$$

on V_1.

Much of the internal structure of the Basic Brill–Noether curve W is directly related to the bi-elliptic representation of $X_1(17)$ mentioned above, so let us return to it with a bit more detail. The diamond operators of $X_1(17)$ acting functorially on W preserve the irreducible component W_1 and we have the following curiously similar sequences of double covers:

- Consider the sequence of double covers:[17]

$$\begin{array}{ccccccc} X \longrightarrow & X/(\langle 3\rangle^4) & \longrightarrow & X/(\langle 3\rangle^2) & \longrightarrow & X/(\langle 3\rangle) \\ & \downarrow \approx & & \downarrow \approx & & \downarrow = \\ & 17a4 & \longrightarrow & 17a2 & \longrightarrow & X_0(17) \end{array}$$

  We easily compute that $X/(\langle 3\rangle^4), X/(\langle 3\rangle^2)$ and $X/(\langle 3\rangle)$ are curves of genus 1, and the automorphism $\langle 3\rangle$ acts freely on them of order $4, 2$ and 1 respectively. In particular, the action of $\langle 3\rangle$ on $X/(\langle 3\rangle^4)$ can be understood as the action of translation by a ($\mathbf{Q}$-rational) point $\mathcal{P}$ of order 4 in the jacobian, $\mathcal{J} := \mathrm{Pic}^0(X/(\langle 3\rangle^4)))$. This pins things down, after consulting Cremona's tables, forcing (the jacobian of) $X/(\langle 3\rangle^4)$ to be $17a4$ (which is the only curve of conductor 17 that has a rational 4-torsion point, the quotient by which is isomorphic to $X_0(17)$) and forcing (the jacobian of) $X/(\langle 3\rangle^2)$ to then be 17a2.

  It is an exercise to see, with no computation at all, that W_0 can be canonically identified as the curve of genus one given as the quotient of the curve $X/(\langle 3\rangle^4)$ by the natural action of the 2-torsion subgroup of its jacobian. It follows then that W_0 is isomorphic to 17a4, and therefore has exactly four rational points. These four points break up into two orbits under the action of the 'diamond operators' Δ contributing to two $\mathbf{Q}$-similarity classes represented by the functions "y" and "z" of our theorem.
- The curve W_1 is a curve of genus 7, but is also directly related to 17a4 and neatly mimics the sequence displayed in the previous bullet as follows. Consider the diamond operators acting on W_1 which can be computed to produce the sequence of double covers:

$$\begin{array}{ccccccccc} W_1 \longrightarrow & W_1/(\langle 3\rangle^4) & \longrightarrow & W_1/(\langle 3\rangle^2) & \longrightarrow & W_1/(\langle 3\rangle) & \longrightarrow & X_0(17) \\ & \downarrow = & & \downarrow \approx & & \downarrow \approx & & \downarrow = \\ & V_1 & \longrightarrow & 17a4 & \longrightarrow & 17a2 & \longrightarrow & X_0(17) \end{array}$$

The curve V_1 has exactly four $\mathbf{Q}$-rational points: $(1, \pm 1, \pm 1)$ and the eight points in W_1 comprising the inverse image of those four points are all $\mathbf{Q}$-rational, and therefore give the full set of $\mathbf{Q}$-rational points of W_1. These eight point comprise a single Δ-orbit. Therefore they give rise to a unique $\mathbf{Q}$-similarity class of rational parametrizations of $X_1(17)$, for which the function "x" of the theorem is a representative.

6.2. Via Fine Siegel Units. As is clear from the account already given, to compute the rational points on the Basic Brill–Noether curve $WX_1(17)$ is not greatly difficult since each of its connected components is a perfectly specified finite

[17]We use Cremona's classification to refer to some of the elliptic curves that occur in these computations.

cover of an elliptic curve possessing only four rational points. Section 4 above offers an utterly independent way of making this computation: by Proposition 4.1 the only very sporadic points on $X_1(17)$ are the eight rational cusps, and therefore any **Q**-rational function ϕ of degree 4 on $X_1(17)$ has the curious property, as discussed in section 4 that

- any of its fibers that contain even a single rational cusp must consist entirely of rational cusps—call such a fiber a **rational cuspidal fiber** and
- there are at least two such rational cuspidal fibers.

It follows that by composing ϕ with an appropriate **Q**-automorphism of $\mathbf{P}^1$ one gets a fine Siegel unit. It follows that the problem of computing the **Q**-rational points on $WX_1(17)$ is essentially equivalent to that of computing fine Siegel units of degree four. As mentioned in Section 4, this is a finite computation.

We will be giving the collection of all fine Siegel units f of geometric degree 4—up to composition by appropriate **Q**-automorphisms $b : \mathbf{P}^1 \to \mathbf{P}^1$. This we do by listing the divisors that describe the cuspidal-fibers for each f and organizing these cuspidal-fiber matrices in Δ-orbits. Each such Δ-orbit describes *one* class of **Q**-parametrizations of $X_1(17)$ (of geometric degree 4); there are three of them.

Each divisor with support on these rational cusps is encoded by 8 integers, where the i-th integer is the multiplicity of the i-th cusp in the ordering:

$$\{2/17, 3/17, 4/17, 5/17, 6/17, 7/17, 8/17, \infty\}.$$

We display this data for a given f as a matrix, with exactly 8 columns, and as many rows as there are cuspidal-fibers for f. This is the **cuspidal-fiber matrix of** f as discussed in section 4 above.

The first two classes factor through the quotient of $X_1(17)$ under the action of the involution $\langle 13\rangle$. That is, they factor through the double cover

$$X_1(17) \xrightarrow{\pi} X_1(17)/\langle 13\rangle.$$

The quotient curve $X_1(17)/\langle 13\rangle$ is isomorphic over **Q** to the elliptic curve $E := 17A4$ in Cremona's classification. The Mordell–Weil group of 17A4 (over **Q**), is cyclic of order four. Make one (of the four possible) identifications—rational over **Q**:

$$X_1(17)/\langle 13\rangle \overset{\iota}{\cong} E$$

The determination of the cuspidal-fiber matrix for each of these two classes uses a minimum of computation; i.e., we work essentially by 'pure thought,' given the fact that $E(\mathbf{Q})$ is cyclic of order four. Since there are eight rational cusps on $X_1(17)$ and $\iota \cdot \pi$ is of degree two, these eight rational cusps are unramified for the mapping $\iota \cdot \pi$, and the set of them map surjectively—by a two-to-one mapping—to $E(\mathbf{Q})$. Now E itself has precisely four **Q**-rational involutions v_a such that $E/\langle v_a\rangle \cong \mathbf{P}^1$. These are given by the formulae $x \mapsto a - x$ for $a \in E(\mathbf{Q})$. Note that

- $|E(\mathbf{Q})/\langle v_a\rangle| = 3$ if a is trivial or of order two, while
- $|E(\mathbf{Q})/\langle v_a\rangle| = 2$ if a is one of the two generators of $E(\mathbf{Q})$.

Denote

$$f_a : E \to E/\langle v_a\rangle \approx \mathbf{P}^1$$

the double cover associated to the involution v_a. Now if t_b is translation by b with b a point of order 4 and if $a' - a$ is in $E(\mathbf{Q})[2] \setminus \{0\}$ then $v_a = t_b \circ v_{a'} \circ t_b^{-1}$, implying that f_a and $f_{a'}$ are in the same parameterization class. So we have (at most) two $\mathbf{Q}$-rational classes of parametrizations of E of degree two coming from the four maps f_a.That these are in fact different equivalence classes can be seen from the bullets above. For more specificity, choose an identification ("$\approx$") of $E/\langle v_a\rangle$ with $\mathbf{P}^1$ so that in the first case above $E(\mathbf{Q})/\langle v_a\rangle$ is identified with the set $\{0,1,\infty\}$ (any order will do) and in the second case it is identified with $\{0,\infty\}$. Fixing such an identification, but composing with $\iota\cdot\pi$ for the four possible choices of ι gives two Δ-orbits of fine Siegel units of degree four on $X_1(17)$.

The cuspidal-fiber matrices for the two $\mathbf{Q}$-parametrizations of $X_1(17)$ (of geometric degree 4) that factor through $X_1(17)/\langle 13\rangle$ are immediately computable from this discussion. In particular they each consist of a single Δ-orbit of order two. We'll call them "$\mathbf{Q}$-similarity classes **(A)** and **(B)**."

- **$\mathbf{Q}$-similarity class of parametrizations (A):**

$$M_1 := \begin{pmatrix} 0 & 0 & 2 & 0 & 0 & 0 & 0 & 2 \\ 0 & 1 & 0 & 1 & 1 & 1 & 0 & 0 \\ 2 & 0 & 0 & 0 & 0 & 0 & 2 & 0 \end{pmatrix}$$

$$M_2 := \begin{pmatrix} 1 & 0 & 1 & 0 & 0 & 0 & 1 & 1 \\ 0 & 2 & 0 & 2 & 0 & 0 & 0 & 0 \\ 0 & 0 & 0 & 0 & 2 & 2 & 0 & 0 \end{pmatrix}$$

- **$\mathbf{Q}$-similarity class of parametrizations (B):**

$$M_3 := \begin{pmatrix} 1 & 0 & 0 & 0 & 1 & 1 & 1 & 0 \\ 0 & 1 & 1 & 1 & 0 & 0 & 0 & 1 \end{pmatrix}$$

$$M_4 := \begin{pmatrix} 1 & 1 & 0 & 1 & 0 & 0 & 1 & 0 \\ 0 & 0 & 1 & 0 & 1 & 1 & 0 & 1 \end{pmatrix}$$

Far less evident is the third (and last) class of $\mathbf{Q}$-parametrization of $X_1(17)$ of degree four. This class ("**(C)**") is given (as shown in the discussion below) by a single Δ-orbit of order eight, described by eight cuspidal-fiber matrices $M_5, M_6, \ldots, M_{12}$ permuted by the action of Δ. These 8 matrices correspond to the 8 rational points of W_1.

- **$\mathbf{Q}$-similarity class of parametrizations (C):**

$$M_5 := \begin{pmatrix} 1 & 0 & 0 & 1 & 0 & 0 & 0 & 2 \\ 0 & 2 & 0 & 0 & 0 & 1 & 1 & 0 \\ 0 & 0 & 3 & 0 & 1 & 0 & 0 & 0 \end{pmatrix}$$

$$M_6 := \langle 3\rangle M_5 \quad \cdots \quad M_{12} := \langle 3\rangle^7 M_5$$

Depending on how you decide which of the three rational cuspidal fibers will be *zeroes* of your function and which *poles* you get different $\mathbf{Q}$-linear parameterizations of $X_1(17)$ and different fine Siegel units. For example E_1E_3/E_5E_6 has the first row of M_5 as zero-divisor and the second row as polar-divisor, while $E_3E_4E_8/E_2E_6E_7$ has the third row of M_5 as zero-divisor and the second row as polar-divisor.

We can summarize as follows. Let

$$\Gamma \subset X_1(17)(\bar{\mathbf{Q}})$$

denote the set of non-cuspidal algebraic points of $X_1(17)$ defined over number fields of degree four. If $\gamma\in\Gamma$ let $\mathbf{Q}(\gamma)$ denote the number field (of degree four) over which γ is defined. Say that γ is of type **(A)**, **(B)** respectively **(C)** if the projection of γ

under one of the $\mathbf{Q}$-parametrization in the equivalence class **(A)**, **(B)** respectively **(C)** is a $\mathbf{Q}$-rational point of $\mathbf{P}^1$. Let $\Gamma_{(\mathbf{A})} \subset \Gamma$ denote the subset of points of type **(A)**; and similarly for $\Gamma_{(\mathbf{B})}$ and $\Gamma_{(\mathbf{C})}$.

THEOREM 6.2. *The set* Γ *(of non-cuspidal algebraic points of* $X_1(17)$ *defined over number fields of degree four) is the disjoint union*

$$\Gamma = \Gamma_{(\mathbf{A})} \bigsqcup \Gamma_{(\mathbf{B})} \bigsqcup \Gamma_{(\mathbf{C})}.$$

PROOF. The above discussion gives us the full list of $\mathbf{Q}$-similar classes of $\mathbf{Q}$-parametrized points of degree 4 on $X_1(17)$. The fact that $W_4^2 X_1(17)$ is empty (because $X_1(17)$ has no functions of degree 3) shows that the union above is a *disjoint union*. It remains to show that $X_1(17)$ has no very sporadic, or sporadic points that are not cusps. For this, see section 6.3 below. □

PROPOSITION 6.3. *Let* x, y, z *be the functions from theorem* 6.1, *then the Galois groups of* $\mathbf{Q}(x) \subset \mathbf{Q}(X_1(17))$, $\mathbf{Q}(y) \subset \mathbf{Q}(X_1(17))$ *and* $\mathbf{Q}(z) \subset \mathbf{Q}(X_1(17))$ *are* S_4, D_4 *and* D_4 *respectively.*

PROOF. Let K_x denote the Galois closure of $\mathbf{Q}(x) \subset \mathbf{Q}(X_1(17))$, then by looking at matrix M_5 one sees that $[K_x : \mathbf{Q}(X_1(17))]$ has to be divisible by 6, implying that 24 divides $[K_x : \mathbf{Q}(x)]$ hence the Galois group has to be S_4.

For the proof that the other two Galois groups are D_4, one can use the following observation: Suppose that $M \subset L \subset K$ is a tower of field extensions with $[L : M] = [K : L] = 2$ and K/M is not Galois, then K/M has Galois group D_4. One can then apply this observation to $\mathbf{Q}(y) \subset \mathbf{Q}(X_1(17)/\langle 3^4 \rangle) \subset \mathbf{Q}(X_1(17))$ and $\mathbf{Q}(z) \subset \mathbf{Q}(X_1(17)/\langle 3^4 \rangle) \subset \mathbf{Q}(X_1(17))$. Now $\mathbf{Q}(X_1(17))/\mathbf{Q}(y)$ and $\mathbf{Q}(X_1(17))/\mathbf{Q}(z)$ are not Galois follows from the fact that there is no subgroup $H \subset \mathrm{Aut}_{\mathbf{Q}}(X_1(17)) = (\mathbf{Z}/17\mathbf{Z})^*/\pm 1$ of order 4 such that $X_1(17)/H \cong \mathbf{P}^1$.

□

6.3. Sporadic and very sporadic points on $X_1(17)$. The computations in the previous section show that the number of g_4^1's on $X_1(17)$ that are defined over $\mathbf{Q}$ is exactly 12. This is actually proved twice, once by proving $12 = 4 + 8 = \#W_0(\mathbf{Q}) + \#W_1(\mathbf{Q})$, and once by using proposition 4.2 and computing that there are exactly 12 cuspidal fiber matrices corresponding to fine Siegel units of degree 4. The main goal of this section is to prove the following theorem:

THEOREM 6.4. *Every point on* $X_1(17)$ *of degree* 4 *over* $\mathbf{Q}$ *is in one of the 12* g_4^1*'s.*

For the proof of this theorem we will use a slight modification of a theorem due to Michael Stoll in [**Kam89**].

Let $C/\mathbf{Q}$ be a curve with jacobian J, and let $d \geq 1$ be an integer. Let C^d be the dth power, and $C_d := \mathrm{Symm}^d(C)$ the dth symmetric power of C. Let

$$C_d^{\{1\}} := C_d \times_J W_d^1(C).$$

So $C_d^{\{1\}} \subset C_d$ is the closed subvariety parametrizing those divisors D of degree d such that $\dim H^0(O_C(D), C) - 1 = \dim |D| \geq 1$.

Denote by $s : C^d \to C_d$ the natural quotient map.

THEOREM 6.5. *Let* ℓ *be a prime of good reduction for* C. *Let* $P_0 \in C(\mathbf{Q})$ *be chosen as base-point for an embedding* $\iota : C \to J$. *This also induces morphisms*

$C^d \to C_d \to J$. If the following assumptions hold:

(1) *$\ell > 2$ or $J(\mathbf{Q})[2]$ injects into $J(\mathbf{F}_\ell)$ (for example, $\#J(\mathbf{Q})$ is odd).*
(2) *$J(\mathbf{Q})$ is finite.*
(3) *The reduction map $C(\mathbf{Q}) \to C(\mathbf{F}_\ell)$ is surjective.*
(4) *The intersection of the image of $C_d(\mathbf{F}_\ell)$ in $J(\mathbf{F}_\ell)$ with the image of $J(\mathbf{Q})$ under reduction mod ℓ is contained in the image of $C^d(\mathbf{F}_\ell)$.*

Then $C_d(\mathbf{Q}) \setminus C_d^{\{1\}}(\mathbf{Q})$ is contained in the image of $C^d(\mathbf{Q}) \to C_d(\mathbf{Q})$.

PROOF. Let ρ_X denote the reduction map $X(\mathbf{Q}) \to X(\mathbf{F}_\ell)$, where X is a smooth projective variety over $\mathbf{Q}$ with good reduction at ℓ.

From assumptions (2) and (1) we can deduce that $\rho_J : J(\mathbf{Q}) \to J(\mathbf{F}_\ell)$ is injective. By the definition of $C_d^{\{1\}}$ it is also clear that $C_d(\mathbf{Q}) \setminus C_d^{\{1\}}(\mathbf{Q}) \to J(\mathbf{Q})$ is injective.

Finally (3) shows that $\rho_C^d : C^d(\mathbf{Q}) \to C^d(\mathbf{F}_\ell)$ is surjective.

$$\begin{array}{ccccc}
 & & C_d(\mathbf{Q}) \setminus C_d^{\{1\}}(\mathbf{Q}) & & \\
 & & \big\downarrow & \searrow^{\iota} & \\
C^d(\mathbf{Q}) & \xrightarrow{s} & C_d(\mathbf{Q}) & \xrightarrow{\iota} & J(\mathbf{Q}) \\
{\scriptstyle \rho_{C^d}}\big\downarrow{\scriptstyle (3)} & & \big\downarrow{\scriptstyle \rho_{C_d}} & & {\scriptstyle (2,1)}\big\downarrow{\scriptstyle \rho_J} \\
C^d(\mathbf{F}_\ell) & \xrightarrow{s} & C_d(\mathbf{F}_\ell) & \xrightarrow{\iota} & J(\mathbf{F}_\ell).
\end{array}$$

Now let $P \in C_d(\mathbf{Q}) \setminus C_d^{\{1\}}(\mathbf{Q}) \to J(\mathbf{Q})$. We want to show that there is a $Q \in C^d(Q)$ such that $s(Q) = P$. Now $\rho_J \circ \iota(P) = \iota \circ \rho_{C_d}(P) \in J(\mathbf{F}_\ell)$ so from assumption (4) it follows that there is a $\overline{Q} \in C^d(\mathbf{F}_\ell)$ such that $\iota \circ s(\overline{Q}) = \rho_J \circ \iota(P)$. Let $Q \in C^d(\mathbf{Q})$ be such that $\rho_{C^d}(Q) = \overline{Q}$ then

$$\rho_J \circ \iota \circ s(Q) = \iota \circ s(\overline{Q}) = \rho_J \circ \iota(P).$$

The injectivity of ρ_J implies $\iota \circ s(Q) = \iota(P)$ and because $P \notin C_d^1(\mathbf{Q})$ we know that $s(Q) = P$. □

COROLLARY 6.6. *If the above hypotheses hold for $d = \gamma_C(C)$ then all sporadic points of C are $\mathbf{Q}$-rational.*

PROPOSITION 6.7. *There are no sporadic non-cuspidal points on $X_1(17)$.*

PROOF. We apply Theorem 6.5 taking $C := X = X_1(17)$. We take $\ell = 3$, so (1) holds. Since $J_1(17)(\mathbf{Q})$ is of finite order[18], condition (2) holds. The Hasse-Weil bound implies that for an elliptic curve E over $\mathbf{F}_3$ we have $\#E(\mathbf{F}_3) \leq 3+1+2\sqrt{3} < 8$ so this E cannot have an $\mathbf{F}_3$-rational point of order 17 showing that $X = X_1(17)(\mathbf{F}_3)$ consists entirely of cusps, which gives (3). Finally we verified with a computation in magma that assumption (4) is also satisfied.

□

[18] $J_1(17)(\mathbf{Q})$ is of order $584 = 8 \cdot 73$ ([**Kam85**] for the prime-to-2 part; and for the 2-torsion: [**Par03**]). Regarding values of N for which $J_1(N)(\mathbf{Q})$ is finite, consult Prop. 6.2.1 in [**CES03**].

7. Appendix: Gonality

If X is a curve over k the k-**gonality** of X is the smallest positive integer $\gamma = \gamma_k(X)$ for which there is a degree γ mapping $f : X \to \mathbf{P}^1$ defined over k, or–equivalently–a k-parametrization of X of degree γ.

LEMMA 7.1. *Let k be algebraically closed. The k-gonality $\gamma = \gamma_k(X)$ is the smallest integer d for which equivalently:*

- *there exists a positive dimensional linear system of effective divisors of degree d on X defined over k;*
- *$W^1_d(X)$ is nonempty. (Here $W^r_d(X)$ denotes the Brill–Noether variety that classifies g^r_d's on X.)*

PROOF. The equivalence of the two bullets above follows from the definition of $W^1_d(X)$ and the assumption that k is algebraically closed. Suppose given a positive dimensional linear system of effective divisors of degree d on X defined over k, and suppose that d is the smallest degree for which there exists such a linear system. Let $D \subset X \times \mathbf{P}^1$ be the (effective) Cartier divisor—which we may assume to be a reduced and irreducible curve in $X \times \mathbf{P}^1$—representing the linear system so that if $\pi_{\mathbf{P}^1}$ and π_X denote projection to the indicated factors, then

- $\pi_{\mathbf{P}^1} : D \to \mathbf{P}^1$ is finite flat of degree d, and
- $\pi_X \circ \pi_{\mathbf{P}^1}^{-1} : t \mapsto D_t$ gives the linear system, with t a parameter of $\mathbf{P}^1$.

Clearly $d \leq \gamma$. Our task is to find a degree d mapping $f : X \to \mathbf{P}^1$ defined over k, proving that $\gamma \leq d$. We will show that for $t_1 \neq t_2 \in \mathbf{P}^1(k)$ the support of D_{t_1} is disjoint from that of D_{t_2}. Suppose it was not. Then write $D_{t_i} = \Delta + \Delta_i$ for $i = 1, 2$, where Δ consists of an effective divisor of positive degree common to both D_{t_1} and D_{t_2}. We have that Δ_i are effective divisors, $\Delta_1 \neq \Delta_2$ and yet $\Delta_1 \equiv \Delta_2$. But this would give us a positive dimensional linear system of degree strictly less than d, contrary to assumption.

Therefore: **(*):** the support of the divisors D_t are disjoint for distinct values of t.

Now view D as an algebraic system of divisors on $\mathbf{P}^1$ parametrized by X. That is, form the correspondence $D^x := \pi_{\mathbf{P}^1} \circ \pi_X^{-1}(x)$ for $x \in X$. By **(*)** the support of each fiber D_x consists of a single point. Since D is reduced, the projection $\pi_X : D \to X$ is an isomorphism. Define $f := \pi_{\mathbf{P}^1} \circ \pi_X^{-1} : X \to \mathbf{P}^1$. $\square$

References

[Abr96] Dan Abramovich, *A linear lower bound on the gonality of modular curves*, Internat. Math. Res. Notices **20** (1996), 1005–1011, DOI 10.1155/S1073792896000621. MR1422373

[AH91] Dan Abramovich and Joe Harris, *Abelian varieties and curves in $W_d(C)$*, Compositio Math. **78** (1991), no. 2, 227–238. MR1104789

[Bos08] J. Bosman, Explicit computations with modular Galois representations, PhD Thesis, 2008 `http://hdl.handle.net/1887/13364`.

[ACGH85] E. Arbarello, M. Cornalba, P. A. Griffiths, and J. Harris, *Geometry of algebraic curves. Vol. I*, Grundlehren der Mathematischen Wissenschaften [Fundamental Principles of Mathematical Sciences], vol. 267, Springer-Verlag, New York, 1985. MR770932

[ACG11] Enrico Arbarello, Maurizio Cornalba, and Pillip A. Griffiths, *Geometry of algebraic curves. Volume II*, Grundlehren der Mathematischen Wissenschaften [Fundamental Principles of Mathematical Sciences], vol. 268, Springer, Heidelberg, 2011. With a contribution by Joseph Daniel Harris. MR2807457

[CCS13] Pete L. Clark, Brian Cook, and James Stankewicz, *Torsion points on elliptic curves with complex multiplication (with an appendix by Alex Rice)*, Int. J. Number Theory **9** (2013), no. 2, 447–479, DOI 10.1142/S1793042112501436. MR3005559

[DvH14] Maarten Derickx and Mark van Hoeij, *Gonality of the modular curve $X_1(N)$*, J. Algebra **417** (2014), 52–71, DOI 10.1016/j.jalgebra.2014.06.026. MR3244637

[Kam89] S. Kamienny, *Torsion points on elliptic curves over fields of low degree*, Manuscripta Math. **65** (1989), no. 3, 349–355, DOI 10.1007/BF01303042. MR1015660

[Dol12] Igor V. Dolgachev, *Classical algebraic geometry*, Cambridge University Press, Cambridge, 2012. A modern view. MR2964027

[CES03] Brian Conrad, Bas Edixhoven, and William Stein, *$J_1(p)$ has connected fibers*, Doc. Math. **8** (2003), 331–408. MR2029169

[Fal94] Gerd Faltings, *The general case of S. Lang's conjecture*, Barsotti Symposium in Algebraic Geometry (Abano Terme, 1991), Perspect. Math., vol. 15, Academic Press, San Diego, CA, 1994, pp. 175–182. MR1307396

[Fre94] Gerhard Frey, *Curves with infinitely many points of fixed degree*, Israel J. Math. **85** (1994), no. 1-3, 79–83, DOI 10.1007/BF02758637. MR1264340

[Hes02] F. Hess, *Computing Riemann-Roch spaces in algebraic function fields and related topics*, J. Symbolic Comput. **33** (2002), no. 4, 425–445, DOI 10.1006/jsco.2001.0513. MR1890579

[Hoe14] M. Hoeij, Low degree places on the modular curve $X_1(N)$ (preprint `http://arxiv.org/abs/1202.4355`)

[JKL11a] Daeyeol Jeon, Chang Heon Kim, and Yoonjin Lee, *Families of elliptic curves over cubic number fields with prescribed torsion subgroups*, Math. Comp. **80** (2011), no. 273, 579–591, DOI 10.1090/S0025-5718-10-02369-0. MR2728995

[JKL11b] Daeyeol Jeon, Chang Heon Kim, and Yoonjin Lee, *Families of elliptic curves over quartic number fields with prescribed torsion subgroups*, Math. Comp. **80** (2011), no. 276, 2395–2410, DOI 10.1090/S0025-5718-2011-02493-2. MR2813367

[Kam82] S. Kamienny, *On $J_1(p)$ and the conjecture of Birch and Swinnerton-Dyer*, Duke Math. J. **49** (1982), no. 2, 329–340. MR659944

[Kam85] S. Kamienny, *Rational points on modular curves and abelian varieties*, J. Reine Angew. Math. **359** (1985), 174–187, DOI 10.1515/crll.1985.359.174. MR794803

[Kam86] S. Kamienny, *Torsion points on elliptic curves over all quadratic fields*, Duke Math. J. **53** (1986), no. 1, 157–162, DOI 10.1215/S0012-7094-86-05310-X. MR835802

[Kam92a] S. Kamienny, *Torsion points on elliptic curves and q-coefficients of modular forms*, Invent. Math. **109** (1992), no. 2, 221–229, DOI 10.1007/BF01232025. MR1172689

[Kam92b] S. Kamienny, *Torsion points on elliptic curves over fields of higher degree*, Internat. Math. Res. Notices **6** (1992), 129–133, DOI 10.1155/S107379289200014X. MR1167117

[KM95] S. Kamienny and B. Mazur, *Rational torsion of prime order in elliptic curves over number fields*, Astérisque **228** (1995), 3, 81–100. With an appendix by A. Granville; Columbia University Number Theory Seminar (New York, 1992). MR1330929

[KN12] Sheldon Kamienny and Filip Najman, *Torsion groups of elliptic curves over quadratic fields*, Acta Arith. **152** (2012), no. 3, 291–305, DOI 10.4064/aa152-3-5. MR2885789

[KM88] M. A. Kenku and F. Momose, *Torsion points on elliptic curves defined over quadratic fields*, Nagoya Math. J. **109** (1988), 125–149. MR931956

[KL81] Daniel S. Kubert and Serge Lang, *Modular units*, Grundlehren der Mathematischen Wissenschaften [Fundamental Principles of Mathematical Science], vol. 244, Springer-Verlag, New York-Berlin, 1981. MR648603

[LL85] Michael Laska and Martin Lorenz, *Rational points on elliptic curves over* $\mathbf{Q}$ *in elementary abelian* 2*-extensions of* $\mathbf{Q}$, J. Reine Angew. Math. **355** (1985), 163–172, DOI 10.1515/crll.1985.355.163. MR772489

[LR13] Álvaro Lozano-Robledo, *On the field of definition of p-torsion points on elliptic curves over the rationals*, Math. Ann. **357** (2013), no. 1, 279–305, DOI 10.1007/s00208-013-0906-5. MR3084348

[Maz77] B. Mazur, *Modular curves and the Eisenstein ideal*, Inst. Hautes Études Sci. Publ. Math. **47** (1977), 33–186 (1978). MR488287

[Maz78] B. Mazur, *Rational isogenies of prime degree (with an appendix by D. Goldfeld)*, Invent. Math. **44** (1978), no. 2, 129–162, DOI 10.1007/BF01390348. MR482230

[MT73] B. Mazur and J. Tate, *Points of order* 13 *on elliptic curves*, Invent. Math. **22** (1973/74), 41–49, DOI 10.1007/BF01425572. MR0347826

[Mer96] Loïc Merel, *Bornes pour la torsion des courbes elliptiques sur les corps de nombres* (French), Invent. Math. **124** (1996), no. 1-3, 437–449, DOI 10.1007/s002220050059. MR1369424

[Mom84] Fumiyuki Momose, *p-torsion points on elliptic curves defined over quadratic fields*, Nagoya Math. J. **96** (1984), 139–165. MR771075

[Naj16] Filip Najman, *Torsion of rational elliptic curves over cubic fields and sporadic points on* $X_1(n)$, Math. Res. Lett. **23** (2016), no. 1, 245–272, DOI 10.4310/MRL.2016.v23.n1.a12. MR3512885

[Par03] Pierre Parent, *No 17-torsion on elliptic curves over cubic number fields* (English, with English and French summaries), J. Théor. Nombres Bordeaux **15** (2003), no. 3, 831–838. MR2142238

[Ste82] Glenn Stevens, *Arithmetic on modular curves*, Progress in Mathematics, vol. 20, Birkhäuser Boston, Inc., Boston, MA, 1982. MR670070

[Sut] A. Sutherland, `http://math.mit.edu/~drew/X1_altcurves.html`

[Yan09] Yifan Yang, *Modular units and cuspidal divisor class groups of* $X_1(N)$, J. Algebra **322** (2009), no. 2, 514–553, DOI 10.1016/j.jalgebra.2009.04.012. MR2529102

Mathematisch Instituut Universiteit Leiden Niels Bohrweg 1 2333 CA Leiden Nederland

Current address: Mathematisches Institut Universität Bayreuth 95440 Bayreuth Deutschland
Email address: `maarten@mderickx.nl`
URL: `http://www.mderickx.nl`

Mathematics Department Harvard University 1 Oxford Street Cambridge, Massachusetts 02138

Email address: `mazur@math.harvard.edu`
URL: `http://www.math.harvard.edu/~mazur/`

Department of Mathematics University of Southern California 3620 South Vermont Ave. Los Angeles, California 90089-2532

Email address: `kamienny@usc.edu`

Contemporary Mathematics
Volume **701**, 2018
http://dx.doi.org/10.1090/conm/701/14143

An explicit integral representation of Siegel-Whittaker functions on $\mathrm{Sp}(2,\mathbb{R})$ for the large discrete series representations

Yasuro Gon and Takayuki Oda

In Memoriam: Fumiyuki Momose

Abstract. We obtain an explicit integral representation of Siegel-Whittaker functions on $\mathrm{Sp}(2,\mathbb{R})$ for the large discrete series representations. We have another integral expression different from that of Miyazaki, 2000.

1. Introduction

In this article, we study Siegel-Whittaker functions on $G = \mathrm{Sp}(2,\mathbb{R})$, the real symplectic group of degree two for the large discrete series representations. Let P_S be the Siegel parabolic subgroup of G, which is a maximal parabolic subgroup with abelian unipotent radical N_S. Let π be an admissible representation of G and ξ be a *definite* unitary character of N_S. Siegel-Whittaker model for an admissible π is a realization of π in the induced module from a certain closed subgroup R which contains N_S. (See (2.2) for definition of R.) We consider the intertwining space

$$SW(\pi;\eta) = \mathrm{Hom}_{(\mathfrak{g}_{\mathbb{C}},K)}(\pi, C_\eta^\infty(R\backslash G)),$$

where η is an irreducible R-module such that $\eta|_{N_S}$ contains ξ, K is an maximal compact subgroup of G and $\mathfrak{g}_{\mathbb{C}}$ is the complexification of the Lie algebra of G. A function in this realization is called a *Siegel-Whittaker function* for π.

Takuya Miyazaki obtained a system of partial differential equations satisfied by Siegel-Whittaker functions for the large discrete representations in [**8**]. He also obtained a multiplicity one property and its formal power series solutions. In this article, we investigate further the system obtained by Miyazaki (Proposition 3.1) and give an explicit integral representation of the Siegel-Whittaker functions, which is of rapid decay, for a large discrete series representation π. (Theorem 5.1.) In other words, we show that the rapidly decreasing Siegel-Whittaker function for the large discrete series representation π is uniquely determined and (up to polynomials)

2010 *Mathematics Subject Classification.* Primary 11F70; Secondary 22E45.

Key words and phrases. large discrete series; Siegel-Whittaker functions.

The first author was also partially supported by JSPS Grant-in-Aid for Scientific Research (C) No. 26400017 and No. 17K05178.

The second author was partially supported by JSPS Grant-in-Aid for Scientific Research (A) No. 23244003.

described by the *partially confluent hypergeometric functions* in the A-radial part $(a_1, a_2) \in (\mathbb{R}_{>0})^2$:
(1.1)

$$\Psi_{\alpha,\beta}(a_1,a_2) = e^{-2\pi(h_1a_1^2+h_2a_2^2)} \int_0^1 F_{SW}\left(2\pi h_1 a_1^2 t + 2\pi h_2 a_2^2(1-t)\right) t^{\alpha-1}(1-t)^{\beta-1}\, dt,$$

with

$$F_{SW}(x) = F(x) = e^x x^{-\gamma} W_{\kappa,\mu}(2x).$$

Here, $W_{\kappa,\mu}(x)$ is the Whittaker's classical confluent hypergeometric function, real numbers $\alpha, \beta, \gamma > 0$ and κ, μ are half integers depending on parameters of representations π and η. This kind of integral already appeared in Gon [**2**], treated the cases for the large discrete series representations and P_J-principal series representations of $\mathrm{SU}(2,2)$.

The reason we begin to believe the validity of similar formula for $\mathrm{Sp}(2,\mathbb{R})$, is the paper Hirano-Ishii-Oda [**3**]. By this new integral expression (1.1), it seems possible to extend the former argument in [**3**] of the confluence from Siegel-Whittaker functions to Whittaker functions for P_J-principal series, to the confluence from Siegel-Whittaker functions to Whittaker functions [**10**] for the large discrete series of $\mathrm{Sp}(2,\mathbb{R})$. Here we recall the template of the integral formulas of Whittaker functions in $(a_1, a_2) \in (\mathbb{R}_{>0})^2$:

$$\mu_W(a_1,a_2) \int_0^\infty F_W(t)\, t^{-C} e^{-A\,t^2/a_2^2 - B\,a_1^2/t^2}\, dt, \tag{1.2}$$

where $F_W(t) = W_{0,\gamma}(t)$ with a positive integral parameter γ, $\mu_W(a_1, a_2)$ is a monomial in a_1, a_2 times $e^{\delta a_2^2}$ $(\delta > 0)$ and A, B, C are positive real parameters.

Probably we can proceed a bit more. Since Iida [**4**], Theorems 8.7 and 8.9 (Formulas (8.8) and (8.10), resp.) give analogous integral expressions for the matrix coefficient of P_J-principal series, and also Oda [**11**] (p.247), Theorem 6.2 similarly gives analogous integral expression for matrix coefficients of the large discrete series. We notice that the template of formulas in Theorem 6.2 of [**11**] is given by

$$\mu_{MC}(a_1,a_2) \int_0^1 F_{MC}\left(tx_1 + (1-t)x_2\right) t^{\alpha-1}(1-t)^{\beta-1}\, dt, \tag{1.3}$$

with $x_i = -(a_i - a_i^{-1})^2/4$ $(i = 1,2)$. Here $F_{MC}(x) = {}_2F_1(A, B; C; x)$ with A, B, C, α, β determined by Harish-Chandra parameters and the components of the minimal K-type. The elementary factor $\mu_{MC}(a_1, a_2)$ is a monomial function of $\left((a_i \pm a_i^{-1})/2\right)^{\pm 1/2}$ $(i = 1, 2)$.

These facts suggests that there will be a deformation of confluence (1.3) $\mapsto$ (1.1) from the matrix coefficients realization to Siegel-Whittaker realization in a very simple natural way. We believe the similar argument is possible for the principal series: but in this case the natures of integrands of Whittaker case Ishii [**6**], Theorem 3.2 and Siegel-Whittaker case [**5**], Theorem 10.1 are still difficult to deform.

For the P_J-principal series of the other group $\mathrm{SU}(2,2)$, we have explicit integral expression of the matrix coefficients (Theorem 5.4 of [**7**]) similar to the above formula (1.3). We can expect the confluence from this formula to the integral expression of the Siegel-Whittaker realization of [**2**] analogous to (1.1), corresponding to the limit of one parameter conjugations of a compact subgroup K of $\mathrm{SU}(2,2)$:

$$\lim_{t\to 0} h_t K h_t^{-1} = R.$$

The deformation of the Siegel-Whittaker function of the P_J-principal series to the Whittaker function could handled in a similar way as in [**3**].

In view of this observation, we may hope that similar phenomenon occurs for more general groups $\mathrm{SO}(2,q)$ $(q>4)$.

Finally, we have a few comments on related works. As kindly pointed out by the referee, the study of the Siegel-Whittaker models is a special case of a more general theory of Bessel models which makes sense for all classical groups (cf. [**1**]). These models were introduced in papers of M. E. Novodvorsky and I. I. Piatetski-Shapiro [**9**] and are among basic cases of general conjectures of W. T. Gan, B. H. Gross and D. Prasad (cf. [**1**]). In particular, D. Prasad and R. Takloo-Bighash proved existence and uniqueness of Bessel models for the discrete series representations of $\mathrm{GSp}(2,\mathbb{R})$ in [**12**]. However, it seems that the detailed study of analytic natures of the Siegel-Whittaker functions as special functions in two variables is quite another problem. Therefore, we believe that our explicit results are important and useful for further investigations.

2. Preliminaries

2.1. Basic notations. Let G be the real symplectic group of degree two:

$$\mathrm{Sp}(2,\mathbb{R})=\left\{g\in \mathrm{SL}(4,\mathbb{R})\;\middle|\; {}^t gJ_2g=J_2=\begin{pmatrix}0_2 & 1_2\\ -1_2 & 0_2\end{pmatrix}\right\},$$

with 1_2 the unit matrix of degree two and 0_2 the zero matrix of degree two.

Fix a maximal compact subgroup K of G by

$$K=\left\{k(A,B)=\begin{pmatrix}A & B\\ -B & A\end{pmatrix}\in G\;\middle|\; A,B\in \mathrm{M}(2,\mathbb{R})\right\}.$$

It is isomorphic to the unitary group $\mathrm{U}(2)$ via the homomorphism

$$K\ni k(A,B)\mapsto A+\sqrt{-1}B\in \mathrm{U}(2).$$

We define a certain spherical subgroup R of G as follows. Let $P_S=L_S\ltimes N_S$ be the Siegel parabolic subgroup with the Levi part L_S and the abelian unipotent radical N_S given by

$$L_S=\left\{\begin{pmatrix}A & 0_2\\ 0_2 & {}^tA^{-1}\end{pmatrix}\;\middle|\; A\in \mathrm{GL}(2,\mathbb{R})\right\},$$

$$N_S=\left\{n(T)=\begin{pmatrix}1_2 & T\\ 0_2 & 1_2\end{pmatrix}\;\middle|\; {}^tT=T\in \mathrm{M}(2,\mathbb{R})\right\}.$$

Fix a non-degenerate unitary character ξ of N_S by

$$\xi(n(T))=\exp\bigl(2\pi\sqrt{-1}\mathrm{tr}(H_\xi T)\bigr)$$

with $H_\xi=\left(\begin{smallmatrix}h_1 & h_3/2\\ h_3/2 & h_2\end{smallmatrix}\right)\in \mathrm{M}(2,\mathbb{R})$ and $\det H_\xi\neq 0$. Consider the action of L_S on N_S by conjugation and the induced action on the character group $\widehat{N_S}$. Define $\mathrm{SO}(\xi)$ to the identity component of the subgroup of L_S which stabilize ξ:

$$\mathrm{SO}(\xi):=\mathrm{Stab}_{L_S}(\xi)^\circ=\left\{\begin{pmatrix}A & 0_2\\ 0_2 & {}^tA^{-1}\end{pmatrix}\in L_S\;\middle|\; {}^tAH_\xi A=H_\xi\right\}.$$

Then $\mathrm{SO}(\xi)$ is isomorphic to $\mathrm{SO}(2)$ if $\det H_\xi>0$ and to $\mathrm{SO}_\circ(1,1)$ if $\det H_\xi<0$. In this article we treat the case that ξ is a 'definite' character, that is $\det H_\xi>0$. So we may assume $h_1,h_2>0$ and $h_3=0$ without loss of generality. We sometimes identify the element of $\mathrm{SO}(\xi)$ with its upper left 2×2 component. Fix a unitary character χ_{m_0} ($m_0\in\mathbb{Z}$) of $\mathrm{SO}(\xi)\cong\mathrm{SO}(2)$ by

$$\chi_{m_0}\left(\begin{pmatrix}\sqrt{h_1} & \\ & \sqrt{h_2}\end{pmatrix}^{-1}\begin{pmatrix}\cos\theta & \sin\theta\\ -\sin\theta & \cos\theta\end{pmatrix}\begin{pmatrix}\sqrt{h_1} & \\ & \sqrt{h_2}\end{pmatrix}\right)=\exp(\sqrt{-1}m_0\theta). \tag{2.1}$$

We define

$$R=\mathrm{SO}(\xi)\ltimes N_S \quad\text{and}\quad \eta=\chi_{m_0}\boxtimes\xi. \tag{2.2}$$

Taking a maximal split torus A of G by

$$A=\{a=(a_1,a_2)=\mathrm{diag}(a_1,a_2,a_1^{-1},a_2^{-1})\mid a_1,\,a_2>0\},$$

we have the decomposition $G=RAK$.

2.2. Siegel-Whittaker functions. We consider the space $C_\eta^\infty(R\backslash G)$ of complex valued C^∞ functions f on G satisfying

$$f(rg)=\eta(r)f(g)\quad\forall(r,g)\in R\times G.$$

By the right translation, $C_\eta^\infty(R\backslash G)$ is a smooth G-module and we denote the same symbol its underlying $(\mathfrak{g}_\mathbb{C},K)$-module. For an irreducible admissible representation (π,H_π) of G and the subspace $H_{\pi,K}$ of K-finite vectors, the intertwining space

$$\mathcal{I}_{\eta,\pi}=\mathrm{Hom}_{(\mathfrak{g}_\mathbb{C},K)}(H_{\pi,K},C_\eta^\infty(R\backslash G))$$

between the $(\mathfrak{g}_\mathbb{C},K)$-modules is called the space of *algebraic Siegel-Whittaker functionals*. For a finite-dimensional K-module (τ,V_τ), denote by $C_{\eta,\tau}^\infty(R\backslash G/K)$ the space

$$\bigl\{\phi\colon G\to V_\tau,\,C^\infty\mid\phi(rgk)=\eta(r)\tau(k^{-1})\phi(g)\quad\forall(r,g,k)\in R\times G\times K\bigr\}.$$

Let (τ^*,V_{τ^*}) be a K-type of π and $\iota\colon V_{\tau^*}\to H_\pi$ be an injection. Here, τ^* means the contragredient representation of τ. Then for $\Phi\in\mathcal{I}_{\eta,\pi}$, we can find an element ϕ_ι in

$$C_{\eta,\tau}^\infty(R\backslash G/K)=C_\eta^\infty(R\backslash G)\otimes V_{\tau^*}\cong\mathrm{Hom}_K(V_{\tau^*},C_\eta^\infty(R\backslash G))$$

via $\Phi(\iota(v^*))(g)=\langle v^*,\phi_\iota(g)\rangle$ with $\langle,\rangle$ the canonical pairing on $V_{\tau^*}\times V_\tau$.

Since there is the decomposition $G=RAK$, our generalized spherical function ϕ_ι is determined by its restriction $\phi_\iota|_A$, which we call the *radial part* of ϕ_ι. For a subspace X of $C_{\eta,\tau}^\infty(R\backslash G/K)$, we denote $X|_A=\{\phi|_A\in C^\infty(A)\mid\phi\in X\}$.

Let us define the space $\mathrm{SW}(\pi,\eta,\tau)$ of Siegel-Whittaker functions and its subspace $\mathrm{SW}(\pi,\eta,\tau)^{\mathrm{rap}}$ as follows:

$$\mathrm{SW}(\pi,\eta,\tau)=\bigcup_{\iota\in\mathrm{Hom}_K(\tau^*,\pi)}\{\phi_\iota\mid\Phi\in\mathcal{I}_{\eta,\pi}\}$$

and

$$\mathrm{SW}(\pi,\eta,\tau)^{\mathrm{rap}}=\bigl\{\phi_\iota\in\mathrm{SW}(\pi,\eta,\tau)\mid\phi_\iota|_A\text{ decays rapidly as }a_1,a_2\to\infty\bigr\}.$$

We call an element in $\mathrm{SW}(\pi,\eta,\tau)$ a *Siegel-Whittaker function* for (π,η,τ).

2.3. Parametrization of the discrete series representations. Let $E_{ij} \in \mathrm{M}_4(\mathbb{R})$ be the matrix unit with 1 as its (i,j)-component and 0 at the other entries. The root system of G with respect to a compact Cartan subgroup

$$T = \exp\big(\mathbb{R}(E_{13} - E_{31}) + \mathbb{R}(E_{24} - E_{42})\big)$$

is given by a set of vectors in the Euclidean plane:

$$\{\pm 2\varepsilon_1, \pm 2\varepsilon_2, \pm\varepsilon_1 \pm \varepsilon_2\}.$$

Here,

$$\begin{aligned}\varepsilon_1\big(r_1(E_{13} - E_{31}) + r_2(E_{24} - E_{42})\big) &= \sqrt{-1}r_1,\\ \varepsilon_2\big(r_1(E_{13} - E_{31}) + r_2(E_{24} - E_{42})\big) &= \sqrt{-1}r_2.\end{aligned}$$

We fix a subset of simple roots and the associated positive roots by

$$\{\varepsilon_1 - \varepsilon_2, \varepsilon_2\}, \quad \{2\varepsilon_1, \varepsilon_1 + \varepsilon_2, \varepsilon_1 - \varepsilon_2, 2\varepsilon_2\}$$

respectively.

Then the set of the unitary characters of T (or their derivatives) is identified naturally with $\mathbb{Z} \oplus \mathbb{Z}$, and the subset consisting of dominant integral weight is

$$\Xi = \{(n_1, n_2) \in \mathbb{Z} \oplus \mathbb{Z} \mid n_1 \geq n_2\}.$$

There is a bijection between $\widehat{K}$ and Ξ by highest weight theory. Because the half-sum of the positive root is integral, the discrete series representations of $G = \mathrm{Sp}(2, \mathbb{R})$ are parametrized by the subset of regular elements in Ξ:

$$\Xi' = \{(n_1, n_2) \in \mathbb{Z} \oplus \mathbb{Z} \mid n_1 > n_2,\ n_1 \neq 0,\ n_2 \neq 0,\ n_1 + n_2 \neq 0\}.$$

Here the condition $n_1 > n_2$ means the positivity of weight (n_1, n_2) with respect to the compact root $\varepsilon_1 - \varepsilon_2 = (1, -1)$.

The subsets $\Xi_{\mathrm{I}} = \{(n_1, n_2) \mid n_1 > n_2 > 0\}$ and $\Xi_{\mathrm{IV}} = \{(n_1, n_2) \mid 0 > n_1 > n_2\}$ parametrize the holomorphic discrete series and the anti-holomorphic discrete series representations, respectively. Set

$$\Xi_{\mathrm{II}} = \{(n_1, n_2) \mid n_1 > 0 > n_2,\ n_1 + n_2 > 0\},$$

and

$$\Xi_{\mathrm{III}} = \{(n_1, n_2) \mid n_1 > 0 > n_2,\ 0 > n_1 + n_2\}.$$

Then the union $\Xi_{\mathrm{II}} \cup \Xi_{\mathrm{III}}$ parametrize the large discrete series representations of G.

3. Miyazaki's results

Miyazaki derived a system of partial differential equations satisfied by Siegel-Whittaker functions for the large discrete representations in [**8**]. He also obtained multiplicity one property. We recall his results in this section.

Let $\tau = \tau_{(\lambda_1, \lambda_2)} = \mathrm{Sym}^{\lambda_1 - \lambda_2} \otimes \det^{\lambda_2}$ be the irreducible K-module with the highest weight (λ_1, λ_2), then the dimension of τ is $d+1$ with $d = \lambda_1 - \lambda_2$. We take the basis $\{v_j\}_{j=0}^{d}$ of V_{τ^*} with $\tau^* = \tau_{(-\lambda_2, -\lambda_1)}$ as in [**8**, Lemma 3.1].

We remark on a compatibility condition satisfied by Siegel-Whittaker functions. For a non-zero function ϕ in $C^\infty_{\eta, \tau^*}(R\backslash G/K)$, we have

$$\phi(a) = \phi(mam^{-1}) = (\chi_{m_0} \boxtimes \xi)(m)\tau_{(-\lambda_2, -\lambda_1)}(m)\,\phi(a),$$

where, $a \in A$ and $m \in \mathrm{SO}(\xi) \cap Z_K(A) = \{\pm 1_4\}$.

If we take $m = -1_4$, $(\chi_{m_0} \boxtimes \xi)(m) = \chi_{m_0}(m) = (-1)^{m_0}$ and $\tau_{(-\lambda_2, -\lambda_1)}(m) = (-1)^d$ imply that $(m_0 + d)/2$ is an integer.

PROPOSITION 3.1 (Miyazaki [**8**]). *Let $\pi = \pi_\Lambda$ be a large discrete series representation of G with the Harish-Chandra parameter $\Lambda = (\lambda_1 - 1, \lambda_2) \in \Xi_{\mathrm{II}}$ and its minimal K-type $\tau = \tau_{(\lambda_1,\lambda_2)}$. Let ξ be a unitary character of N_S associated with a positive definite matrix $H_\xi = \left(\begin{smallmatrix} h_1 & 0 \\ 0 & h_2 \end{smallmatrix}\right)$. Put $\eta = \chi_{m_0} \boxtimes \xi$, as in (2.1). Then we have the following:*

(i) *we have $\dim_{\mathbb{C}} \mathrm{SW}(\pi, \eta, \tau^*) \leq 4$ and a V_{τ^*}-valued function*

$$\phi_{SW}(a) = \sum_{j=0}^{d} \Big\{ (\sqrt{h_1} a_1)^{\lambda_1 - j} (\sqrt{h_2} a_2)^{\lambda_2 + j} e^{-2\pi(h_1 a_1^2 + h_2 a_2^2)} c_j(a) \Big\} v_j$$

is in the space $\mathrm{SW}(\pi_\Lambda, \eta, \tau^*)|_A$ if and only if $\{c_j(a)\}_{j=0}^d$ is a smooth solution of the following system:

(3.1)
$$\left[\partial_1 + j\frac{h_2 a_2^2}{\triangle}\right] c_{j-1}(a) + \sqrt{-1} m_0 \frac{h_2 a_2^2}{\triangle} c_j(a) - (d-j)\frac{h_2 a_2^2}{\triangle} c_{j+1}(a) = 0 \quad (1 \leq j \leq d),$$

(3.2)
$$j\frac{h_1 a_1^2}{\triangle} c_{j-1}(a) + \sqrt{-1} m_0 \frac{h_1 a_1^2}{\triangle} c_j(a) + \left[\partial_2 - (d-j)\frac{h_1 a_1^2}{\triangle}\right] c_{j+1}(a) = 0 \quad (0 \leq j \leq d-1),$$

(3.3)
$$\begin{aligned} h_1 a_1^2 &\left[\partial_2 - 8\pi h_2 a_2^2 - 2j\frac{h_2 a_2^2}{\triangle} + 2\lambda_2 - 2\right] c_{j-1}(a) - 2\sqrt{-1} m_0 \frac{h_1 a_1^2 h_2 a_2^2}{\triangle} c_j(a) \\ &+ h_2 a_2^2 \left[\partial_1 - 8\pi h_1 a_1^2 + 2(d-j)\frac{h_1 a_1^2}{\triangle} + 2\lambda_2 - 2\right] c_{j+1}(a) = 0 \quad (1 \leq j \leq d-1), \end{aligned}$$

with $\partial_i = a_i(\partial/\partial a_i)$ $(i = 1, 2)$ and $\triangle = h_1 a_1^2 - h_2 a_2^2$.

(ii) $\dim_{\mathbb{C}} \mathrm{SW}(\pi, \eta, \tau^*)^{rap} \leq 1$.

This is a paraphrase of Propositions 10.2, 10.7 and Theorem 11.5 of [**8**]. Here (3.1), (3.2), (3.3) are essentially identical equations to (10.4), (10.5), (10.6) of [**8**] deduced from Proposition 10.2. However we replace the symbol $\chi(Y_\eta)$ of [**8**] by its explicit value $\sqrt{-1} m_0 / \sqrt{h_1 h_2}$, and the symbol D of [**8**] by $\triangle$.

4. Partially Confluent hypergeometric functions in two variables

We introduce and study certain partially confluent hypergeometric functions in two variables on $A \simeq (\mathbb{R}_{>0})^2$. These functions play a key role in describing Siegel-Whittaker functions $\phi_{SW}(a)$. We remark that these types of confluent hypergeometric functions have also appeared in [**2**], for the large discrete series representations and P_J-principal series representations of $\mathrm{SU}(2,2)$, and [**3**], for the P_J-principal series representations of $\mathrm{Sp}(2, \mathbb{R})$.

DEFINITION 4.1 (Partially confluent hypergeometric functions). Let $\pi = \pi_\Lambda$ be a large discrete series representation of G with the Harish-Chandra parameter $\Lambda = (\lambda_1 - 1, \lambda_2) \in \Xi_{\mathrm{II}}$ and its minimal K-type $\tau = \tau_{(\lambda_1,\lambda_2)}$. Let ξ be a unitary character of N_S associated with a positive definite matrix $H_\xi = \left(\begin{smallmatrix} h_1 & 0 \\ 0 & h_2 \end{smallmatrix}\right)$. Put $\eta = \chi_{m_0} \boxtimes \xi$, as in (2.1).

We assume that

$$|m_0| \geq d \quad \text{and} \quad |m_0| \equiv d \pmod 2.$$

For $0 \le k \le d$, define

$$\begin{aligned} f_k(a) &= f_k(a;[\pi,\eta,\tau^*]) \\ (4.1) \qquad &= (\sqrt{h_1}a_1)^{2k+1}(\sqrt{h_2}a_2)^{2d+1-2k}(h_1a_1^2-h_2a_2^2)^{\frac{|m_0|-d}{2}} \\ &\times \int_0^1 F\big(2\pi h_1a_1^2t+2\pi h_2a_2^2(1-t)\big)\, t^{\frac{|m_0|-d-1}{2}+k}(1-t)^{\frac{|m_0|+d-1}{2}-k}\,dt, \end{aligned}$$

with

$$F(x) = e^x x^{-\frac{|m_0|+\lambda_1+1}{2}} W_{\frac{\lambda_1-|m_0|-1}{2},\frac{\lambda_2}{2}}(2x).$$

Here, $W_{\kappa,\mu}(z)$ is Whittaker's confluent hypergeometric function. (See [**13**] Chapter 16 for definition.)

Since the indices $\alpha_k - 1 = \frac{|m_0|-d-1}{2}+k$, $\beta_k - 1 = \frac{|m_0|+d-1}{2}-k$ in the integrand of f_k satisfy $\alpha_k, \beta_k > 0$ $(0 \le k \le d)$, we see that $f_k(a)$ is a smooth function on A and of moderate growth when each a_1, a_2 tends to infinity. We have further more,

PROPOSITION 4.2. *Partially confluent hypergeometric functions $\{f_k(a)\}_{k=0}^d$ satisfy the following system of the difference-differential equations.*

$$(4.2)\quad \partial_1 f_k = -(2k+1)\frac{h_2a_2^2}{\triangle} f_k + (|m_0|+d-1-2k)\frac{h_2a_2^2}{\triangle} f_{k+1} \quad (0 \le k \le d-1),$$

$$(4.3)\quad \partial_2 f_k = (2d-2k+1)\frac{h_1a_1^2}{\triangle} f_k - (|m_0|-d-1+2k)\frac{h_1a_1^2}{\triangle} f_{k-1} \quad (1 \le k \le d),$$

(4.4)

$$\big[(\partial_1+\partial_2)^2 + 2(\lambda_2-2)(\partial_1+\partial_2) - 8\pi(h_1a_1^2\partial_1 + h_2a_2^2\partial_2) - 4(\lambda_2-1)\big] f_k(a) = 0$$

$(0 \le k \le d)$.

Here, $\triangle = h_1a_1^2 - h_2a_2^2$.

PROOF. For $0 \le k \le d$, put

$$\check{f}_k(a) = \int_0^1 F\big(2\pi h_1a_1^2t+2\pi h_2a_2^2(1-t)\big)\, t^{\frac{|m_0|-d-1}{2}+k}(1-t)^{\frac{|m_0|+d-1}{2}-k}\,dt.$$

Then we can verify that

$$\partial_1 f_k = (\sqrt{h_1}a_1)^{2k+1}(\sqrt{h_2}a_2)^{2d+1-2k}\triangle^{\frac{|m_0|-d}{2}}\Big[\partial_1 + (2k+1) + (|m_0|-d)\frac{h_1a_1^2}{\triangle}\Big]\check{f}_k$$

and

$$\partial_1 \check{f}_k = -(|m_0|-d+1+2k)\frac{h_1a_1^2}{\triangle}\check{f}_k + (|m_0|+d-1-2k)\frac{h_1a_1^2}{\triangle}\check{f}_{k+1}$$

for $0 \le k \le d-1$. Therefore, we obtain the formula (4.2). Similarly, we have

$$\partial_2 f_k = (\sqrt{h_1}a_1)^{2k+1}(\sqrt{h_2}a_2)^{2d+1-2k}\triangle^{\frac{|m_0|-d}{2}}\Big[\partial_2 + (2d+1-2k) - (|m_0|-d)\frac{h_2a_2^2}{\triangle}\Big]\check{f}_k$$

and

$$\partial_2 \check{f}_k = -(|m_0|-d-1+2k)\frac{h_2a_2^2}{\triangle}\check{f}_{k-1} + (|m_0|+d+1-2k)\frac{h_2a_2^2}{\triangle}\check{f}_k$$

for $1 \le k \le d$. Thus, we have the formula (4.3).

Let us prove (4.4). Put Ω and $\check{\Omega}$ be the partial differential operators defined by

$$\Omega = (\partial_1 + \partial_2)^2 + 2(\lambda_2 - 2)(\partial_1 + \partial_2) - 8\pi(h_1 a_1^2 \partial_1 + h_2 a_2^2 \partial_2) - 4(\lambda_2 - 1)$$

and

$$\begin{aligned}\check{\Omega} =& (\partial_1 + \partial_2)^2 + 2(|m_0| + d + \lambda_2)(\partial_1 + \partial_2) - 8\pi(h_1 a_1^2 \partial_1 + h_2 a_2^2 \partial_2) \\ &- 8\pi(h_1 a_1^2 + h_2 a_2^2)(|m_0| + 1) - 8\pi(h_1 a_1^2 - h_2 a_2^2)(2k - d) \\ &+ (|m_0| + d)(|m_0| + d + 2\lambda_2).\end{aligned}$$

Then we can check that

$$\Omega\, f_k = (\sqrt{h_1} a_1)^{2k+1} (\sqrt{h_2} a_2)^{2d+1-2k} \triangle^{\frac{|m_0|-d}{2}}\, \check{\Omega}\, \check{f}_k.$$

By interchanging differentiation and integration, we have

$$\check{\Omega}\, \check{f}_k(a) = \int_0^1 G\big(2\pi h_1 a_1^2 t + 2\pi h_2 a_2^2 (1-t)\big)\, t^{\frac{|m_0|-d-1}{2}+k} (1-t)^{\frac{|m_0|+d-1}{2}-k}\, dt$$

with

$$G(x) = 4\left[x^2 \frac{d^2}{dx^2} - \{2x - (|m_0| + \lambda_1 + 1)\} x \frac{d}{dx} - 2(|m_0|+1)x + \frac{(|m_0| + \lambda_1)^2 - \lambda_2^2}{4}\right] F(x).$$

Put $F(x) = e^x x^{-(|m_0|+\lambda_1+1)/2} H(x)$. Then we have

$$G(x) = 4e^x x^{-(|m_0|+\lambda_1+1)/2} \cdot x^2 \left[\frac{d^2}{dx^2} + \left\{-1 + \frac{\lambda_1 - |m_0| - 1}{x} + \frac{1 - \lambda_2^2}{4x^2}\right\}\right] H(x).$$

It is known that the differential equation

$$\left[\frac{d^2}{dx^2} + \left\{-1 + \frac{\lambda_1 - |m_0| - 1}{x} + \frac{1 - \lambda_2^2}{4x^2}\right\}\right] H(x) = 0$$

has two linearly independent solutions:

$$W_{\frac{\lambda_1 - |m_0| - 1}{2}, \frac{\lambda_2}{2}}(2x), \quad M_{\frac{\lambda_1 - |m_0| - 1}{2}, \frac{\lambda_2}{2}}(2x).$$

Here, $W_{\kappa,\mu}(z)$ and $M_{\kappa,\mu}(z)$ are Whittaker's confluent hypergeometric functions. (See [**13**] Chapter 16 for definition.) Therefore, we have (4.4). It completes the proof. □

5. Main results

We state our main results on an explicit integral formula of Siegel-Whittaker functions which are of rapid decay for the large discrete series representations of $\mathrm{Sp}(2, \mathbb{R})$.

THEOREM 5.1. *Let $\pi = \pi_\Lambda$ be a large discrete series representation of G with the Harish-Chandra parameter $\Lambda = (\lambda_1 - 1, \lambda_2) \in \Xi_{\mathrm{II}}$ and its minimal K-type $\tau = \tau_{(\lambda_1, \lambda_2)}$. Let ξ be a unitary character of N_S associated with a positive definite matrix $H_\xi = \left(\begin{smallmatrix} h_1 & 0 \\ 0 & h_2 \end{smallmatrix}\right)$. Put $\eta = \chi_{m_0} \boxtimes \xi$, as in (2.1).*

We assume that

$$|m_0| \geq d \quad \textit{and} \quad |m_0| \equiv d \pmod 2.$$

Then we have the following:

(i) $\dim_{\mathbb{C}} \mathrm{SW}(\pi, \eta, \tau^*)^{rap} = 1$.

(ii) *Let* $\{f_k(a)\}_{k=0}^d$ *be the partially confluent hypergeometric functions, defined in Definition* 4.1. *We consider the following* $\mathbb{C}$*-linear combinations* $\{g_j(a)\}_{j=0}^d$ *of elements of* $\{f_k(a)\}_{k=0}^d$*, given by*

$$g_j(a) = \sum_{k=0}^{d} x_{jk}\, f_k(a), \tag{5.1}$$

where the complex numbers $\{x_{jk}\}_{0\le j,k\le d}$ *are given by*

(5.2)

$$x_{jk} = (-1)^{j+k}\left(\mathrm{sgn}(m_0)\sqrt{-1}\right)^j \binom{d}{2k-j} |m_0|^{\delta(j)} \prod_{l=1}^{k} \frac{2l-1}{|m_0|-d+2l-1}$$

$$\times \sum_{r=0}^{[\frac{j}{2}]} \binom{[\frac{j}{2}]}{r} \frac{(d-2r-\delta(j))!}{d!} {}_1\prod_{l=1}^{r} \left\{ \frac{2k-j-2l+2-\delta(j)}{2k-2l+1} \left(|m_0|^2-(d-2l+2)^2\right) \right\}$$

($\delta(j)=1$ *if* j *is odd, otherwise* 0). *Then the* V_{τ^*}*-valued function*

$$\phi_{SW}(a) = \sum_{j=0}^{d} \left\{ (\sqrt{h_1}a_1)^{\lambda_1-j}(\sqrt{h_2}a_2)^{\lambda_2+j} e^{-2\pi(h_1a_1^2+h_2a_2^2)} g_j(a) \right\} v_j \tag{5.3}$$

gives a non-zero element in $\mathrm{SW}(\pi,\eta,\tau^*)^{rap}|_A$ *which is unique up to constant multiple.*

6. Proof of main results

We claim that the coefficient functions $c_0(a), c_1(a), \ldots, c_d(a)$ appearing in the Siegel-Whittaker function $\phi_{SW}(a)$, in Proposition 3.1, are $\mathbb{C}$-linear combination of the confluent hypergeometric functions $f_0(a), f_1(a), \ldots, f_d(a)$ defined in Definition 4.1.

PROPOSITION 6.1. *We assume that*

$$|m_0| \ge d \quad \textit{and} \quad |m_0| \equiv d \pmod 2.$$

Let $\{x_{jk}\}_{0\le j,k\le d}$ *be a sequence of complex numbers, which satisfy*

(6.1) $j\,x_{j-1,k}+\sqrt{-1}m_0\,x_{j,k}+(d-2k+j+1)\,x_{j+1,k}-(|m_0|-d+2k+1)\,x_{j+1,k+1}=0$

$(0\le j\le d-1,\ 0\le k\le d)$,

(6.2)

$(j-2k-1)\,x_{j-1,k}+(|m_0|+d-2k+1)\,x_{j-1,k-1}+\sqrt{-1}m_0\,x_{j,k}-(d-j)\,x_{j+1,k}=0$

$(1\le j\le d,\ 0\le k\le d)$,

$$x_{1,0}=x_{2,0}=\cdots=x_{d,0}=0, \tag{6.3}$$

and

$$x_{0,d}=x_{1,d}=\cdots=x_{d-1,d}=0. \tag{6.4}$$

For $0\le j\le d$*, define*

$$g_j(a) = \sum_{k=0}^{d} x_{jk}\, f_k(a),$$

then $\{g_j(a)\}_{j=0}^d$ *is a smooth solution of the system of the difference-differential equations* (3.1)*,* (3.2) *and* (3.3) *in Proposition* 3.1.

PROOF. By using Proposition 4.2, we see that

$$\begin{aligned}&\frac{\triangle}{h_2a_2^2}\Big[\partial_1+j\frac{h_2a_2^2}{\triangle}\Big]g_{j-1}(a)\\&=\sum_{k=0}^{d}(j-2k-1)x_{j-1,k}\,f_k+\sum_{k=0}^{d}(|m_0|+d-2k-1)x_{j-1,k}\,f_{k+1}.\end{aligned}$$

Therefore, we have

$$\begin{aligned}&\frac{\triangle}{h_2a_2^2}\Big[\partial_1+j\frac{h_2a_2^2}{\triangle}\Big]g_{j-1}(a)+\sqrt{-1}m_0g_j(a)-(d-j)g_{j+1}(a)\\&=\sum_{k=0}^{d}(j-2k-1)x_{j-1,k}\,f_k+\sum_{k=1}^{d}(|m_0|+d-2k+1)x_{j-1,k-1}\,f_k\\&\quad+\sqrt{-1}m_0\sum_{k=0}^{d}x_{j,k}\,f_k-(d-j)\sum_{k=0}^{d}x_{j+1,k}\,f_k\\&=0\quad(1\le j\le d).\end{aligned}$$

This is the desired (3.1) for $\{g_j(a)\}_{j=0}^d$. In the above calculation, we used the relations (6.2) and (6.4) on $\{x_{jk}\}_{0\le j,k\le d}$. Again, by using Proposition 4.2, we see that

$$\begin{aligned}&\frac{\triangle}{h_1a_1^2}\Big[\partial_2-(d-j)\frac{h_1a_1^2}{\triangle}\Big]g_{j+1}(a)\\&=\sum_{k=0}^{d}(d-2k+j+1)x_{j+1,k}\,f_k-\sum_{k=0}^{d}(|m_0|-d+2k-1)x_{j+1,k}\,f_{k-1}.\end{aligned}$$

Therefore, we have

$$\begin{aligned}&jg_{j-1}(a)+\sqrt{-1}m_0g_j(a)+\frac{\triangle}{h_1a_1^2}\Big[\partial_2-(d-j)\frac{h_1a_1^2}{\triangle}\Big]g_{j+1}(a)\\&=j\sum_{k=0}^{d}x_{j-1,k}\,f_k+\sqrt{-1}m_0\sum_{k=0}^{d}x_{j,k}\,f_k+\sum_{k=0}^{d}(d-2k+j+1)x_{j+1,k}\,f_k\\&\quad-\sum_{k=0}^{d-1}(|m_0|-d+2k+1)x_{j+1,k+1}\,f_k\\&=0\quad(0\le j\le d-1).\end{aligned}$$

This is the desired (3.2) for $\{g_j(a)\}_{j=0}^d$. In the above calculation, we used the relations (6.1) and (6.3) on $\{x_{jk}\}_{0\le j,k\le d}$.

By considering (3.1) $\times\, h_1a_1^2$ + (3.2) $\times\, h_2a_2^2$ + (3.3), we have

$$\begin{aligned}&h_1a_1^2[(\partial_1+\partial_2)-8\pi h_2a_2^2+2\lambda_2-2]c_{j-1}(a)\\&+h_2a_2^2[(\partial_1+\partial_2)-8\pi h_1a_1^2+2\lambda_2-2]c_{j+1}(a)=0\quad(1\le j\le d-1).\end{aligned}\tag{6.5}$$

By considering (3.1) $\times\, h_1a_1^2$ − (3.2) $\times\, h_2a_2^2$, we have

$$h_1a_1^2\partial_1\,c_{j-1}(a)-h_2a_2^2\partial_2\,c_{j+1}(a)=0\quad(1\le j\le d-1).\tag{6.6}$$

Operating $\partial_2(h_2a_2^2)^{-1}$ on both hand sides of (6.5), we have

$$\begin{aligned}(6.7)\qquad &\partial_2\frac{h_1a_1^2}{h_2a_2^2}[(\partial_1+\partial_2)-8\pi h_2a_2^2+2\lambda_2-2]c_{j-1}(a)\\ &+[(\partial_1+\partial_2)-8\pi h_1a_1^2+2\lambda_2-2]\partial_2\,c_{j+1}(a)=0\quad(1\le j\le d-1).\end{aligned}$$

Combining (6.7) and (6.6), we have, for $0\le j\le d-2$,

$$(6.8)\quad \big[(\partial_1+\partial_2)^2+2(\lambda_2-2)(\partial_1+\partial_2)\partial_2-8\pi(h_1a_1^2\partial_1+h_2a_2^2\partial_2)-4(\lambda_2-1)\big]c_j(a)=0.$$

Operating $\partial_1(h_1a_1^2)^{-1}$ on both hand sides of (6.5) and combining with (6.6), we have (6.8) for $2\le j\le d$. As a result, we have (6.8) for $0\le j\le d$.

Lastly we prove that $\{g_j(a)\}_{j=0}^d$ satisfy (3.3). By Prop. 4.2, $f_0(a),\ldots,f_d(a)$ satisfy the same differential equation (6.8), therefore their $\mathbb{C}$-linear combinations $\{g_j(a)\}_{j=0}^d$ also satisfy (6.8). Since $\{g_j(a)\}_{j=0}^d$ satisfy (6.6) and (6.8), we see that

$$(6.9)\qquad \partial_2\frac{F_j(a)}{h_2a_2^2}=\partial_1\frac{F_j(a)}{h_1a_1^2}=0\quad(1\le j\le d-1),$$

where we put

$$\begin{aligned}F_j(a)=&h_1a_1^2[(\partial_1+\partial_2)-8\pi h_2a_2^2+2\lambda_2-2]g_{j-1}(a)\\ &+h_2a_2^2[(\partial_1+\partial_2)-8\pi h_1a_1^2+2\lambda_2-2]g_{j+1}(a).\end{aligned}$$

By (6.9), there exist constants β_j $(1\le j\le d-1)$ such that

$$F_j(a)=\beta_j\,h_1a_1^2h_2a_2^2.$$

Let us determine β_j. For $y>0$, define $L=\{(a_1,a_2)\in A\mid h_1a_1^2=h_2a_2^2=y\}$. We show that $(F_j|_L)(y)=0$ to deduce $\beta_j=0$. We can verify that

$$\begin{aligned}(\partial_1+\partial_2)\,f_k&=(\sqrt{h_1}a_1)^{2k+1}(\sqrt{h_2}a_2)^{2d+1-2k}\triangle^{\frac{|m_0|-d}{2}}\big[(\partial_1+\partial_2)+(d+|m_0|+2)\big]\\ &\quad\times\int_0^1F\big(2\pi h_1a_1^2t+2\pi h_2a_2^2(1-t)\big)\,t^{\frac{|m_0|-d-1}{2}+k}(1-t)^{\frac{|m_0|+d-1}{2}-k}\,dt\\ &=(\sqrt{h_1}a_1)^{2k+1}(\sqrt{h_2}a_2)^{2d+1-2k}\triangle^{\frac{|m_0|-d}{2}}\\ &\quad\times\int_0^1F^*\big(2\pi h_1a_1^2t+2\pi h_2a_2^2(1-t)\big)\,t^{\frac{|m_0|-d-1}{2}+k}(1-t)^{\frac{|m_0|+d-1}{2}-k}\,dt.\end{aligned}$$

Here,

$$F^*(x)=\Big(\Big(2x\frac{d}{dx}+(d+|m_0|+2)\Big)F\Big)(x).$$

There are two cases:

(i) If $|m_0|-d\ge 2$, then both $(\partial_1+\partial_2)f_k$ and f_k have zeros at $\{(a_1,a_2)\mid h_1a_1^2-h_2a_2^2=0\}$. Hence, both $(\partial_1+\partial_2)g_j$ and g_j have zeros at there. Therefore we have $(F_j|_L)(y)=0$.

(ii) Suppose that $|m_0|=d$, then we have

$$\big((\partial_1+\partial_2)\,f_k\big)\big|_L(y)=y^{d+1}F^*(2\pi y)\,B\Big(k+\frac{1}{2},d-k+\frac{1}{2}\Big).$$

Here,

$$B(\alpha,\beta)=\int_0^1t^{\alpha-1}(1-t)^{\beta-1}\,dt=\frac{\Gamma(\alpha)\,\Gamma(\beta)}{\Gamma(\alpha+\beta)}$$

is the Beta function. Then we can verify that

$$(6.10)\quad \begin{aligned} &\left((\partial_1+\partial_2)\,g_j\right)\big|_L(y) = \left(\sum_{k=0}^{d} x_{jk}\,B\Big(k+\frac{1}{2},d-k+\frac{1}{2}\Big)\right)y^{d+1}F^*(2\pi y) \\ &= \left(\operatorname{sgn}(m_0)\sqrt{-1}\right)^j 2^{-d}\pi\, y^{d+1}F^*(2\pi y). \end{aligned}$$

To derive (6.10), we used (6.16) in Proposition 6.2: (We will prove later.)

$$x_{jk} = (-1)^{j+k}\left(\operatorname{sgn}(m_0)\sqrt{-1}\right)^j\binom{d}{2k-j} \quad (\text{when } |m_0| = d),$$

under the condition $x_{0,0} = 1$, and the equality:

$$(6.11)\quad \sum_{k=0}^{d}(-1)^k\binom{2k}{j}\binom{2d-2k}{d-j}\binom{d}{k} = (-1)^j 2^d\binom{d}{j}.$$

Let $S_{d,j}$ be the left hand side of (6.11). We remark that the above equality (6.11) is proved by showing that

$$S_{d,j} = -\frac{2d}{j}S_{d-1,j-1} \quad (j \geq 1) \text{ and } S_{d,0} = 2^d.$$

See p.620 no. 63 in [**14**] for the equality $S_{d,0} = 2^d$.

Similarly, we also obtain

$$g_j\big|_L(y) = \left(\operatorname{sgn}(m_0)\sqrt{-1}\right)^j 2^{-d}\pi\, y^{d+1}F(2\pi y).$$

Therefore, we have

$$\left((\partial_1+\partial_2)\,(g_{j-1}+g_{j+1})\right)\big|_L(y) = (g_{j-1}+g_{j+1})\big|_L(y) = 0,$$

for $1 \leq j \leq d-1$. Therefore we have $(F_j|_L)(y) = 0$.

In any cases, $\{g_j(a)\}_{j=0}^d$ satisfy (6.5). Hence, $\{g_j(a)\}_{j=0}^d$ satisfy (3.3) and it completes the proof. □

PROPOSITION 6.2. *We assume that*

$$|m_0| \geq d \quad \textit{and} \quad |m_0| \equiv d \pmod 2.$$

Let $\{x_{jk}\}_{0\leq j,k\leq d}$ be a sequence of complex numbers, which satisfy

$$(6.12)\quad j\,x_{j-1,k}+\sqrt{-1}m_0\,x_{j,k}+(d-2k+j+1)\,x_{j+1,k}-(|m_0|-d+2k+1)\,x_{j+1,k+1} = 0$$

$(0 \leq j \leq d-1,\ 0 \leq k \leq d)$,

(6.13)

$$(j-2k-1)\,x_{j-1,k}+(|m_0|+d-2k+1)\,x_{j-1,k-1}+\sqrt{-1}m_0\,x_{j,k}-(d-j)\,x_{j+1,k} = 0$$

$(1 \leq j \leq d,\ 0 \leq k \leq d)$,

$$(6.14)\quad x_{1,0} = x_{2,0} = \cdots = x_{d,0} = 0,$$

and

$$(6.15)\quad x_{0,d} = x_{1,d} = \cdots = x_{d-1,d} = 0.$$

Then the sequence $\{x_{jk}\}_{0\le j,k\le d}$ *is uniquely determined and given by, up to a constant multiple,*

(6.16)

$$x_{jk} = (-1)^{j+k}\big(\mathrm{sgn}(m_0)\sqrt{-1}\big)^j \binom{d}{2k-j} |m_0|^{\delta(j)} \prod_{l=1}^{k} \frac{2l-1}{|m_0|-d+2l-1}$$

$$\times \sum_{r=0}^{[\frac{j}{2}]} \binom{[\frac{j}{2}]}{r} \frac{(d-2r-\delta(j))!}{d!} \prod_{l=1}^{r} \Big\{ \frac{2k-j-2l+2-\delta(j)}{2k-2l+1} \big(|m_0|^2-(d-2l+2)^2\big)\Big\},$$

where, $\delta(j)=1$ *if* j *is odd, otherwise* 0.

PROOF. We may assume that $x_{0,0}=1$. Let us write down (6.12) with $j=0,1$ and k replaced by $k-1,k$, and (6.13) with $j=1$ and k replaced by $k-1,k,k+1$. Then we have the following system of seven difference equations:

(6.17)

$$\sqrt{-1}m_0\, x_{0,k-1} + (d-2k+3)\, x_{1,k-1} - (|m_0|-d+2k-1)\, x_{1,k} = 0,$$
$$\sqrt{-1}m_0\, x_{0,k} + (d-2k+1)\, x_{1,k} - (|m_0|-d+2k+1)\, x_{1,k+1} = 0,$$
$$x_{0,k-1} + \sqrt{-1}m_0\, x_{1,k-1} + (d-2k+4)\, x_{2,k-1} - (|m_0|-d+2k-1)\, x_{2,k} = 0,$$
$$x_{0,k} + \sqrt{-1}m_0\, x_{1,k} + (d-2k+2)\, x_{2,k} - (|m_0|-d+2k+1)\, x_{2,k+1} = 0,$$
$$-(2k-2)\, x_{0,k-1} + (|m_0|+d-2k+3)\, x_{0,k-2} + \sqrt{-1}m_0\, x_{1,k-1} - (d-1)\, x_{2,k-1} = 0,$$
$$-2k\, x_{0,k} + (|m_0|+d-2k+1)\, x_{0,k-1} + \sqrt{-1}m_0\, x_{1,k} - (d-1)\, x_{2,k} = 0,$$
$$-(2k+2)\, x_{0,k+1} + (|m_0|+d-2k-1)\, x_{0,k} + \sqrt{-1}m_0\, x_{1,k+1} - (d-1)\, x_{2,k+1} = 0.$$

We eliminate $x_{j,k}$, $x_{j,k\pm1}$ $(j=1,2)$ in the above system. Then we have

$$\begin{aligned}
&-2(k+1)(|m_0|-d+2k+1)(|m_0|-d+2k-1)\, x_{0,k+1} \\
&+ \{2k(4d-6k+3)-d(d-1)\}(|m_0|-d+2k-1)\, x_{0,k} \\
&+ \Big[2(d-2k+2)\{d(d-1)-(2k-2)(2d-2k+3)\} \\
&\quad + (2k-2)(|m_0|+d-2k+1)(|m_0|-d+2k-1)\Big]\, x_{0,k-1} \\
&+ (d-2k+4)(d-2k+3)(|m_0|+d-2k+3)\, x_{0,k-2} \\
&= 0.
\end{aligned} \tag{6.18}$$

By solving the difference equation (6.18) for $\{x_{0,k}\}$ with $x_{0,0}=1$, we obtain

$$x_{0,k} = (-1)^k \binom{d}{2k} \prod_{l=1}^{k} \frac{2l-1}{|m_0|-d+2l-1}. \tag{6.19}$$

From the second equation of (6.17), (6.19) and (6.14), we have the following difference equation for $\{x_{1,k}\}$:

$$(|m_0|-d+2k+1)\, x_{1,k+1} = (d-2k+1)\, x_{1,k} + \sqrt{-1}m_0\, (-1)^k \binom{d}{2k} \prod_{l=1}^{k} \frac{2l-1}{|m_0|-d+2l-1}$$

with $x_{1,0} = 0$. Thus, we obtain

$$x_{1,k} = (-1)^{k+1}(\operatorname{sgn}(m_0)\sqrt{-1})\binom{d}{2k-1}\frac{|m_0|}{d}\prod_{l=1}^{k}\frac{2l-1}{|m_0|-d+2l-1}. \tag{6.20}$$

We prove the formula (6.16) for general x_{jk} by induction on the index j. For $0 \le 2p, 2p+1, k \le d$, let us prove

(6.21)

$$\begin{aligned} x_{2p,k} =& (-1)^{k+p}\binom{d}{2k-2p}\prod_{l=1}^{k}\frac{2l-1}{|m_0|-d+2l-1} \\ &\times \sum_{r=0}^{p}\binom{p}{r}\frac{(d-2r)!}{d!}\prod_{l=1}^{r}\left\{\frac{2k-2p-2l+2}{2k-2l+1}(|m_0|^2-(d-2l+2)^2)\right\}, \end{aligned}$$

(6.22)

$$\begin{aligned} x_{2p+1,k} =& (-1)^{k+p+1}(\operatorname{sgn}(m_0)\sqrt{-1})\binom{d}{2k-2p-1}|m_0|\prod_{l=1}^{k}\frac{2l-1}{|m_0|-d+2l-1} \\ &\times \sum_{r=0}^{p}\binom{p}{r}\frac{(d-2r-1)!}{d!}\prod_{l=1}^{r}\left\{\frac{2k-2p-2l}{2k-2l+1}(|m_0|^2-(d-2l+2)^2)\right\}. \end{aligned}$$

Suppose that $2p+1 \le d$ and the above formulas are true for $x_{2p-1,k}$, $x_{2p,k}$ $(0 \le k \le d)$, then substitute them into (6.13):

(6.23)

$$x_{2p+1,k} = \frac{2p-2k-1}{d-2p}x_{2p-1,k} + \frac{|m_0|+d-2k+1}{d-2p}x_{2p-1,k-1} + \frac{\sqrt{-1}\,m_0}{d-2p}x_{2p,k}.$$

Put

$$a(k) = \prod_{l=1}^{k}\frac{2l-1}{|m_0|-d+2l-1}, \quad b(r;k,p) = \prod_{l=1}^{r}\left\{\frac{2k-2p-2l}{2k-2l+1}(|m_0|^2-(d-2l+2)^2)\right\}.$$

Then, we have

(6.24)

$$\begin{aligned} &\frac{(-1)^{k+p+1}(2p-2k-1)}{\sqrt{-1}\,m_0(d-2p)}x_{2p-1,k} \\ &= -\frac{2p-2k-1}{d-2p}\binom{d}{2k-2p+1}a(k)\sum_{r=0}^{p-1}\binom{p-1}{r}\frac{(d-2r-1)!}{d!}b(r;k,p-1) \\ &= \frac{a(k)}{d-2p}\binom{d}{2k-2p-1}(d-2k+2p+1)(d-2k+2p) \\ &\quad \times \sum_{r=0}^{p-1}\binom{p-1}{r}\frac{(d-2r-1)!}{d!}\frac{b(r;k,p)}{2k-2p-2r}, \end{aligned}$$

and

(6.25)

$$
\begin{aligned}
&\frac{(-1)^{k+p+1}(|m_0|+d-2k+1)}{\sqrt{-1}\,m_0(d-2p)}\,x_{2p-1,k-1}\\
&=\frac{|m_0|+d-2k+1}{d-2p}\binom{d}{2k-2p-1}a(k-1)\sum_{r=0}^{p-1}\binom{p-1}{r}\frac{(d-2r-1)!}{d!}\,b(r;k-1,p-1)\\
&=\frac{a(k)}{d-2p}\binom{d}{2k-2p-1}\{|m_0|^2-(d-2k+1)^2\}\sum_{r=0}^{p-1}\binom{p-1}{r}\frac{(d-2r-1)!}{d!}\,\frac{b(r;k,p)}{2k-2r-1}\\
&=\frac{a(k)}{d-2p}\binom{d}{2k-2p-1}\sum_{r=0}^{p-1}\binom{p-1}{r}\frac{(d-2r-1)!}{d!}\,\frac{b(r+1;k,p)}{2k-2p-2r-2}\\
&\quad+\frac{a(k)}{d-2p}\binom{d}{2k-2p-1}\sum_{r=0}^{p-1}\binom{p-1}{r}\frac{(d-2r-1)!}{d!}\,(2d-2r-2k+1)\,b(r;k,p).
\end{aligned}
$$

Here, we used the equality:

$$
\left\{|m_0|^2-(d-2k+1)^2\right\}=\left\{|m_0|^2-(d-2r)^2\right\}+(2d-2r-2k+1)(2k-2r-1)
$$

in the above calculation. By noting the equality:

$$
\frac{(d-2k+2p+1)(d-2k+2p)}{2k-2p-2r}+(2d-2r-2k+1)=\frac{(d-2r)(d-2r+1)}{2k-2p-2r}-2p,
$$

we have

$$
\begin{aligned}
&(6.24)+(6.25)\\
&=\frac{a(k)}{d-2p}\binom{d}{2k-2p-1}\sum_{r=0}^{p-1}\binom{p-1}{r}\frac{(d-2r-1)!}{d!}\left\{\frac{(d-2r)(d-2r+1)}{2k-2p-2r}-2p\right\}b(r;k,p)\\
&\quad+\frac{a(k)}{d-2p}\binom{d}{2k-2p-1}\sum_{r=1}^{p}\binom{p-1}{r-1}\frac{(d-2r+1)!}{d!}\,\frac{b(r;k,p)}{2k-2p-2r}\\
&=\frac{a(k)}{d-2p}\binom{d}{2k-2p-1}\sum_{r=0}^{p}\binom{p}{r}\frac{(d-2r+1)!}{d!}\,\frac{b(r;k,p)}{2k-2p-2r}\\
&\quad-2p\,\frac{a(k)}{d-2p}\binom{d}{2k-2p-1}\sum_{r=0}^{p-1}\binom{p-1}{r}\frac{(d-2r-1)!}{d!}\,b(r;k,p).
\end{aligned}
$$

On the other hand, we have

(6.26)

$$\begin{aligned}
&\frac{(-1)^{k+p+1}}{\sqrt{-1}\,m_0(d-2p)}x_{2p,k}\\
&=-\frac{1}{d-2p}\binom{d}{2k-2p}a(k)\sum_{r=0}^{p}\binom{p}{r}\frac{(d-2r)!}{d!}b(r;k,p-1)\\
&=-\frac{a(k)}{d-2p}\binom{d}{2k-2p-1}(d-2k+2p+1)\sum_{r=0}^{p}\binom{p}{r}\frac{(d-2r)!}{d!}\frac{b(r;k,p)}{2k-2p-2r}\\
&=\frac{a(k)}{d-2p}\binom{d}{2k-2p-1}\sum_{r=0}^{p}\binom{p}{r}\frac{(d-2r)!}{d!}b(r;k,p)\\
&\quad-\frac{a(k)}{d-2p}\binom{d}{2k-2p-1}\sum_{r=0}^{p}\binom{p}{r}\frac{(d-2r+1)!}{d!}\frac{b(r;k,p)}{2k-2p-2r}.
\end{aligned}$$

Therefore, we have

$$\begin{aligned}
&(6.24)+(6.25)+(6.26)\\
&=\frac{a(k)}{d-2p}\binom{d}{2k-2p-1}\sum_{r=0}^{p}\binom{p}{r}\frac{(d-2r)!}{d!}b(r;k,p)\\
&\quad-2p\,\frac{a(k)}{d-2p}\binom{d}{2k-2p-1}\sum_{r=0}^{p-1}\binom{p-1}{r}\frac{(d-2r-1)!}{d!}b(r;k,p)\\
&=a(k)\binom{d}{2k-2p-1}\sum_{r=0}^{p}\binom{p}{r}\frac{(d-2r-1)!}{d!}b(r;k,p).
\end{aligned}$$

By (6.23) and the above formula, we obtain

$$x_{2p+1,k}=(-1)^{k+p+1}(m_0\sqrt{-1})a(k)\binom{d}{2k-2p-1}\sum_{r=0}^{p}\binom{p}{r}\frac{(d-2r-1)!}{d!}b(r;k,p).$$

Therefore, the formula is valid for $x_{2p+1,k}$ $(0\le k\le d)$. Similarly, if we suppose that $2p\le d$ and the formulas (6.21) and (6.22) are true for $x_{2p-1,k}$, $x_{2p-2,k}$ $(0\le k\le d)$, then we can prove the formula is also valid for $x_{2p,k}$ $(0\le k\le d)$. It completes the proof. □

Let us complete the proof of Theorem 5.1.

PROOF OF 5.1. By the condition $|m_0|\ge d$, $|m_0|\equiv d \pmod 2$, Propositions 6.1 and 6.2, $\{g_j(a)\}_{j=0}^{d}$ is a *non-zero* smooth solution of the system (3.1), (3.2) and (3.3) in Proposition 3.1. Furthermore, we can check that all of $\{e^{-2\pi(h_1a_1^2+h_2a_2^2)}\,g_j(a)\}_{j=0}^{d}$ are rapidly decreasing when each a_1, a_2 tends to infinity, by Definition 4.1. We complete the proof. □

7. Remarks on $\{x_{jk}\}$

We have some remarks on the sequence of complex numbers $\{x_{jk}\}_{0\le j,k\le d}$ appearing in Theorem 5.1 and Proposition 6.2. Let $X_d=(x_{jk})_{0\le j,k\le d}\in M_{d+1}(\mathbb{C})$ be the square matrix of size $(d+1)$, defined by the sequence $\{x_{jk}\}_{0\le j,k\le d}$.

Put $\varepsilon = \operatorname{sgn}(m_0)\sqrt{-1}$ and $t = |m_0|$. Then x_{jk} is given by

(7.1)

$$x_{jk} = (-1)^{j+k}\varepsilon^j \binom{d}{2k-j} t^{\delta(j)} \prod_{l=1}^{k} \frac{2l-1}{t-d+2l-1}$$
$$\times \sum_{r=0}^{[\frac{j}{2}]} \binom{[\frac{j}{2}]}{r} \frac{(d-2r-\delta(j))!}{d!} \prod_{l=1}^{r} \left\{ \frac{2k-j-2l+2-\delta(j)}{2k-2l+1} \left(t^2-(d-2l+2)^2\right) \right\},$$

where, $\delta(j) = 1$ if j is odd, otherwise 0.

Furthermore, we set

$$Z_d = (z_{jk})_{0\le j,k\le d} \in M_{d+1}(\mathbb{C}) \text{ with } z_{jk} = \varepsilon^{-j}x_{jk}. \tag{7.2}$$

Though the Harish-Chandra parameter $\Lambda = (\lambda_1 - 1, \lambda_2) \in \Xi_{\mathrm{II}}$ implies that $d = \lambda_1 - \lambda_2 \ge 4$, we formally write down matrices Z_d for $1 \le d \le 7$.

EXAMPLE 7.1 (Z_d for $1 \le d \le 7$).

$$Z_1 = \begin{bmatrix} 1 & 0 \\ 0 & 1 \end{bmatrix}, \quad Z_2 = \begin{bmatrix} 1 & -\frac{1}{t-1} & 0 \\ 0 & \frac{t}{t-1} & 0 \\ 0 & -\frac{1}{t-1} & 1 \end{bmatrix}, \quad Z_3 = \begin{bmatrix} 1 & -\frac{3}{t-2} & 0 & 0 \\ 0 & \frac{t}{t-2} & -\frac{1}{t-2} & 0 \\ 0 & -\frac{1}{t-2} & \frac{t}{t-2} & 0 \\ 0 & 0 & -\frac{3}{t-2} & 1 \end{bmatrix},$$

$$Z_4 = \begin{bmatrix} 1 & -\frac{6}{t-3} & \frac{3}{(t-3)(t-1)} & 0 & 0 \\ 0 & \frac{t}{t-3} & -\frac{3t}{(t-3)(t-1)} & 0 & 0 \\ 0 & -\frac{1}{t-3} & \frac{t^2+2}{(t-3)(t-1)} & -\frac{1}{t-3} & 0 \\ 0 & 0 & -\frac{3t}{(t-3)(t-1)} & \frac{t}{t-3} & 0 \\ 0 & 0 & \frac{3}{(t-3)(t-1)} & -\frac{6}{t-3} & 1 \end{bmatrix},$$

$$Z_5 = \begin{bmatrix} 1 & -\frac{10}{t-4} & \frac{15}{(t-4)(t-2)} & 0 & 0 & 0 \\ 0 & \frac{t}{t-4} & -\frac{6t}{(t-4)(t-2)} & \frac{3}{(t-4)(t-2)} & 0 & 0 \\ 0 & -\frac{1}{t-4} & \frac{t^2+5}{(t-4)(t-2)} & -\frac{3t}{(t-4)(t-2)} & 0 & 0 \\ 0 & 0 & -\frac{3t}{(t-4)(t-2)} & \frac{t^2+5}{(t-4)(t-2)} & -\frac{1}{t-4} & 0 \\ 0 & 0 & \frac{3}{(t-4)(t-2)} & -\frac{6t}{(t-4)(t-2)} & \frac{t}{t-4} & 0 \\ 0 & 0 & 0 & \frac{15}{(t-4)(t-2)} & -\frac{10}{t-4} & 1 \end{bmatrix},$$

$$Z_6 = \begin{bmatrix} 1 & -\frac{15}{t-5} & \frac{45}{(t-5)(t-3)} & -\frac{15}{(t-5)(t-3)(t-1)} & 0 & 0 & 0 \\ 0 & \frac{t}{t-5} & -\frac{10t}{(t-5)(t-3)} & \frac{15t}{(t-5)(t-3)(t-1)} & 0 & 0 & 0 \\ 0 & -\frac{1}{t-5} & \frac{t^2+9}{(t-5)(t-3)} & -\frac{3(2t^2+3)}{(t-5)(t-3)(t-1)} & \frac{3}{(t-5)(t-3)} & 0 & 0 \\ 0 & 0 & -\frac{3t}{(t-5)(t-3)} & \frac{t(t^2+14)}{(t-5)(t-3)(t-1)} & -\frac{3t}{(t-5)(t-3)} & 0 & 0 \\ 0 & 0 & \frac{3}{(t-5)(t-3)} & -\frac{3(2t^2+3)}{(t-5)(t-3)(t-1)} & \frac{t^2+9}{(t-5)(t-3)} & -\frac{1}{t-5} & 0 \\ 0 & 0 & 0 & \frac{15t}{(t-5)(t-3)(t-1)} & -\frac{10t}{(t-5)(t-3)} & \frac{t}{t-5} & 0 \\ 0 & 0 & 0 & -\frac{15}{(t-5)(t-3)(t-1)} & \frac{45}{(t-5)(t-3)} & -\frac{15}{t-5} & 1 \end{bmatrix},$$

$Z_7 =$

$$\begin{bmatrix} 1 & -\frac{21}{t-6} & \frac{105}{(t-6)(t-4)} & -\frac{105}{(t-6)(t-4)(t-2)} & 0 & 0 & 0 & 0 \\ 0 & \frac{t}{t-6} & -\frac{15t}{(t-6)(t-4)} & \frac{45t}{(t-6)(t-4)(t-2)} & -\frac{15}{(t-6)(t-4)(t-2)} & 0 & 0 & 0 \\ 0 & -\frac{1}{t-6} & \frac{t^2+14}{(t-6)(t-4)} & -\frac{5(2t^2+7)}{(t-6)(t-4)(t-2)} & \frac{15t}{(t-6)(t-4)(t-2)} & 0 & 0 & 0 \\ 0 & 0 & -\frac{3t}{(t-6)(t-4)} & \frac{t(t^2+26)}{(t-6)(t-4)(t-2)} & -\frac{3(2t^2+7)}{(t-6)(t-4)(t-2)} & \frac{3}{(t-6)(t-4)} & 0 & 0 \\ 0 & 0 & \frac{3}{(t-6)(t-4)} & -\frac{3(2t^2+7)}{(t-6)(t-4)(t-2)} & \frac{t(t^2+26)}{(t-6)(t-4)(t-2)} & -\frac{3t}{(t-6)(t-4)} & 0 & 0 \\ 0 & 0 & 0 & \frac{15t}{(t-6)(t-4)(t-2)} & -\frac{5(2t^2+7)}{(t-6)(t-4)(t-2)} & \frac{t^2+14}{(t-6)(t-4)} & -\frac{1}{t-6} & 0 \\ 0 & 0 & 0 & -\frac{15}{(t-6)(t-4)(t-2)} & \frac{45t}{(t-6)(t-4)(t-2)} & -\frac{15t}{(t-6)(t-4)} & \frac{t}{t-6} & 0 \\ 0 & 0 & 0 & 0 & -\frac{105}{(t-6)(t-4)(t-2)} & \frac{105}{(t-6)(t-4)} & -\frac{21}{t-6} & 1 \end{bmatrix}.$$

By direct calculation, we have

PROPOSITION 7.2 ($\det Z_d$ for $1 \leq d \leq 7$).

$$\det Z_1 = 1, \quad \det Z_2 = \frac{t}{t-1}, \quad \det Z_3 = \frac{t^2-1}{(t-2)^2}, \quad \det Z_4 = \frac{t^2(t^2-4)}{(t-1)^1(t-3)^3},$$

$$\det Z_5 = \frac{(t^2-1)^2(t^2-9)^1}{(t-2)^2(t-4)^4}, \quad \det Z_6 = \frac{t^3(t^2-4)^2(t^2-16)^1}{(t-1)^1(t-3)^3(t-5)^5},$$

$$\det Z_7 = \frac{(t^2-1)^3(t^2-9)^2(t^2-25)^1}{(t-2)^2(t-4)^4(t-6)^6}.$$

Since $\det X_d = \varepsilon^{d(d+1)/2} \det Z_d$, we have the following conjecture on $\det X_d$.

CONJECTURE 7.3. *For a natural number q, we conjecture that*

$$\det X_{2q-1} = \left(\operatorname{sgn}(m_0)\sqrt{-1}\right)^{q(2q-1)} \frac{\prod_{l=1}^{q-1} \left(t^2 - (2l-1)^2\right)^{q-l}}{\prod_{l=1}^{q-1} (t-2l)^{2l}},$$

$$\det X_{2q} = \left(\operatorname{sgn}(m_0)\sqrt{-1}\right)^{q(2q+1)} \frac{t^q \prod_{l=1}^{q-1} \left(t^2 - (2l)^2\right)^{q-l}}{\prod_{l=1}^{q} (t-2l+1)^{2l-1}}.$$

In this article, we have shown that the Siegel-Whittaker functions of rapid decay for the large discrete series representations of $\mathrm{Sp}(2,\mathbb{R})$ are described by the partially confluent hypergeometric functions $\{f_k(a)\}_{k=0}^d$. Conversely, if the above conjecture is true, we have,

COROLLARY 7.4. *We assume that*

$$|m_0| \geq d \quad \text{and} \quad |m_0| \equiv d \pmod 2.$$

Let $\{c_j(a)\}_{j=0}^d$ be a smooth solution of the system (3.1), (3.2) and (3.3) in Proposition 3.1, and suppose that all of $\{e^{-2\pi(h_1a_1^2+h_2a_2^2)}c_j(a)\}_{j=0}^d$ are rapidly decreasing.

If Conjecture 7.3 is true, then all of $\{f_k(a)\}_{k=0}^d$ are $\mathbb{C}$-linear combinations of elements of $\{c_j(a)\}_{j=0}^d$.

PROOF. Since $|m_0| \geq d$ and $|m_0| \equiv d \pmod 2$, we can check that

$$\det X_d \neq 0,$$

and X_d^{-1} exists. It completes the proof. $\square$

References

[1] Wee Teck Gan, Benedict H. Gross, and Dipendra Prasad, *Symplectic local root numbers, central critical L values, and restriction problems in the representation theory of classical groups* (English, with English and French summaries), Astérisque **346** (2012), 1–109. Sur les conjectures de Gross et Prasad. I. MR3202556

[2] Yasuro Gon, *Generalized Whittaker functions on* $\mathrm{SU}(2,2)$ *with respect to the Siegel parabolic subgroup*, Mem. Amer. Math. Soc. **155** (2002), no. 738, viii+116, DOI 10.1090/memo/0738. MR1878340

[3] Miki Hirano, Taku Ishii, and Takayuki Oda, *Confluence from Siegel-Whittaker functions to Whittaker functions on* $\mathrm{Sp}(2,\mathbb{R})$, Math. Proc. Cambridge Philos. Soc. **141** (2006), no. 1, 15–31, DOI 10.1017/S0305004106009224. MR2238640

[4] Masatoshi Iida, *Errata: "Spherical functions of the principal series representations of* $\mathrm{Sp}(2,\mathbf{R})$ *as hypergeometric functions of* C_2*-type" [Publ. Res. Inst. Math. Sci.* **32** *(1996), no. 4, 689–727; MR1421998]*, Publ. Res. Inst. Math. Sci. **43** (2007), no. 2, 521–524. MR2341023

[5] Taku Ishii, *Siegel-Whittaker functions on* $\mathrm{Sp}(2,\mathbf{R})$ *for principal series representations*, J. Math. Sci. Univ. Tokyo **9** (2002), no. 2, 303–346. MR1904934

[6] Taku Ishii, *On principal series Whittaker functions on* $\mathrm{Sp}(2,\mathbf{R})$, J. Funct. Anal. **225** (2005), no. 1, 1–32, DOI 10.1016/j.jfa.2005.03.023. MR2149916

[7] Harutaka Koseki and Takayuki Oda, *Matrix coefficients of representations of* $\mathrm{SU}(2,2)$*: the case of* P_J*-principal series*, Internat. J. Math. **15** (2004), no. 10, 1033–1064, DOI 10.1142/S0129167X04002648. MR2106262

[8] Takuya Miyazaki, *The generalized Whittaker functions for* $\mathrm{Sp}(2,\mathbf{R})$ *and the gamma factor of the Andrianov L-function*, J. Math. Sci. Univ. Tokyo **7** (2000), no. 2, 241–295. MR1768466

[9] M. E. Novodvorskiĭ and I. I. Pjateckiĭ-Šapiro, *Generalized Bessel models for the symplectic group of rank* 2 (Russian), Mat. Sb. (N.S.) **90(132)** (1973), 246–256, 326. MR0338279

[10] Takayuki Oda, *An explicit integral representation of Whittaker functions on* $\mathrm{Sp}(2;\mathbf{R})$ *for the large discrete series representations*, Tohoku Math. J. (2) **46** (1994), no. 2, 261–279, DOI 10.2748/tmj/1178225761. MR1272882

[11] Takayuki Oda, *Matrix coefficients of the large discrete series representations of* $\mathrm{Sp}(2;\mathbf{R})$ *as hypergeometric series of two variables*, Nagoya Math. J. **208** (2012), 201–263. MR3006701

[12] Dipendra Prasad and Ramin Takloo-Bighash, *Bessel models for GSp(4)*, J. Reine Angew. Math. **655** (2011), 189–243, DOI 10.1515/CRELLE.2011.045. MR2806111

[13] E. T. Whittaker and G. N. Watson, *A course of modern analysis*, Cambridge Mathematical Library, Cambridge University Press, Cambridge, 1996. An introduction to the general theory of infinite processes and of analytic functions; with an account of the principal transcendental functions; Reprint of the fourth (1927) edition. MR1424469

[14] Y. Otsuki (eds.), New mathematical formulas I. Elementary functions (in Japanese), Maruzen (1991).

Faculty of Mathematics, Kyushu University, 744 Motooka, Nishi-ku, Fukuoka 819-0395, Japan

Email address: `ygon@math.kyushu-u.ac.jp`

Graduate School of Mathematical Sciences, The University of Tokyo, 3-8-1 Komaba, Meguro-ku, Tokyo 153-8914, Japan

Email address: `takayuki@ms.u-tokyo.ac.jp`

Contemporary Mathematics
Volume **701**, 2018
http://dx.doi.org/10.1090/conm/701/14144

On implementation of GHS attack against elliptic curve cryptosystems over cubic extension fields of odd characteristic

Naoki Hashizume, Fumiyuki Momose, and Jinhui Chao

ABSTRACT. In this paper, we present algorithms implementing the GHS attack against Elliptic curve cryptosystems (ECC). In particular, we consider two large classes of elliptic curves over cubic extension fields of odd characteristic which have weak covering curves against GHS attack, whose existence have been shown recently [**17**], [**18**], [**19**], [**20**]. We give algorithms to compute the defining equation of the covering curve and to transfer the DLP from the elliptic curve to the Jacobian of the covering curve. An algorithm to test if the covering curve is hyperelliptic is also given in the appendix.

1. Introduction

Elliptic curve cryptography is known as a rich source of secure and efficient cryptosystems. In particular, it can provide the same level of security as RSA and ElGamal cryptosystems while using much shorter key length. This property is also desirable in implementation of compact and low cost cryptosystems. Against cryptosystems based on low genus algebraic curves, the fastest known attacks (in general) are "square-root" attacks such as the baby-step giant-step attack, Pollard's rho and lambda algorithms. Recently, index calculus attacks have been proposed by Gaudry, Nagao, Gaudry-Thériault-Thomé-Diem [**1**], [**2**], [**3**] for hyperelliptic curves of genera larger than 3 and by Diem [**4**] for non-hyperelliptic curves of genera larger than or equal to 3.

A relatively new attack called GHS attack, which is based on the idea of Weil descent suggested by Frey [**5**], was proposed by Gaudry, Hess, and Smart in 2000 [**6**]. The GHS attack transfers the discrete logarithm problem (DLP) in the group of rational points of an elliptic curve E over an extension k_d of a finite field k to the DLP in the Jacobian variety of a new curve C of higher genus over the smaller definition field k.

The GHS attack has already been investigated extensively. However, although theoretically interesting, its analysis seemed nontrivial [**7**], [**8**] [**9**], [**10**], [**11**], [**12**], [**13**], [**14**], [**15**], [**16**]. The classes of the weak elliptic curves or curves for which the GHS attack efficiently works have not yet been fully understood. At the beginning,

2010 *Mathematics Subject Classification.* Primary 14G50,11T71, Secondary 11G25, 94A60.
Key words and phrases. Elliptic curve cryptosystems, Discrete logarithm problem, GHS attack.

it seemed that the class of curves subjected to the GHS attack must be special so the number of such curves will not be very large. Recently, the existence of certain large classes of elliptic and hyperelliptic curves which are weak against GHS was shown [**17**], [**18**], [**19**], [**20**]. Further results on the subject can be found in [**21**], [**22**], [**23**].

In modern cryptography, one of the most efficient and reliable approaches for the security analysis of a particular cryptosystem is (particularly if the security is not theoretically provable) to apply every possible attacks to it in order to find its weak point. Only systems which have resisted all such attacks can be trusted in practical usage. Thus it is both important and interesting to implement GHS attack to these weak curves.

A GHS attack consists of three steps: finding a covering curve C/k of an elliptic curve E/k_d where k_d is the degree d extension of a finite field k; transferring the discrete logarithm on E/k_d to the Jacobian $J(C)/k$; and finally, applying an index calculus algorithm to solve the discrete logarithm in $J(C)/k$. As to the first step, it seems to be nontrivial to find the defining equation of a weak curve E/k_d. Certain cases were discussed in [**14**], [**15**], [**17**]. For the second step, although a general strategy using norm-conorm map is well known, efficient and explicit implementation of the strategy does not seem to be available and appear to be nontrivial.

In this paper, we show explicit procedures for the first two steps of the GHS attack against two large classes of the elliptic curves over cubic extension fields of odd characteristic. These two classes, called Type I and Type II curves have been obtained in [**17**][**18**][**19**][**20**], both of them have non-hyperelliptic covering curves of genus three, which are subjected to Diem's double-large-prime attack. We present an algorithm to explicitly construct these covering curves C over k from the elliptic curves E over the cubic extension of k with odd characteristic. Then an algorithm is given to map the rational point on the elliptic curve E to the divisor of the covering curve C, in order to transfer the DLP. In the appendix, we also present an algorithm to test if a Type I or II curve is hyperellipic. These algorithms are implemented and examples are shown.

The first and third authors would like to dedicate the publication of this work to their friend, colleague, and collaborator, Fumiyuki Momose.

2. Weak Covering C over k_3, char$k \neq 2$

Let $k = \mathbb{F}_q$ be a finite field of odd characteristic, and $k_d = \mathbb{F}_{q^d}$.

We consider the GHS attack against an algebraic curve C_0/k_d with genus $g_0 = g(C_0)$. A special case is when $g_0 = 1$ and $C_0 = E/k_d$ is an elliptic curve.

Assume there exists an algebraic curve C/k such that

$$\pi/k_d : C \longrightarrow C_0 \tag{2.1}$$

is a covering defined over k_d, which induces the map

$$\pi_*/k_d : \mathrm{Jac}(C) \longrightarrow \mathrm{Jac}(C_0). \tag{2.2}$$

Also assume the restriction of π_* onto k

$$\mathrm{Re}(\pi_*)/k : \mathrm{Jac}(C) \longrightarrow \mathrm{Re}_{k_d/k}(\mathrm{Jac}(C_0)) \tag{2.3}$$

defines an isogeny over k. Then C has genus $g(C) = dg_0$. Here, $\mathrm{Re}_{k_d/k}(\mathrm{Jac}(C_0))$ is the Weil restriction of $\mathrm{Jac}(C_0)$ with respect to extension field k_d/k.

Throughout this paper, we will assume that $g_0 = 1,\ d = 3,\ \mathrm{char}(k) \neq 2$.

According to [**17**][**18**][**19**][**20**], the elliptic curves C_0 which have weak covering C as genus three nonhyperelliptic curves can be divided into two types.

$$C_0/k_3: \ y^2 = (x-\alpha)(x-\alpha^q)(x-\beta)(x-\beta^q) \tag{2.4}$$

$$\text{Type I}: \ \alpha,\ \beta \in k_3\backslash k,\ \#\{\alpha,\ \alpha^q, \beta,\ \beta^q\} = 4 \tag{2.5}$$

$$\text{Type II}: \ \alpha \in k_6\backslash(k_2 \cup k_3),\ \beta = \alpha^{q^3} \tag{2.6}$$

These elliptic curves can be transformed to the following Legendre canonical forms:

- Type I:

$$C_0/k_3: \ y^2 = x(x-1)(x-\lambda),\ \lambda = \frac{(\beta-\alpha^q)(\beta^q-\alpha)}{(\beta-\alpha)(\beta^q-\alpha^q)} \tag{2.7}$$

- Type II:

$$C_0/k_3: \ y^2 = \mathrm{N}_{k_6/k_3}(\beta-\alpha^q)x(x-1)(x-\lambda),\ \lambda = \mathrm{N}_{k_6/k_3}\left(\frac{\alpha^q-\alpha}{\alpha^q-\beta}\right) \tag{2.8}$$

And $\#\{\lambda\} \approx \frac{1}{2}q^3$.

The discrete logarithm on C_0/k_3 has a complexity of $\tilde{O}(q^{4/3})$ against the Pollard's rho method. On the other hand, apply Diem's algorithm to nonhyperellitic C, the complexity of discrete logarithm reduces to $\tilde{O}(q)$.

Here we use $M \cdot \gamma$ to denote a PGL_2-action as follows.

$$M := \begin{pmatrix} a & b \\ c & d \end{pmatrix} \in PGL_2(k),\ \ \gamma \in \overline{k} \qquad M \cdot \gamma := \frac{a\gamma + b}{c\gamma + d}. \tag{2.9}$$

Now, define

$$\mu \ := \ \begin{pmatrix} \alpha^q & -\alpha \\ 1 & -1 \end{pmatrix} \cdot \lambda, \tag{2.10}$$

$$A \ := \ \begin{pmatrix} -\mu + \alpha + \alpha^q & -\alpha^{1+q} \\ 1 & -\mu \end{pmatrix}, \tag{2.11}$$

$$B \ := \ {}^{\sigma^2}\!A\ {}^{\sigma}\!A\ A. \tag{2.12}$$

According to Lemma 7, 1,2 [**20**], the necessary and sufficient condition for C_0 to be Type I is that the quadratic equation

$$B \cdot \beta = \beta \tag{2.13}$$

has a solution β.

Besides, by Lemma 4 [**20**], the covering curve C of such a curve C_0 is hyperelliptic if and only if

$$\beta = A \cdot \alpha,\ {}^{\exists}A \in \mathrm{GL}_2(k),\ \mathrm{Tr}A = 0. \tag{2.14}$$

Hereafter we assume that α and β do not satisfy Condition (2.14). Then, the curve C is a nonhyperelliptic curve over k of genus three. We show in the appendix an algorithm to test if C is hyperelliptic.

In this paper, we describe the following two algorithms:

(i) To construct the curve C/k, or to find the defining equation explicitly from the given curve C_0/k_d.

(ii) To transfer the DLP over C_0/k_d to the DLP over $J(C/k)$.

3. How to construct C/k from C_0/k_d

Assume C is a nonhyperelliptic curve of genus $g = dg_0 = 3$. Thus, its canonical embedding is a quartic curve in $\mathbb{P}^2$. Let σ be a q-th power Frobenius map satisfying

$$l(x) = \sum_{i=1}^{n} a_i x^i \quad \longmapsto \quad {}^{\sigma}l(x) = \sum_{i=1}^{n} {a_i}^q x^i \qquad ({}^{\forall}l(x) \in k_d[x]). \tag{3.1}$$

The embedding map is

$$C \quad \hookrightarrow \quad \mathbb{P}^2 \tag{3.2}$$

$$P \quad \longmapsto \quad \left(\omega(P) : {}^{\sigma}\omega(P) : {}^{\sigma^2}\omega(P)\right) \tag{3.3}$$

where $\omega = \frac{\mathrm{d}x}{y}$ and its conjugates generate the first cohomology group

$$H^0(C/k_3, \Omega^1) = \langle\ \omega,\ \ {}^{\sigma}\omega,\ \ {}^{\sigma^2}\omega\ \rangle. \tag{3.4}$$

We use hereafter the correspondence

$$\omega \longleftrightarrow X, \qquad {}^{\sigma}\omega \longleftrightarrow Y, \qquad {}^{\sigma^2}\omega \longleftrightarrow Z. \tag{3.5}$$

The Galois action on $H^0(C/k_3, \Omega^1)$ is a cyclic shift.

Now we consider the automorphism group of the first coholomogy group

$$Aut(H^0(C/k_3, \Omega^1)) = \{id,\ \phi,\ {}^{\sigma}\phi,\ {}^{\sigma^2}\phi\}. \tag{3.6}$$

The identity on $H^0(C/k_3, \Omega^1)$ is

$$id : \begin{cases} X & \longmapsto & X \\ Y & \longmapsto & Y \\ Z & \longmapsto & Z \end{cases}. \tag{3.7}$$

The bi-elliptic involution changes the signs of both Y and Z

$$\phi : \begin{cases} X & \mapsto & X \\ Y & \longmapsto & -Y \\ Z & \longmapsto & -Z \end{cases}. \tag{3.8}$$

Then the bi-elliptic involution under the Galois action of σ has the following form

$${}^{\sigma}\phi : \begin{cases} X & \longmapsto & -X \\ Y & \longmapsto & Y \\ Z & \longmapsto & -Z \end{cases}, \tag{3.9}$$

and the bi-elliptic involution under the action of σ^2 has the following form

$${}^{\sigma^2}\phi : \begin{cases} X & \longmapsto & -X \\ Y & \longmapsto & -Y \\ Z & \longmapsto & Z \end{cases}. \tag{3.10}$$

3.1. Defining equation of C/k_3. The quartic curve C/k_3 has its defining equation invariant under $\mathrm{Gal}(k_3/k)$, thus in the following symmetric form.

$$\begin{aligned} C/k_3 \quad : \quad & aX^4 + a^qY^4 + a^{q^2}Z^4 \\ & +bX^3Y + b^qY^3Z + b^{q^2}Z^3X \\ & +cX^3Z + c^qY^3X + c^{q^2}Z^3Y \\ & +dX^2Y^2 + d^qY^2Z^2 + d^{q^2}Z^2X^2 \\ & +eX^2YZ + e^qXY^2Z + e^{q^2}XYZ^2 = 0. \end{aligned} \tag{3.11}$$

Since the defining equation $F = 0$ of C is invariant under the action of automorphisms of $\mathrm{Aut}(H^0(C, \Omega^1))$, C will also be defined by $F + \phi(F) + {}^{\sigma}\phi(F) + {}^{\sigma^2}\phi(F) = 0$.

On the other hand, since ϕ, ${}^{\sigma}\phi$, ${}^{\sigma^2}\phi$ change the signs of two variables, the terms with odd degrees of variables are canceled each other.

Thus the equation of the curve C/k_3 is in the following form.

$$\begin{aligned} C/k_3 : \quad & aX^4 + a^qY^4 + a^{q^2}Z^4 + bX^2Y^2 + b^qY^2Z^2 + b^{q^2}Z^2X^2 = 0. \\ & a, b \in k_3 \end{aligned} \tag{3.12}$$

3.2. Evaluation of a and b. To find the coefficients a and b in (3.12), we substitute into it $X = \frac{dx}{y}$, $Y = \frac{dx}{{}^{\sigma}y}$, $Z = \frac{dx}{{}^{\sigma^2}y}$.

Since

$$\frac{1}{y^2} = \frac{(x - \alpha^{q^2})(x - \beta^{q^2})}{\mathrm{N}_{k_3/k}((x-\alpha)(x-\beta))}, \tag{3.13}$$

$$\frac{1}{({}^{\sigma}y)^2} = \frac{(x - \alpha)(x - \beta)}{\mathrm{N}_{k_3/k}((x-\alpha)(x-\beta))}, \tag{3.14}$$

we substitute these into (3.12) to obtain

$$\begin{aligned} & \mathrm{Tr}_{k_3/k}(a(x - \alpha^{q^2})^2(x - \beta^{q^2})^2) + \\ & \qquad \mathrm{Tr}_{k_3/k}(b(x-\alpha)(x - \alpha^{q^2})(x-\beta)(x - \beta^{q^2})) = 0. \end{aligned} \tag{3.15}$$

3.2.1. *Type I.* From expansion of (3.15) we can express the coefficients of each x^i as

$$\begin{aligned} x^4 &: \ \mathrm{Tr}(a) + \mathrm{Tr}(b) \\ x^3 &: \ -2\mathrm{Tr}(a(\alpha^{q^2} + \beta^{q^2})) - \mathrm{Tr}(b(\alpha + \beta + \alpha^{q^2} + \beta^{q^2})) \\ x^2 &: \ \mathrm{Tr}(a(\alpha^{2q^2} + 4\alpha^{q^2}\beta^{q^2} + \beta^{2q^2})) + \mathrm{Tr}(b\{\alpha^{q^2+1} + (\alpha + \alpha^{q^2})(\beta + \beta^{q^2}) + \beta^{q^2+1}\}) \\ x &: \ -2\mathrm{Tr}(a(\alpha^{2q^2}\beta^{q^2} + \alpha^{q^2}\beta^{2q^2})) - \mathrm{Tr}(b\{\alpha^{q^2+1}(\beta + \beta^{q^2}) + \beta^{q^2+1}(\alpha + \alpha^{q^2})\}) \\ 1 &: \ \mathrm{Tr}(a\alpha^{2q^2}\beta^{2q^2}) + \mathrm{Tr}(b\alpha^{q^2+1}\beta^{q^2+1}) \end{aligned}$$

which are identically zeros.

In order to calculate a, b explicitly, we express $a, b \in k_3$ on a k-basis of k_3.

$$a = a_0 + a_1\epsilon + a_2\epsilon^2 \ (a_0,\ a_1,\ a_2 \in k) \tag{3.16}$$

$$b = b_0 + b_1\epsilon + b_2\epsilon^2 \ (b_0,\ b_1,\ b_2 \in k) \tag{3.17}$$

where ϵ generates $k_3 = k(\epsilon)$.

Belows, we express the coefficients of x^i in (3.15) in terms of a_i, b_i.

First, in the coefficient of x^4, $\mathrm{Tr}(a)$ is given by

$$\mathrm{Tr}(a) = 3a_0 + \mathrm{Tr}(\epsilon)a_1 + \mathrm{Tr}(\epsilon^2)a_2. \tag{3.18}$$

Similarly,

$$\mathrm{Tr}(b) = 3b_0 + \mathrm{Tr}(\epsilon)b_1 + \mathrm{Tr}(\epsilon^2)b_2. \tag{3.19}$$

Next, in the coefficient of x^3, $\mathrm{Tr}(a(\alpha^{q^2} + \beta^{q^2}))$ is given by

$$\begin{aligned}\mathrm{Tr}(a(\alpha^{q^2} + \beta^{q^2})) &= (\alpha^{q^2} + \beta^{q^2})(a_0 + a_1\epsilon + a_2\epsilon^2)\\ &\quad +(\alpha + \beta)(a_0 + a_1\epsilon^q + a_2\epsilon^{2q})\\ &\quad +(\alpha^q + \beta^q)(a_0 + a_1\epsilon^{q^2} + a_2\epsilon^{2q^2})\\ &= \mathrm{Tr}(\alpha + \beta)a_0 + \mathrm{Tr}((\alpha + \beta)\epsilon^q)a_1 + \mathrm{Tr}((\alpha + \beta)\epsilon^{2q})a_2.\end{aligned} \tag{3.20}$$

$\mathrm{Tr}(b(\alpha + \beta + \alpha^{q^2} + \beta^{q^2}))$ is given by

$$\begin{aligned}\mathrm{Tr}(b(\alpha + \beta + \alpha^{q^2} + \beta^{q^2})) &= (\alpha + \beta + \alpha^{q^2} + \beta^{q^2})(b_0 + b_1\epsilon + b_2\epsilon^2)\\ &\quad +(\alpha^q + \beta^q + \alpha + \beta)(b_0 + b_1\epsilon^q + b_2\epsilon^{2q})\\ &\quad +(\alpha^{q^2} + \beta^{q^2} + \alpha^q + \beta^q)(b_0 + b_1\epsilon^{q^2} + b_2\epsilon^{2q^2})\\ &= \mathrm{Tr}(\alpha^q + \beta^q + \alpha + \beta)b_0 + \mathrm{Tr}((\alpha^q + \beta^q + \alpha + \beta)\epsilon^q)b_1\\ &\quad +\mathrm{Tr}((\alpha^q + \beta^q + \alpha + \beta)\epsilon^{2q})b_2.\end{aligned} \tag{3.21}$$

In the coefficient of x^2, $\mathrm{Tr}(a(\alpha^{2q^2} + 4\alpha^{q^2}\beta^{q^2} + \beta^{2q^2}))$ is given by

$$\begin{aligned}\mathrm{Tr}(a(\alpha^{2q^2} + 4\alpha^{q^2}\beta^{q^2} + \beta^{2q^2})) &= \mathrm{Tr}(\alpha^2 + 4\alpha\beta + \beta^2)a_0 +\\ &\quad \mathrm{Tr}((\alpha^2 + 4\alpha\beta + \beta^2)\epsilon^q)a_1 + \mathrm{Tr}((\alpha^2 + 4\alpha\beta + \beta^2)\epsilon^{2q})a_2.\end{aligned} \tag{3.22}$$

and $\mathrm{Tr}(b\{\alpha^{q^2+1} + (\alpha + \alpha^{q^2})(\beta + \beta^{q^2}) + \beta^{q^2+1}\})$ is given by

$$\begin{aligned}&\mathrm{Tr}(b\{\alpha^{q^2+1} + (\alpha + \alpha^{q^2})(\beta + \beta^{q^2}) + \beta^{q^2+1}\}) =\\ &\quad \mathrm{Tr}(\alpha^{q+1} + (\alpha^q + \alpha)(\beta^q + \beta) + \beta^{q+1})b_0\\ &\quad +\mathrm{Tr}(\{\alpha^{q+1} + (\alpha^q + \alpha)(\beta^q + \beta) + \beta^{q+1}\}\epsilon^q)b_1\\ &\quad +\mathrm{Tr}(\{\alpha^{q+1} + (\alpha^q + \alpha)(\beta^q + \beta) + \beta^{q+1}\}\epsilon^{2q})b_2.\end{aligned} \tag{3.23}$$

In the coefficient of x, $\mathrm{Tr}(a(\alpha^{2q^2}\beta^{q^2} + \alpha^{q^2}\beta^{2q^2}))$ is given by

$$\begin{aligned}\mathrm{Tr}(a(\alpha^{2q^2}\beta^{q^2} + \alpha^{q^2}\beta^{2q^2})) &= \mathrm{Tr}(\alpha^2\beta + \alpha\beta^2)a_0\\ &\quad +\mathrm{Tr}((\alpha^2\beta + \alpha\beta^2)\epsilon^q)a_1\\ &\quad +\mathrm{Tr}((\alpha^2\beta + \alpha\beta^2)\epsilon^{2q})a_2.\end{aligned} \tag{3.24}$$

and $\mathrm{Tr}(b\{\alpha^{q^2+1}(\beta + \beta^{q^2}) + \beta^{q^2+1}(\alpha + \alpha^{q^2})\})$ is given by

$$\begin{aligned}&\mathrm{Tr}(b\{\alpha^{q^2+1}(\beta + \beta^{q^2}) + \beta^{q^2+1}(\alpha + \alpha^{q^2})\}) =\\ &\quad \mathrm{Tr}(\alpha^q\beta^q(\alpha + \beta) + \alpha\beta(\alpha^q + \beta^q))b_0\\ &\quad +\mathrm{Tr}(\{\alpha^q\beta^q(\alpha + \beta) + \alpha\beta(\alpha^q + \beta^q)\}\epsilon^q)b_1\\ &\quad +\mathrm{Tr}(\{\alpha^q\beta^q(\alpha + \beta) + \alpha\beta(\alpha^q + \beta^q)\}\epsilon^{2q})b_2.\end{aligned} \tag{3.25}$$

In the constant term of (3.15), $\mathrm{Tr}(a\alpha^{2q^2}\beta^{2q^2})$ is given by

$$\mathrm{Tr}(a\alpha^{2q^2}\beta^{2q^2}) = \mathrm{Tr}(\alpha^2\beta^2)a_0 + \mathrm{Tr}(\alpha^2\beta^2\epsilon^q)a_1 + \mathrm{Tr}(\alpha^2\beta^2\epsilon^{2q})a_2. \tag{3.26}$$

and $\mathrm{Tr}(b\alpha^{q^2+1}\beta^{q^2+1})$ is given by

$$\text{(3.27)}\quad \mathrm{Tr}(b\alpha^{q^2+1}\beta^{q^2+1}) = \mathrm{Tr}(\alpha^{q+1}\beta^{q+1})b_0 + \mathrm{Tr}(\alpha^{q+1}\beta^{q+1}\epsilon^q)b_1 + \mathrm{Tr}(\alpha^{q+1}\beta^{q+1}\epsilon^{2q})b_2.$$

Combining the equations above yields the following system of simultaneous linear equations.

$$\begin{cases} 3a_0 + \mathrm{Tr}(\epsilon)a_1 + \mathrm{Tr}(\epsilon^2)a_2 + 3b_0 + \mathrm{Tr}(\epsilon)b_1 + \mathrm{Tr}(\epsilon^2)b_2 = 0 \\[2ex] 2\mathrm{Tr}(\alpha+\beta)a_0 + 2\mathrm{Tr}((\alpha+\beta)\epsilon^q)a_1 + 2\mathrm{Tr}((\alpha+\beta)\epsilon^{2q})a_2 \\ \quad +\mathrm{Tr}(\alpha^q+\beta^q+\alpha+\beta)b_0 \\ \quad\quad +\mathrm{Tr}((\alpha^q+\beta^q+\alpha+\beta)\epsilon^q)b_1 \\ \quad\quad\quad +\mathrm{Tr}((\alpha^q+\beta^q+\alpha+\beta)\epsilon^{2q})b_2 = 0 \\[2ex] \mathrm{Tr}(\alpha^2+4\alpha\beta+\beta^2)a_0 + \mathrm{Tr}((\alpha^2+4\alpha\beta+\beta^2)\epsilon^q)a_1 \\ \quad +\mathrm{Tr}((\alpha^2+4\alpha\beta+\beta^2)\epsilon^{2q})a_2 \\ \quad\quad +\mathrm{Tr}(\alpha^{q+1}+(\alpha^q+\alpha)(\beta^q+\beta)+\beta^{q+1})b_0 \\ \quad\quad\quad +\mathrm{Tr}(\{\alpha^{q+1}+(\alpha^q+\alpha)(\beta^q+\beta)+\beta^{q+1}\}\epsilon^q)b_1 \\ \quad\quad\quad\quad +\mathrm{Tr}(\{\alpha^{q+1}+(\alpha^q+\alpha)(\beta^q+\beta)+\beta^{q+1}\}\epsilon^{2q})b_2 = 0 \\[2ex] 2\mathrm{Tr}(\alpha^2\beta+\alpha\beta^2)a_0 + 2\mathrm{Tr}((\alpha^2\beta+\alpha\beta^2)\epsilon^q)a_1 + 2\mathrm{Tr}((\alpha^2\beta+\alpha\beta^2)\epsilon^{2q})a_2 \\ \quad +\mathrm{Tr}(\alpha^q\beta^q(\alpha+\beta)+\alpha\beta(\alpha^q+\beta^q))b_0 \\ \quad\quad +\mathrm{Tr}(\{\alpha^q\beta^q(\alpha+\beta)+\alpha\beta(\alpha^q+\beta^q)\}\epsilon^q)b_1 \\ \quad\quad\quad +\mathrm{Tr}(\{\alpha^q\beta^q(\alpha+\beta)+\alpha\beta(\alpha^q+\beta^q)\}\epsilon^{2q})b_2 = 0 \\[2ex] \mathrm{Tr}(\alpha^2\beta^2)a_0 + \mathrm{Tr}(\alpha^2\beta^2\epsilon^q)a_1 + \mathrm{Tr}(\alpha^2\beta^2\epsilon^{2q})a_2 \\ \quad +\mathrm{Tr}(\alpha^{q+1}\beta^{q+1})b_0 + \mathrm{Tr}(\alpha^{q+1}\beta^{q+1}\epsilon^q)b_1 + \mathrm{Tr}(\alpha^{q+1}\beta^{q+1}\epsilon^{2q})b_2 = 0 \end{cases}$$

From the equation (3.12), we can assume $a_0 = 1$. Accordingly, the simultaneous linear equations can be written as

$$\text{(3.28)}\qquad \begin{pmatrix} d_{11} & d_{12} & d_{13} & d_{14} & d_{15} \\ d_{21} & d_{22} & d_{23} & d_{24} & d_{25} \\ d_{31} & d_{32} & d_{33} & d_{34} & d_{35} \\ d_{41} & d_{42} & d_{43} & d_{44} & d_{45} \\ d_{51} & d_{52} & d_{53} & d_{54} & d_{55} \end{pmatrix} \begin{pmatrix} a_1 \\ a_2 \\ b_0 \\ b_1 \\ b_2 \end{pmatrix} = \begin{pmatrix} e_1 \\ e_2 \\ e_3 \\ e_4 \\ e_5 \end{pmatrix}.$$

where d_{ij} are the coefficients of a_1, a_2, b_0, b_1, b_2 in each equation and e_i are the negations of the coefficients of a_0.

Thus a_1, a_2, b_0, b_1, b_2 can be obtained by solving the linear system of equations for the given set of α, β and ϵ.

3.2.2. *Type II.* For Type II curves, the coefficients of x^i in Equation (3.15) are as follows.

First, the coefficient of x^4 is

$$\text{(3.29)}\quad \mathrm{Tr}(a) + \mathrm{Tr}(b) = 3a_0 + \mathrm{Tr}(\epsilon)a_1 + \mathrm{Tr}(\epsilon^2)a_2 + 3b_0 + \mathrm{Tr}(\epsilon)b_1 + \mathrm{Tr}(\epsilon^2)b_2 = 0.$$

Next, the coefficient of x^3 is as follows:

$$\begin{aligned}(3.30)\quad 2\mathrm{Tr}(a(\alpha^{q^2}+\beta^{q^2}))+\mathrm{Tr}(b(\alpha+\beta+\alpha^{q^2}+\beta^{q^2})) &= 2\mathrm{Tr}(\mathrm{Tr}_{k_6/k_3}(\alpha))a_0\\ &+2\mathrm{Tr}(\mathrm{Tr}_{k_6/k_3}(\alpha)\epsilon^q)a_1\\ &+2\mathrm{Tr}(\mathrm{Tr}_{k_6/k_3}(\alpha)\epsilon^{2q})a_2\\ &+\mathrm{Tr}(\{\mathrm{Tr}_{k_6/k_3}(\alpha)\}^q+\mathrm{Tr}_{k_6/k_3}(\alpha))b_0\\ &+\mathrm{Tr}(\left[\{\mathrm{Tr}_{k_6/k_3}(\alpha)\}^q+\mathrm{Tr}_{k_6/k_3}(\alpha)\right]\epsilon^q)b_1\\ &+\mathrm{Tr}(\left[\{\mathrm{Tr}_{k_6/k_3}(\alpha)\}^q+\mathrm{Tr}_{k_6/k_3}(\alpha)\right]\epsilon^{2q})b_2=0.\end{aligned}$$

The coefficient of x^2 is

$$\begin{aligned}(3.31)\quad &\mathrm{Tr}(a(\alpha^{2q^2}+4\alpha^{q^2}\beta^{q^2}+\beta^{2q^2}))+\\ &\qquad\mathrm{Tr}(b\{\alpha^{q^2+1}+(\alpha+\alpha^{q^2})(\beta+\beta^{q^2})+\beta^{q^2+1}\})\\ &=\mathrm{Tr}(\{\mathrm{Tr}_{k_6/k_3}(\alpha)\}^2+2\mathrm{N}_{k_6/k_3}(\alpha))a_0\\ &+\mathrm{Tr}(\left[\{\mathrm{Tr}_{k_6/k_3}(\alpha)\}^2+2\mathrm{N}_{k_6/k_3}(\alpha)\right]\epsilon^q)a_1\\ &+\mathrm{Tr}(\left[\mathrm{Tr}_{k_6/k_3}(\alpha)^2+2\mathrm{N}_{k_6/k_3}(\alpha)\right]\epsilon^{2q})a_2\\ &+\mathrm{Tr}(\{\mathrm{Tr}_{k_6/k_3}(\alpha)\}^{q+1}+\{\mathrm{N}_{k_6/k_3}(\alpha)\}^q+\mathrm{N}_{k_6/k_3}(\alpha))b_0\\ &+\mathrm{Tr}(\left[\{\mathrm{Tr}_{k_6/k_3}(\alpha)\}^{q+1}+\{\mathrm{N}_{k_6/k_3}(\alpha)\}^q+\mathrm{N}_{k_6/k_3}(\alpha)\right]\epsilon^q)b_1\\ &+\mathrm{Tr}(\left[\{\mathrm{Tr}_{k_6/k_3}(\alpha)\}^{q+1}+\{\mathrm{N}_{k_6/k_3}(\alpha)\}^q+\mathrm{N}_{k_6/k_3}(\alpha)\right]\epsilon^{2q})b_2=0.\end{aligned}$$

The coefficient of x is

$$\begin{aligned}(3.32)\quad &2\mathrm{Tr}(a(\alpha^{2q^2}\beta^{q^2}+\alpha^{q^2}\beta^{2q^2}))+\mathrm{Tr}(b\{\alpha^{q^2+1}(\beta+\beta^{q^2})+\beta^{q^2+1}(\alpha+\alpha^{q^2})\})\\ &=2\mathrm{Tr}(\mathrm{Tr}_{k_6/k_3}(\alpha)\mathrm{N}_{k_6/k_3}(\alpha))a_0+2\mathrm{Tr}(\mathrm{Tr}_{k_6/k_3}(\alpha)\mathrm{N}_{k_6/k_3}(\alpha)\epsilon^q)a_1\\ &+2\mathrm{Tr}(\mathrm{Tr}_{k_6/k_3}(\alpha)\mathrm{N}_{k_6/k_3}(\alpha)\epsilon^{2q})a_2\\ &+\mathrm{Tr}(\mathrm{Tr}_{k_6/k_3}(\alpha)\{\mathrm{N}_{k_6/k_3}(\alpha)\}^q+\{\mathrm{Tr}_{k_6/k_3}(\alpha)\}^q\mathrm{N}_{k_6/k_3}(\alpha))b_0\\ &+\mathrm{Tr}(\left[\mathrm{Tr}_{k_6/k_3}(\alpha)\{\mathrm{N}_{k_6/k_3}(\alpha)\}^q+\{\mathrm{Tr}_{k_6/k_3}(\alpha)\}^q\mathrm{N}_{k_6/k_3}(\alpha)\right]\epsilon^q)b_1\\ &+\mathrm{Tr}(\left[\mathrm{Tr}_{k_6/k_3}(\alpha)\{\mathrm{N}_{k_6/k_3}(\alpha)\}^q+\{\mathrm{Tr}_{k_6/k_3}(\alpha)\}^q\mathrm{N}_{k_6/k_3}(\alpha)\right]\epsilon^{2q})b_2=0.\end{aligned}$$

The constant term of (3.15) for Type II curves is

$$\begin{aligned}(3.33)\qquad &\mathrm{Tr}(a\alpha^{2q^2}\beta^{2q^2})+\mathrm{Tr}(b\alpha^{q^2+1}\beta^{q^2+1})=\\ &\mathrm{Tr}(\{\mathrm{N}_{k_6/k_3}(\alpha)\}^2)a_0+\mathrm{Tr}(\{\mathrm{N}_{k_6/k_3}(\alpha)\}^2\epsilon^q)a_1\\ &+\mathrm{Tr}(\{\mathrm{N}_{k_6/k_3}(\alpha)\}^2\epsilon^{2q})a_2+\mathrm{Tr}(\{\mathrm{N}_{k_6/k_3}(\alpha)\}^{q+1})b_0\\ &+\mathrm{Tr}(\{\mathrm{N}_{k_6/k_3}(\alpha)\}^{q+1}\epsilon^q)b_1+\mathrm{Tr}(\{\mathrm{N}_{k_6/k_3}(\alpha)\}^{q+1}\epsilon^{2q})b_2=0.\end{aligned}$$

Then one can also build and solve a system of simultaneous linear equations, as in the case of Type I, in $a_1,\ a_2,\ b_0,\ b_1,\ b_2$.

Hereafter, we assume that a, b are known.

3.3. Definition equation of C/k. Notice that X, Y, Z correspond to a basis $\omega,\ {}^{\sigma}\omega,\ {}^{\sigma^2}\omega$ of $H^0(C/k_3, \Omega^1)$. Since C is defined over k, the next step is to find a basis of $H^0(C/k, \Omega^1)$.

The necessary and sufficient condition for $\{\omega_1, \omega_2, \omega_3\}$ to be such a basis, i.e. $H^0(C/k, \Omega^1) = \langle \omega_1, \omega_2, \omega_3 \rangle$ is

$$\omega = \gamma\omega_1 + \delta\omega_2 + \psi\omega_3, \ {}^{\exists}\gamma,\ \delta,\ \psi \in k_3 \tag{3.34}$$

$$\text{such that} \quad \det(U) \neq 0. \quad \text{where} \quad U := \begin{pmatrix} \gamma & \delta & \psi \\ \gamma^q & \delta^q & \psi^q \\ \gamma^{q^2} & \delta^{q^2} & \psi^{q^2} \end{pmatrix}. \tag{3.35}$$

We will use the following correspondence.

$$\omega_1 \longleftrightarrow \underline{x}, \quad \omega_2 \longleftrightarrow \underline{y}, \quad \omega_3 \longleftrightarrow \underline{z} \tag{3.36}$$

Then $X,\ Y,\ Z$ are expressed as

$$\begin{cases} X = \gamma\underline{x} + \delta\underline{y} + \psi\underline{z} \\ Y = \gamma^q\underline{x} + \delta^q\underline{y} + \psi^q\underline{z} \\ Z = \gamma^{q^2}\underline{x} + \delta^{q^2}\underline{y} + \psi^{q^2}\underline{z} \end{cases}. \tag{3.37}$$

or

$$\begin{pmatrix} X \\ Y \\ Z \end{pmatrix} = U \begin{pmatrix} \underline{x} \\ \underline{y} \\ \underline{z} \end{pmatrix}. \tag{3.38}$$

Given $\gamma,\ \delta,\ \psi$, one substitutes (3.37) into (3.12) to obtain a definition equation of the curve C/k as

$$\begin{aligned}
(3.39)\ C/k:\quad & \mathrm{Tr}(a\gamma^4 + b\gamma^{2q+2})\underline{x}^4 \\
& + \mathrm{Tr}(4a\gamma^3\delta + \{2\gamma^{q+2}\delta^q + 2\gamma^{2q+1}\delta\}b)\underline{x}^3\underline{y} \\
& + \mathrm{Tr}(4a\gamma^3\psi + \{2\gamma^{q+2}\psi^q + 2\gamma^{2q+1}\psi\}b)\underline{x}^3\underline{z} \\
& + \mathrm{Tr}(6a\gamma^2\delta^2 + \{\gamma^2\delta^{2q} + \gamma^{2q}\delta^2 + 4\gamma^{q+1}\delta^{q+1}\}b)\underline{x}^2\underline{y}^2 \\
& + \mathrm{Tr}(12a\gamma^2\delta\psi + \{2\gamma^2\delta^q\psi^q + 4\gamma^{q+1}\delta\psi^q + 2\gamma^{2q}\delta\psi + 4\gamma^{q+1}\delta^q\psi\}b)\underline{x}^2\underline{y}\underline{z} \\
& + \mathrm{Tr}(6a\gamma^2\psi^2 + \{\gamma^2\psi^{2q} + \gamma^{2q}\psi^2 + 4\gamma^{q+1}\psi^{q+1}\}b)\underline{x}^2\underline{z}^2 \\
& + \mathrm{Tr}(4a\gamma\delta^3 + \{2\gamma^q\delta^{q+2} + 2\gamma\delta^{2q+1}\}b)\underline{x}\underline{y}^3 \\
& + \mathrm{Tr}(12a\gamma\delta^2\psi + \{2\gamma^q\delta^2\psi^q + 4\gamma\delta^{q+1}\psi^q + 4\gamma^q\delta^{q+1}\psi + 2\gamma\delta^{2q}\psi\}b)\underline{x}\underline{y}^2\underline{z} \\
& + \mathrm{Tr}(12a\gamma\delta\psi^2 + \{2\gamma^q\delta^q\psi^2 + 2\gamma\delta\psi^{2q} + 4\gamma^q\delta\psi^{q+1} + 4\gamma\delta^q\psi^{q+1}\}b)\underline{x}\underline{y}\underline{z}^2 \\
& + \mathrm{Tr}(4a\gamma\psi^3 + \{2\gamma^q\psi^{q+2} + 2\gamma\psi^{2q+1}\}b)\underline{x}\underline{z}^3 \\
& + \mathrm{Tr}(a\delta^4 + b\delta^{2q+2})\underline{y}^4 \\
& + \mathrm{Tr}(4a\delta^3\psi + \{2\delta^{q+2}\psi^q + 2\delta^{2q+1}\psi\}b)\underline{y}^3\underline{z} \\
& + \mathrm{Tr}(6a\delta^2\psi^2 + \{\delta^2\psi^{2q} + \delta^{2q}\psi^2 + 4\delta^{q+1}\psi^{q+1}\}b)\underline{y}^2\underline{z}^2 \\
& + \mathrm{Tr}(4a\delta\psi^3 + \{2\delta^q\psi^{q+2} + 2\delta\psi^{2q+1}\}b)\underline{y}\underline{z}^3 \\
& + \mathrm{Tr}(a\psi^4 + b\psi^{2q+2})\underline{z}^4 \\
& = 0.
\end{aligned}$$

3.4. Find a basis of $H^0(C/k, \Omega^1)$ to determine γ, δ and ψ. In this section, we give explicitly a basis of $H^0(C/k, \Omega^1)$ and determine $\gamma,\ \delta$ and ψ.

Define

$$\omega_1 = \omega + {}^{\sigma}\omega + {}^{\sigma^2}\omega \tag{3.40}$$

$$\omega_2 = \epsilon\omega + \epsilon^q\, {}^{\sigma}\omega + \epsilon^{q^2}\, {}^{\sigma^2}\omega \tag{3.41}$$

$$\omega_3 = \epsilon^2\omega + \epsilon^{2q}\, {}^{\sigma}\omega + \epsilon^{2q^2}\, {}^{\sigma^2}\omega. \tag{3.42}$$

Then

$$\begin{pmatrix} \underline{x} \\ \underline{y} \\ \underline{z} \end{pmatrix} = V \begin{pmatrix} X \\ Y \\ Z \end{pmatrix}. \tag{3.43}$$

The Vandermonde's matrix

$$V = \begin{pmatrix} 1 & 1 & 1 \\ \epsilon & \epsilon^q & \epsilon^{q^2} \\ \epsilon^2 & \epsilon^{2q} & \epsilon^{2q^2} \end{pmatrix} \tag{3.44}$$

has its determinant as

(3.45) $\det(V) = N(\epsilon - \epsilon^q) = (\epsilon - \epsilon^q)(\epsilon^q - \epsilon^{q^2})(\epsilon^{q^2} - \epsilon) = N(\epsilon - \epsilon^q) \neq 0$

then $\{\omega_i\}$ is a basis of $H^0(C/k, \Omega^1)$. We can take $U = V^{-1}$ or

$$\begin{pmatrix} X \\ Y \\ Z \end{pmatrix} = U \begin{pmatrix} \underline{x} \\ \underline{y} \\ \underline{z} \end{pmatrix} \tag{3.46}$$

and the inverse matrix can be expressed by

$$U = V^{-1} = \begin{pmatrix} \gamma & \delta & \psi \\ \gamma^q & \delta^q & \psi^q \\ \gamma^{q^2} & \delta^{q^2} & \psi^{q^2} \end{pmatrix}. \tag{3.47}$$

Thus, one has

$$\gamma = \frac{\epsilon^{2q^2+q} - \epsilon^{q^2+2q}}{\det(V)}, \quad \delta = \frac{\epsilon^{2q} - \epsilon^{2q^2}}{\det(V)} \text{ and } \psi = \frac{\epsilon^{q^2} - \epsilon^q}{\det(V)}. \tag{3.48}$$

Now we have a, b, $\underline{x}, \underline{y}, \underline{z}$ and γ, δ, ψ explicitly thus the definition equation of C/k.

4. Transfer DLP from C_0/k_3 to C/k

The transfer of DLP from C_0/k_d to C/k is usually assumed to follow the norm-conorm map. However, previous works on the subject do not give this map explicitly and its description is not trivial. Here we use the language of divisors instead of function fields to give an explicit map from $Jac(C_0/k_3)$ to $Jac(C/k)$.

The transfer map consists of a trace and a pullback map.

Denote by π^* the pullback map induced by the cover map $\pi/k_3 : C \to C_0$. i.e.,

$$\begin{aligned} \pi^* : \mathrm{Jac}(C_0/k_3) &\to \mathrm{Jac}(C/k_3) \\ P - P_0 &\mapsto D_P - D_{P_0} \end{aligned} \tag{4.1}$$

where $P - P_0$ is a divisor of $\mathrm{Jac}(C_0/k_3)$ and $D_P = \sum_i e_i Q_i$ a divisor of $\mathrm{Jac}(C/k_3)$ s.t. $\pi(Q_i) = P$, e_i is the ramification index at Q_i.

This map corresponds to the conorm map of the function fields.

Denote the trace map of divisor groups as (Here the trace is not on k_3/k as before but on the divisor group)

$$\begin{array}{llcl} (4.2) & \mathrm{Tr}_{k_3/k}: \mathrm{Jac}(C/k_3) & \to & \mathrm{Jac}(C/k) \\ & D_P & \mapsto & D_P + {}^{\sigma}D_P + {}^{\sigma^2}D_P \end{array}$$

which corresponds to the norm map of the function fields.

Then the transfer map is a homomorphism defined by the composition of π^* with the trace map

$$\chi := \mathrm{Tr}_{k_3/k} \circ \pi^* : \mathrm{Jac}(C_0/k_3) \longrightarrow \mathrm{Jac}(C/k). \tag{4.3}$$

Given P_1, P_2, two points on C_0 such that $P_2 \in \langle P_1 \rangle$, the elliptic curve discrete logarithm problem consists in finding an integer λ suchthat $P_2 = \lambda P_1$. Since the group of points on C_0 and the group $\mathrm{Jac}(C_0)$ are isomorphic, we can transfer from $P_2 = \lambda P_1$ to

$$(P_2 - P_\infty) = \lambda(P_1 - P_\infty) \tag{4.4}$$

on $\mathrm{Jac}(C_0)$ where P_∞ is the point at infinity.

Finally, the homomorphism χ transfers the above discrete logarithm to the discrete logarithm on $\mathrm{Jac}(C/k)$ which is to find λ such that

$$(\chi(P_2) - \chi(P_\infty)) = \lambda(\chi(P_1) - \chi(P_\infty)). \tag{4.5}$$

So, it suffices to find π.

In fact, π can be factored into

$$\pi/k_3 = \pi_1 \circ \pi_2 \tag{4.6}$$

where π_1/k_3 is the map from C/k_3 defined by (3.12) to C_0/k_3 and π_2/k_3 is an isomorphism from C/k_3 defined by the equation (3.39) of C/k to C/k_3 defined by (3.12), which can be represented by (3.46) where the matrix U is known.

We find π_1 as follows.

Let s, t be $s = \frac{Y}{X}$, $t = \frac{Z}{X}$ then (3.12) becomes

$$C: \ a + a^q s^4 + a^{q^2} t^4 + bs^2 + b^q s^2 t^2 + b^{q^2} t^2 = 0. \tag{4.7}$$

Additionally let u, v be $u = s^2$, $v = t^2$ then (4.7) becomes

$$a + a^q u^2 + a^{q^2} v^2 + bu + b^q uv + b^{q^2} v = 0 \tag{4.8}$$

which can be identified with $\mathbb{P}^1(k_3)$, while C is its $(2,2)$-covering.

Below, we first consider the case of Type I curves.

4.1. Type I. Since (4.8) is a genus zero curve, we choose the point on it (u_0, v_0), $u_0 = (\alpha\beta)^{-q^2+1}$, $v_0 = (\alpha\beta)^{-q^2+q}$ by letting $x = 0$ in u and v.

Then a point (u, v) of (4.8) are uniquely determined by a line which has slope l and passes through the point $(u_0, v_0) = ((\alpha\beta)^{-q^2+1}, (\alpha\beta)^{-q^2+q})$ and the point $(u,\ v)$.

The equation of the line is

$$v - (\alpha\beta)^{-q^2+q} = l(u - (\alpha\beta)^{-q^2+1}). \tag{4.9}$$

The slope l can be written as

$$l = \frac{v - (\alpha\beta)^{-q^2+q}}{u - (\alpha\beta)^{-q^2+1}}. \tag{4.10}$$

Substituting $u = \dfrac{(x-\alpha)(x-\beta)}{(x-\alpha^{q^2})(x-\beta^{q^2})}$, $v = \dfrac{(x-\alpha^q)(x-\beta^q)}{(x-\alpha^{q^2})(x-\beta^{q^2})}$ into (4.9), the denominator of l becomes

$$u-(\alpha\beta)^{-q^2+1} = \frac{\{1-(\alpha\beta)^{-q^2+1}\}x^2+(-\alpha-\beta+\alpha\beta^{-q^2+1}+\alpha^{-q^2+1}\beta)x}{(x-\alpha^{q^2})(x-\beta^{q^2})}. \tag{4.11}$$

The numerator of l becomes

$$v-(\alpha\beta)^{-q^2+q} = \frac{\{1-(\alpha\beta)^{-q^2+q}\}x^2+(-\alpha^q-\beta^q+\alpha^q\beta^{-q^2+q}+\alpha^{-q^2+q}\beta^q)x}{(x-\alpha^{q^2})(x-\beta^{q^2})}. \tag{4.12}$$

In the sequal,

$$l = \frac{\{1-(\alpha\beta)^{-q^2+q}\}x+(-\alpha^q-\beta^q+\alpha^q\beta^{-q^2+q}+\alpha^{-q^2+q}\beta^q)}{\{1-(\alpha\beta)^{-q^2+1}\}x+(-\alpha-\beta+\alpha\beta^{-q^2+1}+\alpha^{-q^2+1}\beta)}. \tag{4.13}$$

Define G_{11}, G_{12}, G_{21}, $G_{22} \in k_3$

$$G_{11} := 1-(\alpha\beta)^{-q^2+q} \tag{4.14}$$

$$G_{12} := -\alpha^q-\beta^q+\alpha^q\beta^{-q^2+q}+\alpha^{-q^2+q}\beta^q \tag{4.15}$$

$$G_{21} := 1-(\alpha\beta)^{-q^2+1} \tag{4.16}$$

$$G_{22} := -\alpha-\beta+\alpha\beta^{-q^2+1}+\alpha^{-q^2+1}\beta. \tag{4.17}$$

Then l can be expressed by the action of the matrix G on x. Indeed, rewrite (4.13) as

$$l = G\cdot x \quad \text{s.t.} \quad G := \begin{pmatrix} G_{11} & G_{12} \\ G_{21} & G_{22} \end{pmatrix} \in \mathrm{GL}_2(k_3). \tag{4.18}$$

In particular, x is now the image of l under the action of G^{-1}:

$$x = G^{-1}l = \frac{G_{22}l-G_{12}}{-G_{21}l+G_{11}}. \tag{4.19}$$

Now that we expressed x in terms of l, we try to express x directly in terms of X, Y and Z.

Substituting $s = \dfrac{Y}{X}$, $t = \dfrac{Z}{X}$ into l, one has

$$l = \frac{Z^2-(\alpha\beta)^{-q^2+q}X^2}{Y^2-(\alpha\beta)^{-q^2+1}X^2}. \tag{4.20}$$

Therefore

$$x = G^{-1}\cdot l = \frac{G_{22}Z^2-G_{22}(\alpha\beta)^{-q^2+q}X^2-G_{12}Y^2+G_{12}(\alpha\beta)^{-q^2+1}X^2}{-G_{21}Z^2+G_{21}(\alpha\beta)^{-q^2+q}X^2+G_{11}Y^2-G_{11}(\alpha\beta)^{-q^2+1}X^2}. \tag{4.21}$$

To find y, one can use the defining equation of Type I curve $C_0 : y^2 = (x-\alpha)(x-\alpha^q)(x-\beta)(x-\beta^q)$,

$$\frac{(x-\alpha)(x-\alpha^q)(x-\beta)(x-\beta^q)}{{}^{\sigma}y^{\sigma^2}y} = st. \tag{4.22}$$

Then

$$y = \frac{st\mathrm{N}_{k_3/k}(y)}{(x-\alpha)(x-\alpha^q)(x-\beta)(x-\beta^q)}. \tag{4.23}$$

To find $\mathrm{N}_{k_3/k}(y)$, use the definition of C_0 again

$$\mathrm{N}_{k_3/k}(y^2) \quad = \quad \mathrm{N}_{k_3/k}(x-\alpha)^2\mathrm{N}_{k_3/k}(x-\beta)^2. \tag{4.24}$$

Now $\mathrm{N}_{k_3/k}(y)$ is expressed in terms of x as

$$\mathrm{N}_{k_3/k}(y) = \pm\mathrm{N}_{k_3/k}(x-\alpha)\mathrm{N}_{k_3/k}(x-\beta). \tag{4.25}$$

Hence, y can be written as

$$y \quad = \quad \pm st(x-\alpha^{q^2})(x-\beta^{q^2}) \tag{4.26}$$

and we use $y = st(x-\alpha^{q^2})(x-\beta^{q^2})$ hereafter.

Similar to x, y can also be expressed in terms of X, Y and Z.

$$\begin{aligned} y &= st(x-\alpha^{q^2})(x-\beta^{q^2}) \\ &= \frac{YZ}{X^2}(x-\alpha^{q^2})(x-\beta^{q^2}). \end{aligned} \tag{4.27}$$

From the coordinates x, y of the affine curve C_0, one can obtain projective coordinates of C_0 as follows.

First, denote x as a fraction $x = \dfrac{x_2}{x_1}$. Then x, y and z can be expressed as

$$x = \frac{x_2}{x_1},\ y = \frac{YZ}{X^2}\left(\frac{x_2}{x_1} - \alpha^{q^2}\right)\left(\frac{x_2}{x_1} - \beta^{q^2}\right),\quad z = 1. \tag{4.28}$$

Thus one obtains the projective coordinates of C_0 as

$$x = x_1x_2X^2,\ y = YZ(x_2 - \alpha^{q^2}x_1)(x_2 - \beta^{q^2}x_1),\ z = {x_1}^2X^2. \tag{4.29}$$

Now π_1 can be expressed as

$$\begin{aligned} \pi_1 : C \quad &\rightarrow \quad C_0 \\ (X,\ Y,\ Z) \quad &\mapsto \quad (x,\ y,\ z) \end{aligned}$$

such that

$$\begin{aligned} x = \{&-(\alpha\beta)^{-2q^2+2}G_{11}G_{12} + (\alpha\beta)^{-2q^2+q+1}G_{11}G_{22} \\ &+(\alpha\beta)^{-2q^2+q+1}G_{12}G_{21} - (\alpha\beta)^{-2q^2+2q}G_{21}G_{22}\}X^6 \\ &+\{2(\alpha\beta)^{-q^2+1}G_{11}G_{12} - (\alpha\beta)^{-q^2+q}G_{11}G_{22} - (\alpha\beta)^{-q^2+q}G_{12}G_{21}\}X^4Y^2 \\ &+\{-(\alpha\beta)^{-q^2+1}G_{11}G_{22} - (\alpha\beta)^{-q^2+1}G_{12}G_{21} + 2(\alpha\beta)^{-q^2+q}G_{21}G_{22}\}X^4Z^2 \\ &-G_{11}G_{12}X^2Y^4 + (G_{11}G_{22} + G_{12}G_{21})X^2Y^2Z^2 - G_{21}G_{22}X^2Z^4, \end{aligned} \tag{4.30}$$

$$
\begin{aligned}
(4.31)\quad y = \{&(\alpha\beta)^{-q^2+2}{G_{11}}^2 + (\alpha\beta)^{-2q^2+2}(\alpha^{q^2}+\beta^{q^2})G_{11}G_{12} \\
&-2(\alpha\beta)^{-q^2+q+1}G_{11}G_{21} - (\alpha\beta)^{-2q^2+q+1}(\alpha^{q^2}+\beta^{q^2})G_{11}G_{22} \\
&+(\alpha\beta)^{-2q^2+2}{G_{12}}^2 - (\alpha\beta)^{-2q^2+q+1}(\alpha^{q^2}+\beta^{q^2})G_{12}G_{21} \\
&-2(\alpha\beta)^{-2q^2+q+1}G_{12}G_{22} + (\alpha\beta)^{-q^2+2q}{G_{21}}^2 \\
&+(\alpha\beta)^{-2q^2+2q}(\alpha^{q^2}+\beta^{q^2})G_{21}G_{22} \\
&+(\alpha\beta)^{-2q^2+2q}{G_{22}}^2\}X^4YZ \\
+\{&-2\alpha\beta G_{11}^2 - 2(\alpha\beta)^{-q^2+1}(\alpha^{q^2}+\beta^{q^2})G_{11}G_{12} + 2(\alpha\beta)^{q}G_{11}G_{21} \\
&+(\alpha\beta)^{-q^2+q}(\alpha^{q^2}+\beta^{q^2})G_{11}G_{22} - 2(\alpha\beta)^{-q^2+1}{G_{12}}^2 \\
&+(\alpha\beta)^{-q^2+q}(\alpha^{q^2}+\beta^{q^2})G_{12}G_{21} + 2(\alpha\beta)^{-q^2+q}G_{12}G_{22}\}X^2Y^3Z \\
+\{&2\alpha\beta G_{11}G_{21} + (\alpha\beta)^{-q^2+1}(\alpha^{q^2}+\beta^{q^2})G_{11}G_{22} \\
&+(\alpha\beta)^{-q^2+1}(\alpha^{q^2}+\beta^{q^2})G_{12}G_{21} + 2(\alpha\beta)^{-q^2+1}G_{12}G_{22} - 2(\alpha\beta)^{q}{G_{21}}^2 \\
&-2(\alpha\beta)^{-q^2+q}(\alpha^{q^2}+\beta^{q^2})G_{21}G_{22} - 2(\alpha\beta)^{-q^2+q}{G_{22}}^2\}X^2YZ^3 \\
&+\{(\alpha\beta)^{q^2}{G_{11}}^2 + (\alpha^{q^2}+\beta^{q^2})G_{11}G_{12} + {G_{12}}^2\}Y^5Z \\
&-\{2(\alpha\beta)^{q^2}G_{11}G_{21} + (\alpha^{q^2}+\beta^{q^2})G_{11}G_{22} \\
&+(\alpha^{q^2}+\beta^{q^2})G_{12}G_{21} + 2G_{12}G_{22}\}Y^3Z^3 \\
&+\{(\alpha\beta)^{q^2}{G_{21}}^2 + (\alpha^{q^2}+\beta^{q^2})G_{21}G_{22} + {G_{22}}^2\}YZ^5,
\end{aligned}
$$

$$
\begin{aligned}
(4.32)\quad z = \{&(\alpha\beta)^{-2q^2+2}{G_{11}}^2 - 2(\alpha\beta)^{-2q^2+q+1}G_{11}G_{21} + (\alpha\beta)^{-2q^2+2q}{G_{21}}^2\}X^6 \\
&+\{-2(\alpha\beta)^{-q^2+1}{G_{11}}^2 + 2(\alpha\beta)^{-q^2+q}G_{11}G_{21}\}X^4Y^2 \\
&+\{2(\alpha\beta)^{-q^2+1}G_{11}G_{21} - 2(\alpha\beta)^{-q^2+q}{G_{21}}^2\}X^4Z^2 \\
&+{G_{11}}^2X^2Y^4 - 2G_{11}G_{21}X^2Y^2Z^2 + {G_{21}}^2X^2Z^4.
\end{aligned}
$$

4.2. Type II. Calculation for Type II curves is similar to Type I, what we need is to confirm that (4.21), (4.27) are defined over k_3.

For (4.21), first the entries of the matrix G, G_{11}, G_{12}, G_{21}, G_{22} become

$$
\begin{aligned}
(4.33)\quad G_{11} &= 1 - \{\mathrm{N}_{k_6/k_3}(\alpha)\}^{-q^2+q} \\
(4.34)\quad G_{12} &= -\{\mathrm{Tr}_{k_6/k_3}(\alpha)\}^{q} + \{\mathrm{N}_{k_6/k_3}(\alpha)\}^{q}\{\mathrm{Tr}_{k_6/k_3}(\alpha)\}^{-q^2} \\
(4.35)\quad G_{21} &= 1 - \{\mathrm{N}_{k_6/k_3}(\alpha)\}^{-q^2+1} \\
(4.36)\quad G_{22} &= -\mathrm{Tr}_{k_6/k_3}(\alpha) + \mathrm{N}_{k_6/k_3}(\alpha)\{\mathrm{Tr}_{k_6/k_3}(\alpha)\}^{-q^2}.
\end{aligned}
$$

Thus x can be expressed as

(4.37)
$$
x = \frac{G_{22}Z^2 - G_{22}\{\mathrm{N}_{k_6/k_3}(\alpha)\}^{-q^2+q}X^2 - G_{12}Y^2 + G_{12}\{\mathrm{N}_{k_6/k_3}(\alpha)\}^{-q^2+1}X^2}{-G_{21}Z^2 + G_{21}\{\mathrm{N}_{k_6/k_3}(\alpha)\}^{-q^2+q}X^2 + G_{11}Y^2 - G_{11}\{\mathrm{N}_{k_6/k_3}(\alpha)\}^{-q^2+1}X^2}
$$

which has only coefficients in k_3.

Next, (4.27) becomes

$$\begin{aligned} y &= \frac{YZ}{X^2}(x-\alpha^{q^2})(x-\beta^{q^2}) \\ &= \frac{YZ}{X^2}(x^2-\{\mathrm{Tr}_{k_6/k_3}(\alpha)\}^{q^2}x+\{\mathrm{N}_{k_6/k_3}(\alpha)\}^{q^2}) \end{aligned} \tag{4.38}$$

which also has coefficients in k_3. Thus we are done.

5. Computer experiments

The computation environment as follows.

- OS: Windows XP Professional SP2
- CPU: Pentium4 3.2GHz
- Memory: 1.5GB
- Programming language: Magma ver.2.13-14

We start with an elliptic curve E in Legendre form and a base point P_E of E. P_E and its m-multiple mP_E are mapped to points P and mP on an elliptic curve C_0 which is isomorphic to E. Then we find the associated $\chi(P)$ and $\chi(mP)$ in $\mathrm{Jac}(C)$.

5.1. Type I.

$$\begin{gathered} q = 1152921504606851053, \\ k=\mathbb{F}_q,\ k_3 = k[x]/\langle x^3-2\rangle,\ {}^{\exists}\epsilon\ \in k_3 \text{ s.t. } \epsilon^3-2=0 \\ \lambda = 685592167687491848\epsilon^2+685592167687491847\epsilon+3 \end{gathered}$$

The elliptic curve E is in projective Legendre form.

$$E/k_3:\ y^2z = x(x-z)(x-\lambda z)$$

5.1.1. *Testing for Type I curves.* Let $\alpha=\epsilon+1$, then

$$\begin{aligned} A &= \begin{pmatrix} a_{11} & a_{12} \\ a_{21} & a_{22} \end{pmatrix} \\ a_{11} &= 238798614356861922\epsilon+457061445124994566 \\ a_{12} &= 685592167687491848\epsilon^2+685592167687491847\epsilon+ \\ &\quad\ 1152921504606851052 \\ a_{21} &= 1 \\ a_{22} &= 924390782044353769\epsilon+457061445124994564 \end{aligned}$$

$$\begin{aligned} B &= \begin{pmatrix} b_{11} & b_{12} \\ b_{21} & b_{22} \end{pmatrix} \\ b_{11} &= 2\epsilon^2+\epsilon+477597228713723848 \\ b_{12} &= 1152921504606851050\epsilon^2+1152921504606851050\epsilon+ \\ &\quad\ 1152921504606851044 \\ b_{21} &= \epsilon^2+\epsilon+1152921504606851052 \\ b_{22} &= 1152921504606851051\epsilon^2+1152921504606851052\epsilon+ \\ &\quad\ 477597228713723844 \end{aligned}$$

The quadratic equation $b_{21}x^2+(b_{22}-b_{11})x-b_{12}=0$ has two solutions:
$\{\epsilon^2+2\epsilon+1,\ 733677321113450670\epsilon^2+524055229366750479\epsilon+209622091746700193\}$

Therefore, E is Type I. Take $\beta = \epsilon^2 + 2\epsilon + 1 = \alpha^2$, we know that E is k_3-isomorphic to

$$C_0/k_3 : \; y^2z^2 = (x - \alpha z)(x - \alpha^q z)(x - \beta z)(x - \beta^q z).$$

In fact, to test for Type I curves, we chose $\lambda = 2, \ldots, 10001$, the average time to test each curve was 0.0356858 second. Among these curves, 5018 were of Type I.

5.1.2. *Finding the defining equation of the covering curve C/k.* The covering C/k of C_0/k_3 is found using the algorithm shown in Section 4.

$$\begin{aligned} C/k \quad : \quad & 997145058967064651\underline{x}^3\underline{y} + 588586465123877340\underline{x}^3\underline{z} \\ & +907131123326719637\underline{x}^2\underline{y}^2 + 896716725805328597\underline{x}^2\underline{y}\underline{z} \\ & +973749290975691411\underline{x}^2\underline{z}^2 + 1024819115206089825\underline{x}\underline{y}^3 \\ & +280456204442426083\underline{x}\underline{y}^2\underline{z} + 318544658202842297\underline{x}\underline{y}\underline{z}^2 \\ & +1088870309906470439\underline{x}\underline{z}^3 + 973749290975691411\underline{y}^4 \\ & +294293232561938670\underline{y}^3\underline{z} + 1120895907256660746\underline{y}^2\underline{z}^2 \\ & +537516640893478926\underline{y}\underline{z}^3 + 975051090665865291\underline{z}^4 = 0 \end{aligned}$$

To find the C/k from E takes 0.500 second, where 0.063 second is used to test if E is of Type I, the remaining 0.437 is used to build C/k.

5.1.3. *Transferring the DLP.* The isomorphism from E to C_0, $\iota : \; E \to C_0$ is

$$\begin{aligned} \iota : \; E \quad & \to \quad C_0 \\ (x : y : z) \quad & \mapsto \quad (x_{C_0} : y_{C_0} : z_{C_0}) \\ x_{C_0} \quad & = \quad (364080906763379389\epsilon^2 + 963836771592621382\epsilon \\ & \qquad +451137459017005 24)x^2 + (697163568297605614\epsilon^2 \\ & \qquad +434818842429256188\epsilon + 651968585745464837)xz \\ & \qquad +(1110165463009250121\epsilon^2 + 1594118053277349 98\epsilon \\ & \qquad +1139314830835562614)z^2, \\ y_{C_0} \quad & = \quad (1032765162513052 35\epsilon^2 + 814915306056127686\epsilon \\ & \qquad +861572657639767622)yz, \\ z_{C_0} \quad & = \quad (883436713213250245\epsilon^2 + 387404862777293 03\epsilon \\ & \qquad +1108413203079573589)x^2 + (614045874632256899\epsilon^2 \\ & \qquad +476034365815665715\epsilon + 725151688441932395)xz \\ & \qquad +(1080996664374642930\epsilon^2 + 291687986346071 91\epsilon \\ & \qquad +130243006693127807)z^2. \end{aligned}$$

The inverse map ι^{-1} is

$$\begin{aligned}\iota^{-1}:\ C_0 &\rightarrow E\\ (x:y:z) &\mapsto (x_E:y_E:z_E)\\ x_E &= (228530722562497283\epsilon^2+924390782044353770\epsilon\\ &\quad +228530722562497284)x^2+(467329336919359205\epsilon^2\\ &\quad +218262830768132642\epsilon+1152921504606851049)xz\\ &\quad +(467329336919359205\epsilon^2+249066506151226564\epsilon\\ &\quad +685592167687491850)z^2,\\ y_E &= (1098530568356793848\epsilon^2+364091151918511417\epsilon\\ &\quad +156909573516618064)yz,\\ z_E &= x^2+(218262830768132643\epsilon+1152921504606851051)xz\\ &\quad +(685592167687491847\epsilon^2+934658673838718410\epsilon+1)z^2.\end{aligned}$$

For example, take a base point on E

$$\begin{aligned}P_E=(&326484750616207568\epsilon^2+398950984132538563\epsilon\\ &+1105635074365709877:155216221479156187\epsilon^2\\ &+496624914529310471\epsilon+708459555015860335:1)\end{aligned}$$

which has a prime order :

$$ord(P_E)=38312388521647627903649086812540666587976816396877 4759.$$

Under the isomorphism ι, P_E is mapped to $P=\iota(P_E)$ on C_0.

$$\begin{aligned}P=(&382583549840633528\epsilon^2+1049745021810473522\epsilon\\ &+527223886793925136:297304679459601150\epsilon^2\\ &+626540460794459518\epsilon+906489884274840212:1).\end{aligned}$$

From P one obtaines D_P and $\chi(P)$ as follows:

$$D_P=Q_1+Q_2$$

$$\begin{aligned}q_1 &= 712456629299217053\epsilon^2+953676660329800786\epsilon+707524424701837646\\ q_2 &= 666557349447958527\epsilon^2+352353429259986813\epsilon+1073895093206451353\\ q_3 &= 805061362249374584\epsilon^2+1042799979746437227\epsilon+880598497458186947\\ q_4 &= 527740077639497471\epsilon^2+947552956030900685\epsilon+390269122338929978\\ Q_1 &= (q_1:\ q_2:\ 1)\in C/k_3,\qquad Q_2=(q_3:\ q_4:\ 1)\in C/k_3\end{aligned}$$

$$\chi(P)=D_P+{}^{\sigma}D_P+{}^{\sigma^2}D_P$$

$$\begin{aligned}{}^{\sigma}D_P &= {}^{\sigma}Q_1+{}^{\sigma}Q_2, & {}^{\sigma^2}D_P&={}^{\sigma^2}Q_1+{}^{\sigma^2}Q_2\\ {}^{\sigma}Q_1 &= ({q_1}^q:\ {q_2}^q:\ 1), & {}^{\sigma}Q_2&=({q_3}^q:\ {q_4}^q:\ 1)\\ {}^{\sigma^2}Q_1 &= ({q_1}^{q^2}:\ {q_2}^{q^2}:\ 1), & {}^{\sigma^2}Q_2&=({q_3}^{q^2}:\ {q_4}^{q^2}:\ 1).\end{aligned}$$

The time needed to map P_E to $\chi(P)$ is 17.578 seconds.

Now let

$$m=323265910321268664514129224009489670151908972955376519.$$

$$\begin{aligned} E \ni mP_E = (&792310221862816838\epsilon^2 + 180893695299760122\epsilon \\ &+952490131358998041 : 669346193997384009\epsilon^2 \\ &+488209130112427093\epsilon + 787028498315590410 : 1). \end{aligned}$$

This mP_E is also mapped to $C_0 \ni mP = \iota(mP_E)$,

$$\begin{aligned} mP = (&306607799499267855\epsilon^2 + 445518833785785499\epsilon + 141583952331989134 \\ &: 585481570718467983\epsilon^2 + 205882509018091440\epsilon + 573359644129055255 : 1). \end{aligned}$$

One then maps mP to D_{mP} and $\chi(mP)$ as follows.

$$D_{mP} = Q_1 + Q_2$$

$$\begin{aligned} q_1 &= 1062802094539799458\epsilon^2 + 296237055839945308\epsilon + 1057758671244525799 \\ q_2 &= 344189168181796656\epsilon^2 + 529982675029763103\epsilon + 1134629167237810190 \\ q_3 &= 666903385786606500\epsilon^2 + 44288219254827598\epsilon + 362073667770795536 \\ q_4 &= 8690116147489311\epsilon^2 + 330243703134573774\epsilon + 1048131323955608138 \\ Q_1 &= (q_1 : q_2 : 1) \in C/k_3, \qquad Q_2 = (q_3 : q_4 : 1) \in C/k_3 \end{aligned}$$

$$\chi(mP) = D_{mP} + {}^{\sigma}D_{mP} + {}^{\sigma^2}D_{mP}$$

$$\begin{aligned} {}^{\sigma}D_{mP} &= {}^{\sigma}Q_1 + {}^{\sigma}Q_2, & {}^{\sigma^2}D_{mP} &= {}^{\sigma^2}Q_1 + {}^{\sigma^2}Q_2 \\ {}^{\sigma}Q_1 &= ({q_1}^q : {q_2}^q : 1), & {}^{\sigma}Q_2 &= ({q_3}^q : {q_4}^q : 1) \\ {}^{\sigma^2}Q_1 &= ({q_1}^{q^2} : {q_2}^{q^2} : 1), & {}^{\sigma^2}Q_2 &= ({q_3}^{q^2} : {q_4}^{q^2} : 1) \end{aligned}$$

The time taken to compute $\chi(mP)$ from mP_E is 9.859 seconds.

In fact, given $\{2^i P_E | 0 \le i \le 999\}$, the average time to compute $\chi(2^i P)$ is 17.8545 seconds.

5.2. Type II. Assume

$$k = \mathbb{F}_q,\ q = 1152921504606850871$$

$$\begin{aligned} k[x] &\ni a(x) = x^3 + 943550857826445658x^2 + 1018916892242739535x \\ &\qquad +475736851389393367 \\ k_3 &= k[x]/\langle a(x)\rangle,\ {}^{\exists}\epsilon \in k_3 \text{ s.t. } a(\epsilon) = 0 \end{aligned}$$

$$\begin{aligned} k_3[x] &\ni b(x) = x^2 + (595455718590278195\epsilon^2 + 926100813892756385\epsilon \\ &\qquad +508785546940475093)x + 463189347482206220\epsilon^2 \\ &\qquad +936329421988414364\epsilon + 172788951250122324 \\ k_6 &= k_3[x]/\langle b(x)\rangle,\ {}^{\exists}\eta \in k_6 \text{ s.t. } b(\eta) = 0 \end{aligned}$$

$$\alpha = \eta + \epsilon,\ \beta = \alpha^{q^3}$$

and consider the three isomorphic elliptic curves:

$$\begin{aligned} C_0/k_3 &: y^2z^2 = (x - \alpha z)(x - \alpha^q z)(x - \beta z)(x - \beta^q z) \\ E_\lambda/k_3 &: y^2 z = \mathrm{N}_{k_6/k_3}(\beta - \alpha^q)x(x - z)(x - \lambda z),\ \lambda = \mathrm{N}_{k_6/k_3}\left(\frac{\alpha^q - \alpha}{\alpha^q - \beta}\right) \\ E/k_3 &: y^2 z = x(x - z)(x - \lambda z),\ \lambda = \mathrm{N}_{k_6/k_3}\left(\frac{\alpha^q - \alpha}{\alpha^q - \beta}\right) \end{aligned}$$

5.2.1. *Finding defining equation of the covering curve C/k.* Using the algorithm in Section 4, one finds the defining equation of C/k as follows.

$$\begin{aligned} C/k \ : \ & 2619665386729300 61\underline{x}^4 + 719520632819288417\underline{x}^3\underline{y} \\ & +711206123750751637\underline{x}^3\underline{z} + 556061188891864603\underline{x}^2\underline{y}^2 \\ & +311605282877609 88\underline{x}^2\underline{y}\underline{z} + 775851849086806 38\underline{x}^2\underline{z}^2 \\ & +982040544271606073\underline{x}\underline{y}^3 + 860780141350083361\underline{x}\underline{y}^2\underline{z} \\ & +853202732103761301\underline{x}\underline{y}\underline{z}^2 + 953674572673705028\underline{x}\underline{z}^3 \\ & +1020431679265907920\underline{y}^4 + 609659296596817935\underline{y}^3\underline{z} \\ & +954717973652630225\underline{y}^2\underline{z}^2 + 717468332466366860\underline{y}\underline{z}^3 \\ & +1023160869085822939\underline{z}^4 = 0 \end{aligned}$$

Computing C/k takes 0.500 second.

5.2.2. *Transferring the DLP.* We first find the isomorphism from E to E_λ, $\xi : E \to E_\lambda$ as follows.

$$\begin{aligned} \xi : \ E \quad & \to \quad E_\lambda \\ (x : y : z) \quad & \mapsto \quad (x_{E_\lambda} : y_{E_\lambda} : z_{E_\lambda}) \\ x_{E_\lambda} \quad & = \quad (508394311291495279\epsilon^2 + 644802231052062119\epsilon \\ & \qquad +115125795437003532)x, \\ y_{E_\lambda} \quad & = \quad (177549366635458744\epsilon^2 + 533904715816049699\epsilon \\ & \qquad +115337281084752855)y, \\ z_{E_\lambda} \quad & = \quad (508394311291495279\epsilon^2 + 644802231052062119\epsilon \\ & \qquad +115125795437003532)z \end{aligned}$$

Its inverse map ξ^{-1} is

$$\begin{aligned} \xi^{-1} : \ E_\lambda \quad & \to \quad E \\ (x : y : z) \quad & \mapsto \quad (x_E : y_E : z_E) \\ x_E \quad & = \quad (953930729849692988\epsilon^2 + 810853815288336082\epsilon \\ & \qquad +251110930387145558)x, \\ y_E \quad & = \quad (113867255224414650 0\epsilon^2 + 823850992582405 19\epsilon \\ & \qquad +134969511359100 11)y, \\ z_E \quad & = \quad (953930729849692988\epsilon^2 + 810853815288336082\epsilon \\ & \qquad +251110930387145558)z \end{aligned}$$

Next we compute the isomorphism from E_λ to C_0, $\tau: E_\lambda \to C_0$ as follows.

$$
\begin{array}{rcl}
\tau: E_\lambda & \to & C_0 \\
(x:y:z) & \mapsto & (x_{C_0} : y_{C_0} : z_{C_0}) \\
x_{C_0} & = & (510834712742882221\epsilon^2 + 459409699423611549\epsilon \\
 & & +472370343629151306)x^2z + (23471605822501754\epsilon^2 \\
 & & +309377569878570651\epsilon + 7799912042878324)xyz \\
 & & +(931076450504798462\epsilon^2 + 525743454321773525\epsilon \\
 & & +30041499258217822)xz^2 + (977818514557529265\epsilon^2 \\
 & & +765506242357294185\epsilon + 252827041845239982)yz^2 \\
 & & +(1000370112565854753\epsilon^2 + 328209714163922360\epsilon \\
 & & +293352898935549091)z^3, \\
y_{C_0} & = & (1102768582695395466\epsilon^2 + 801656811370788382\epsilon \\
 & & +1017012503317150212)x^3 + (162397320242107152\epsilon^2 \\
 & & +559604911348892417\epsilon + 312861297828079035)x^2z \\
 & & +(558782202587610802\epsilon^2 + 590994009401290871\epsilon \\
 & & +1152361677914957201)xz^2 + (117358029112508 77\epsilon^2 \\
 & & +731149537242710761\epsilon + 3899956021439162)y^2z \\
 & & +(764240535732840601\epsilon^2 + 875626294947314353\epsilon \\
 & & +1076372293311177227)yz^2 + (48504428759686342\epsilon^2 \\
 & & +341476326696745685\epsilon + 96595209872171953)z^3, \\
z_{C_0} & = & (1105978292961847363\epsilon^2 + 534166364849709569\epsilon \\
 & & +1137321680521094223)x^2z + (700411960197286424\epsilon^2 \\
 & & +396739544391375873\epsilon + 141613337225890943)xz^2 \\
 & & +(1019981124724128614\epsilon^2 + 858207083874918419\epsilon \\
 & & +885871207426547152)z^3
\end{array}
$$

The inverse map τ^{-1} is

$$\begin{array}{rcl}
\tau^{-1}: C_0 & \to & E_\lambda \\
(x:y:z) & \mapsto & (x_{E_\lambda}:y_{E_\lambda}:z_{E_\lambda}) \\
x_{E_\lambda} & = & (1180317244173094434\epsilon^2+3507245180500046294\epsilon \\
& & +1076063691653845190)x^2z+(6704052422793404424\epsilon^2 \\
& & +8459489624753854288\epsilon+764269400807635885)xz^2 \\
& & +(1180317244173094434\epsilon^2+3507245180500046294\epsilon \\
& & +1076063691653845190)yz^2+(335044387858599910\epsilon^2 \\
& & +6830302878326106661\epsilon+6177050163273702655)z^3, \\
y_{E_\lambda} & = & (9168580557722320003\epsilon^2+451472468506758283\epsilon \\
& & +1537156259060113622)x^3+(2946272823756804470\epsilon^2 \\
& & +9209176263943963299\epsilon+1303480679079404087)x^2z \\
& & +(9168580557722320003\epsilon^2+451472468506758283\epsilon \\
& & +1537156259060113622)xyz+(4100751878387255568\epsilon^2 \\
& & +2802277467621471644\epsilon+8415193229597810788)xz^2 \\
& & +(4825162623275104447\epsilon^2+3069725421314654443\epsilon \\
& & +3886521037992149866)yz^2+(5749423048423693599\epsilon^2 \\
& & +1073906081772340197\epsilon+2409677446117922259)z^3, \\
z_{E_\lambda} & = & (9796136308903917377\epsilon^2+8733899343624536645\epsilon \\
& & +483213384487444277)z^3
\end{array}$$

For example, a base point on E is chosen as

$$\begin{array}{rcl}
E \ni P_E & = & (8323384416724395277\epsilon^2+3691462625282721400\epsilon \\
& & +7885950516864382000:9164925464481941211\epsilon^2 \\
& & +8053870008812365877\epsilon+2443438155297211559:1)
\end{array}$$

P_E has a prime order :

$$ord(P_E)=383123885216476097596869443538990953306902164540505859.$$

This base point is mapped by ξ, τ to a point on C_0.

First, P_E is mapped to $E_\lambda \ni P_{E_\lambda}=\xi(P_E)$ as follows.

$$\begin{array}{rcl}
P_{E_\lambda} & = & (8323384416724395277\epsilon^2+3691462625282721400\epsilon \\
& & +7885950516864382000:4185534049919400477\epsilon^2 \\
& & +5886066263776092344\epsilon+1115855807315016888:1)
\end{array}$$

Next, it is mapped to $P=\tau(P_{E_\lambda})\in C_0$

$$\begin{array}{rcl}
P & = & (1003935588241243168\epsilon^2+8950662170579869555\epsilon \\
& & +3827737229935504399:6781872062002843533\epsilon^2 \\
& & +1916392135843210088\epsilon+6739556183069205622:1)
\end{array}$$

Now we find D_P and $\chi(P)$ as follows.

$$D_P=Q_1+Q_2$$

$$\begin{aligned}
q_1 &= 1117937506258149424\epsilon^2 + 644917233207069268\epsilon + 165251471146963260 \\
q_2 &= 403047038883440000\epsilon^2 + 653044510390728782\epsilon + 817374729039765305 \\
q_3 &= 994819008370064408\epsilon^2 + 979271450995116569\epsilon + 737452330843672573 \\
q_4 &= 154176739126340404\epsilon^2 + 1152026966659272902\epsilon + 1072497119895785670 \\
Q_1 &= (q_1 : q_2 : 1) \in C/k_3, \qquad Q_2 = (q_3 : q_4 : 1) \in C/k_3
\end{aligned}$$

$$\chi(P) = D_P + {}^{\sigma}D_P + {}^{\sigma^2}D_P$$

$$\begin{aligned}
{}^{\sigma}D_P &= {}^{\sigma}Q_1 + {}^{\sigma}Q_2, \qquad {}^{\sigma^2}D_P = {}^{\sigma^2}Q_1 + {}^{\sigma^2}Q_2 \\
{}^{\sigma}Q_1 &= ({q_1}^q : {q_2}^q : 1), \qquad {}^{\sigma}Q_2 = ({q_3}^q : {q_4}^q : 1) \\
{}^{\sigma^2}Q_1 &= ({q_1}^{q^2} : {q_2}^{q^2} : 1), \qquad {}^{\sigma^2}Q_2 = ({q_3}^{q^2} : {q_4}^{q^2} : 1)
\end{aligned}$$

Computing $\chi(P)$ from P_E takes 21.062 seconds.

Now take $m = 182096100370109847529739170552459116709626522690507709$, $mP_E \in E$ is

$$\begin{aligned}
mP_E = (&5225217305998205366\epsilon^2 + 443211485181667680\epsilon \\
&+408033332463290588 : 191091537075096495\epsilon^2 \\
&+622369471011935091\epsilon + 865873192897372210 : 1)
\end{aligned}$$

mP_E is also mapped first to $E_\lambda \ni mP_{E_\lambda} = \xi(mP_E)$,

$$\begin{aligned}
mP_{E_\lambda} = (&522521730599820536\epsilon^2 \\
&+443211485181667680\epsilon + 408033332463290588 \\
&: 872463812381179496\epsilon^2 + 234010666736627778\epsilon \\
&+346552211766968750 : 1)
\end{aligned}$$

and then to $mP = \tau(mP_{E_\lambda}) \in C_0$:

$$\begin{aligned}
mP = (&457134269332727797\epsilon^2 + 1093275824725039274\epsilon \\
&+664447513560384851 : 955617022224051997\epsilon^2 \\
&+777335844438891994\epsilon + 420110831598890971 : 1)
\end{aligned}$$

From mP, one can find D_{mP} and $\chi(mP)$ as follows.

$$D_{mP} = Q_1 + Q_2$$

$$\begin{aligned}
q_1 &= 300783147327828788\epsilon^2 + 988992501393194153\epsilon + 673404688332712109 \\
q_2 &= 1148714815680333640\epsilon^2 + 423917326839288390\epsilon + 503765461488992377 \\
q_3 &= 734788579677917913\epsilon^2 + 68926008534553154\epsilon + 77740516941101348 \\
q_4 &= 750968410676713515\epsilon^2 + 683426730428696431\epsilon + 823046869633863637 \\
Q_1 &= (q_1 : q_2 : 1) \in C/k_3, \ Q_2 = (q_3 : q_4 : 1) \in C/k_3
\end{aligned}$$

$$\chi(mP) = D_{mP} + {}^{\sigma}D_{mP} + {}^{\sigma^2}D_{mP}$$

$$\begin{aligned}
{}^{\sigma}D_{mP} &= {}^{\sigma}Q_1 + {}^{\sigma}Q_2, \ {}^{\sigma^2}D_{mP} = {}^{\sigma^2}Q_1 + {}^{\sigma^2}Q_2 \\
{}^{\sigma}Q_1 &= ({q_1}^q : {q_2}^q : 1), \ {}^{\sigma}Q_2 = ({q_3}^q : {q_4}^q : 1) \\
{}^{\sigma^2}Q_1 &= ({q_1}^{q^2} : {q_2}^{q^2} : 1), \ {}^{\sigma^2}Q_2 = ({q_3}^{q^2} : {q_4}^{q^2} : 1)
\end{aligned}$$

Computing $\chi(mP)$ from mP_E takes 11.281 seconds.

In fact, given $\{2^i P_E | 0 \le i \le 999\}$, the average time to find $\chi(2^i P)$ is 23.155937 seconds.

6. Conclusion

We presented two algorithms to implement the GHS attack against elliptic curve cryptosystems over cubic extension fields of odd characteristic and the results of the computer simulation. The first algorithm obtains the defining equation for the nonhyperelliptic covering C/k of the elliptic curve C_0/k_3. The second algorithm transfers explicitly the DLP over C_0/k to the DLP over $\mathrm{Jac}(C/k)$. These DLP over $\mathrm{Jac}(C/k)$ can then be solved using Diem's double-large-prime algorithm.

References

[1] Pierrick Gaudry, *An algorithm for solving the discrete log problem on hyperelliptic curves*, Advances in cryptology—EUROCRYPT 2000 (Bruges), Lecture Notes in Comput. Sci., vol. 1807, Springer, Berlin, 2000, pp. 19–34, DOI 10.1007/3-540-45539-6_2. MR1772021

[2] Nicolas Thériault, *Index calculus attack for hyperelliptic curves of small genus*, Advances in cryptology—ASIACRYPT 2003, Lecture Notes in Comput. Sci., vol. 2894, Springer, Berlin, 2003, pp. 75–92, DOI 10.1007/978-3-540-40061-5_5. MR2093253

[3] P. Gaudry, E. Thomé, N. Thériault, and C. Diem, *A double large prime variation for small genus hyperelliptic index calculus*, Math. Comp. **76** (2007), no. 257, 475–492, DOI 10.1090/S0025-5718-06-01900-4. MR2261032

[4] Claus Diem, *An index calculus algorithm for plane curves of small degree*, Algorithmic number theory, Lecture Notes in Comput. Sci., vol. 4076, Springer, Berlin, 2006, pp. 543–557, DOI 10.1007/11792086_38. MR2282948

[5] G. Frey,"How to disguise an elliptic curve," Talk at the 2nd Elliptic Curve Cryptography Workshop, 1998.

[6] P. Gaudry, F. Hess, and N. P. Smart, *Constructive and destructive facets of Weil descent on elliptic curves*, J. Cryptology **15** (2002), no. 1, 19–46, DOI 10.1007/s00145-001-0011-x. MR1880933

[7] Alfred Menezes and Minghua Qu, *Analysis of the Weil descent attack of Gaudry, Hess and Smart*, Topics in cryptology—CT-RSA 2001 (San Francisco, CA), Lecture Notes in Comput. Sci., vol. 2020, Springer, Berlin, 2001, pp. 308–318, DOI 10.1007/3-540-45353-9_23. MR1907106

[8] Alfred Menezes, Edlyn Teske, and Annegret Weng, *Weak fields for ECC*, Topics in cryptology—CT-RSA 2004, Lecture Notes in Comput. Sci., vol. 2964, Springer, Berlin, 2004, pp. 366–386, DOI 10.1007/978-3-540-24660-2_28. MR2092257

[9] Steven D. Galbraith, *Weil descent of Jacobians*, Discrete Appl. Math. **128** (2003), no. 1, 165–180, DOI 10.1016/S0166-218X(02)00443-2. International Workshop on Coding and Cryptography (WCC 2001) (Paris). MR1991424

[10] Nicolas Thériault, *Weil descent attack for Kummer extensions*, J. Ramanujan Math. Soc. **18** (2003), no. 3, 281–312. MR2007146

[11] Nicolas Thériault, *Weil descent attack for Kummer extensions*, J. Ramanujan Math. Soc. **18** (2003), no. 3, 281–312. MR2007146

[12] Florian Hess, *The GHS attack revisited*, Advances in cryptology—EUROCRYPT 2003, Lecture Notes in Comput. Sci., vol. 2656, Springer, Berlin, 2003, pp. 374–387, DOI 10.1007/3-540-39200-9_23. MR2090430

[13] F. Hess, *Generalising the GHS attack on the elliptic curve discrete logarithm problem*, LMS J. Comput. Math. **7** (2004), 167–192, DOI 10.1112/S146115700000108X. MR2087095

[14] Claus Diem, *The GHS attack in odd characteristic*, J. Ramanujan Math. Soc. **18** (2003), no. 1, 1–32. MR1966526

[15] C. Diem and J. Sholten, "Cover attack". Preprint, 2003. Available at http://www.math.uni-leipzig.de/ diem/preprints/english.html

[16] H. Cohen, G. Frey(ed), *Handbook of elliptic and hyperelliptic curve cryptography*, Chapman & Hall/CRC, 2005.

[17] F. Momose, J. Chao, "Scholten forms and elliptic/hyperelliptic curves with weak Weil restrictions". Preprint, 2005. Available at http://eprint.iacr.org/2005/277
[18] F. Momose and J. Chao "Classification of Weil restrictions obtained by $(2, \dots, 2)$ coverings of $\mathbb{P}^1$," Preprint, 2006. Available at http://eprint.iacr.org/2006/347
[19] J. Chao,"Elliptic and hyperelliptic curves with weak covering against Weil descent attacks", 2007 International Workshop on Elliptic Curve Cryptosystems, ECC2007, Sept., 2007.
[20] Fumiyuki Momose and Jinhui Chao, *Elliptic curves with weak coverings over cubic extensions of finite fields with odd characteristic*, J. Ramanujan Math. Soc. **28** (2013), no. 3, 299–357. MR3113387
[21] T. Iijima, F. Momose, and J. Chao "Classification of elliptic/hyperelliptic curves with weak coverings against GHS attack without isogeny condition," Preprint, 2009. Available at http://eprint.iacr.org/2009/613.
[22] T. Iijima, F. Momose and J. Chao "Classification of elliptic/hyperelliptic curves with weak coverings against GHS attack under an isogeny condition," Preprint, 2013. Available at http://eprint.iacr.org/2013/487.
[23] T. Iijima, F. Momose, and J. Chao "A classification of elliptic curves with respect to the GHS attack in odd characteristic," Preprint, 2015. Available at http://eprint.iacr.org/2015/805.

Appendix: On Condition (2.14) of hyperellipticity

Type I. By (2.14), $\beta = A \cdot \alpha = \dfrac{a\alpha + b}{c\alpha + d}$ $(a,\ b,\ c,\ d\ \in k)$. Combining with $\mathrm{Tr}A = 0$, one has the following variation of Condition (2.14)

$$\text{(6.1)}\quad C \text{ is hyperelliptic} \iff \beta = A\cdot\alpha,\ A \in \mathrm{GL}_2(k), \mathrm{Tr}A = 0$$

$$\iff \text{Either (i) or (ii) is true.}$$

$$\text{(6.2)}\qquad \begin{cases} \text{(i)}\ \ A = \begin{pmatrix} a & b \\ 0 & -a \end{pmatrix}, \\ \beta = A\cdot\alpha = \frac{a\alpha+b}{-a} = -\alpha - b', \\ \qquad \textit{or} \qquad \alpha+\beta = -b' \in k \\ \text{(ii)}\ \ A = \begin{pmatrix} a & b \\ 1 & -a \end{pmatrix},\ \beta = A\cdot\alpha = \frac{a\alpha+b}{\alpha-a} \end{cases}$$

In particular, Condition (ii) means $\beta = \dfrac{a\alpha + b}{\alpha - a}$, or

$$\alpha\beta - (\alpha+\beta)a - b \ = \ 0 \tag{6.3}$$

Since any element $l \in k_3$ can be expressed, using the basis $\{1,\ \epsilon,\ \epsilon^2\}$ as

$$l = l_0 + l_1\epsilon + l_2\epsilon^2 \qquad l_0,\ l_1,\ l_2 \in k$$

assume

$$\alpha \ = \ \alpha_0 + \alpha_1\epsilon + \alpha_2\epsilon^2, \tag{6.4}$$

$$\beta \ = \ \beta_0 + \beta_1\epsilon + \beta_2\epsilon^2 \tag{6.5}$$

Then

$$\alpha\beta \ = \ (\alpha\beta)_0 + (\alpha\beta)_1\epsilon + (\alpha\beta)_2\epsilon^2 \tag{6.6}$$

$$-(\alpha+\beta)a \ = \ -(\alpha_0+\beta_0)a - (\alpha_1+\beta_1)a\epsilon - (\alpha_2+\beta_2)a\epsilon^2 \tag{6.7}$$

(6.3) becomes

$$\begin{aligned} & \alpha\beta - (\alpha+\beta)a - b \\ = \ & \{(\alpha\beta)_0 - (\alpha_0+\beta_0)a - b\} + \{(\alpha\beta)_1 - (\alpha_1+\beta_1)a\}\epsilon \\ & +\{(\alpha\beta)_2 - (\alpha_2+\beta_2)a\}\epsilon^2 \end{aligned} \tag{6.8}$$

$$= \ 0 \tag{6.9}$$

Therefore Condition (ii) can be replaced by the existence of solutions in the following linear equations in a, b

$$\begin{cases} -(\alpha_0+\beta_0)a - b + (\alpha\beta)_0 = 0 \\ -(\alpha_1+\beta_1)a + (\alpha\beta)_1 = 0 \\ -(\alpha_2+\beta_2)a + (\alpha\beta)_2 = 0 \end{cases} \tag{6.10}$$

When one wishes to find a nonhyperelliptic curve, Condition (2.14) has to be avoided. Therefore neither (i) nor (ii) should hold for α and β. This means

(6.11) $\overline{\text{(i)}}$ $\alpha+\beta \notin k$

(6.12) $\overline{\text{(ii)}}$ The system of equations: $\begin{cases} -(\alpha_0+\beta_0)a - b + (\alpha\beta)_0 = 0 \\ -(\alpha_1+\beta_1)a + (\alpha\beta)_1 = 0 \\ -(\alpha_2+\beta_2)a + (\alpha\beta)_2 = 0 \end{cases}$

has no solution.

Define

$$B := \begin{pmatrix} -(\alpha_0+\beta_0) & -1 \\ -(\alpha_1+\beta_1) & 0 \\ -(\alpha_2+\beta_2) & 0 \end{pmatrix}, \ B' := \begin{pmatrix} -(\alpha_0+\beta_0) & -1 & -(\alpha\beta)_0 \\ -(\alpha_1+\beta_1) & 0 & -(\alpha\beta)_1 \\ -(\alpha_2+\beta_2) & 0 & -(\alpha\beta)_2 \end{pmatrix} \tag{6.13}$$

then $\overline{\text{(ii)}}$ holds if and only if rank $B \neq$ rank B'.

In other words, to obtain a nonhyperelliptic covering curve C/k, one only needs to choose α and β such that $\alpha+\beta \notin k$ and rank $B \neq$ rank B'.

Type II. For the Type II case, since $\alpha+\beta = \mathrm{Tr}_{k_6/k_3}(\alpha)$, $\alpha\beta = \mathrm{N}_{k_6/k_3}(\alpha)$, $\overline{\text{(i)}}$ and $\overline{\text{(ii)}}$ in Type I can be replaced by

$\overline{\text{(i)}}$ $\mathrm{Tr}_{k_6/k_3}(\alpha) \notin k$

$\overline{\text{(ii)}}$ The system of equations: $\begin{cases} -\{\mathrm{Tr}_{k_6/k_3}(\alpha)\}_0 a - b + \{\mathrm{N}_{k_6/k_3}(\alpha)\}_0 = 0 \\ -\{\mathrm{Tr}_{k_6/k_3}(\alpha)\}_1 a + \{\mathrm{N}_{k_6/k_3}(\alpha)\}_1 = 0 \\ -\{\mathrm{Tr}_{k_6/k_3}(\alpha)\}_2 a + \{\mathrm{N}_{k_6/k_3}(\alpha)\}_2 = 0 \end{cases}$

has no solution.

Define

(6.14)

$$B := \begin{pmatrix} -\{\mathrm{Tr}_{k_6/k_3}(\alpha)\}_0 & -1 \\ -\{\mathrm{Tr}_{k_6/k_3}(\alpha)\}_1 & 0 \\ -\{\mathrm{Tr}_{k_6/k_3}(\alpha)\}_2 & 0 \end{pmatrix}, \ B' := \begin{pmatrix} -\{\mathrm{Tr}_{k_6/k_3}(\alpha)\}_0 & -1 & -\{\mathrm{N}_{k_6/k_3}(\alpha)\}_0 \\ -\{\mathrm{Tr}_{k_6/k_3}(\alpha)\}_1 & 0 & -\{\mathrm{N}_{k_6/k_3}(\alpha)\}_1 \\ -\{\mathrm{Tr}_{k_6/k_3}(\alpha)\}_2 & 0 & -\{\mathrm{N}_{k_6/k_3}(\alpha)\}_2 \end{pmatrix}$$

then $\overline{\text{(ii)}}$ holds if and only if rank $B \neq$ rank B'.

Thus, to obtain a nonhyperelliptic covering for a Type II curve, one needs to choose α and β such that $\mathrm{Tr}_{k_6/k_3}(\alpha) \notin k$ and rank $B \neq$ rank B'.

Graduate School of Science and Engineering, Course of Information and System Engineering, Chuo University, 1-13-27 Kasuga, Bunkyo-ku, Tokyo 112-8551, Japan

Department of Mathematics, Faculty of Science and Engineering, Chuo University, 1-13-27 Kasuga, Bunkyo-ku, Tokyo 112-8551, Japan

Department of Information and System Engineering, Faculty of Science and Engineering, Chuo University, 1-13-27 Kasuga, Bunkyo-ku, Tokyo 112-8551, Japan
Email address: jchao@ise.chuo-u.ac.jp

Contemporary Mathematics
Volume **701**, 2018
http://dx.doi.org/10.1090/conm/701/14151

The Sato-Tate conjecture for a Picard curve with complex multiplication (with an appendix by Francesc Fité)

Joan-Carles Lario and Anna Somoza

To the memory of Fumiyuki Momose

ABSTRACT. Let $C/\mathbb{Q}$ be the genus 3 Picard curve given by the affine model $y^3 = x^4 - x$. In this paper we compute its Sato-Tate group, show the generalized Sato-Tate conjecture for C, and compute the statistical moments for the limiting distribution of the normalized local factors of C.

1. Introduction

Serre [**Ser12**] provides a vast generalization of the Sato-Tate conjecture, which is known to be true for varieties with complex multiplication [**Joh13**]. As a down-to-earth example, in this paper we consider the Picard curve defined over $\mathbb{Q}$ given by the affine model

$$C\colon y^3 = x^4 - x\,.$$

One easily checks that $[0 : 1 : 0]$ is the unique point of C at infinity, and that C has good reduction at all primes different from 3. The Jacobian variety of C is absolutely simple and it has complex multiplication by the cyclotomic field $K = \mathbb{Q}(\zeta)$ where ζ is a primitive 9th root of unity.

With the help of Sage, we compile information on the number of points that the reduction of C has over finite fields of small characteristic in Table 1.

TABLE 1. Number of points $|C(\mathbb{F}_{p^i})|$.

p	$\lvert C(\mathbb{F}_p)\rvert$	$\lvert C(\mathbb{F}_{p^2})\rvert$	$\lvert C(\mathbb{F}_{p^3})\rvert$
2	3	5	9
5	6	26	126
7	8	50	365
11	12	122	1332
13	14	170	2003
17	18	392	4914
19	14	302	6935

2010 *Mathematics Subject Classification.* Primary 11G10; Secondary 11G15, 11G30 and 11G40.

Both authors were partially funded by MINECO grant MTM2015-63829-P.

For every prime p of good reduction, we consider the local zeta function

$$\zeta(C/\mathbb{F}_p; s) = \exp\left(\sum_{k\geq 1} |C(\mathbb{F}_{p^k})| \frac{p^{-ks}}{k}\right).$$

It follows from Weil's conjectures [**Wei49**] that the zeta function is a rational function of $T = p^{-s}$. That is,

$$\zeta(C/\mathbb{F}_p; T) = \exp\left(\sum_{k\geq 1} |C(\mathbb{F}_{p^k})| \frac{T^k}{k}\right) = \frac{L_p(C,T)}{(1-T)(1-pT)}$$

where the so-called *local factor* of C at p

$$L_p(C,T) = \sum_{i=0}^{6} b_i T^i = \prod_{i=1}^{6}(1-\alpha_i T)$$

is a polynomial of degree 6 with integral coefficients and the complex numbers α_i satisfy $|\alpha_i| = \sqrt{p}$. In particular, it is determined by the three numbers $|C(\mathbb{F}_p)|$, $|C(\mathbb{F}_{p^2})|$, $|C(\mathbb{F}_{p^3})|$ according to:

$$\begin{aligned}
b_0 &= 1\\
b_1 &= |C(\mathbb{F}_p)| - (p+1)\\
b_2 &= (|C(\mathbb{F}_{p^2})| - (p^2+1) + b_1^2)/2\\
b_3 &= (|C(\mathbb{F}_{p^3})| - (p^3+1) - b_1^3 + 3b_2b_1)/3\\
b_4 &= pb_2\\
b_5 &= p^2 b_1\\
b_6 &= p^3\,.
\end{aligned}$$

For all $m \geq 1$ we have

$$|C(\mathbb{F}_{p^m})| = 1 + p^m - \sum_{i=1}^{6} \alpha_i^m\,.$$

The local factors for small good primes are:

p	$L_p(C,T)$
2	$(1+2T^2)(1-2T^2+4T^4)$
5	$(1+5T^2)(1-5T^2+25T^4)$
7	$1+7T^3+343T^6$
11	$(1+11T^2)(1-11T^2+121T^4)$
13	$1-65T^3+2197T^6$
17	$(1+17T^2)^3$
19	$1-6T-12T^2+169T^3-228T^4-2166T^5+6859T^6$

Even if for every such prime p all terms of the sequence

$$|C(\mathbb{F}_p)|, |C(\mathbb{F}_{p^2})|, |C(\mathbb{F}_{p^3})|, \ldots, |C(\mathbb{F}_{p^m})|, \ldots \quad (m \geq 1)$$

are determined by the first three, obtaining these first three can be a hard computational task as soon as the prime p gets large. However, the presence of complex multiplication enables the fast computation of the local factors $L_p(C,T)$ (see Section 4.2).

For future use, we introduce some notation. The ring of integers of K will be denoted by $\mathcal{O} = \mathbb{Z}[\zeta]$, and the unit group $\mathcal{O}^* \simeq \mathbb{Z}/18\mathbb{Z} \times \mathbb{Z} \times \mathbb{Z}$ has generators $\epsilon_0 = -\zeta^2$, $\epsilon_1 = \zeta^4 - \zeta^3 + \zeta$, $\epsilon_2 = \zeta^5 + \zeta^2 - \zeta$. Let σ_i denote the automorphism of $\mathrm{Gal}(K/\mathbb{Q})$ determined by $\sigma_i(\zeta) = \zeta^i$; one has that σ_2 generates the Galois group $\mathrm{Gal}(K/\mathbb{Q}) \simeq (\mathbb{Z}/9\mathbb{Z})^*$. The unique ramified prime in $K/\mathbb{Q}$ is $3\mathcal{O} = (1+\zeta+\zeta^4)^6$.

Since the Jacobian variety $\mathrm{Jac}(C)$ has complex multiplication, the work of Shimura and Taniyama [**ST61**] ensures the existence of an ideal $\mathfrak{m}$ of the ring of integers $\mathcal{O}$ and a Grassencharakter $\psi\colon I_K(\mathfrak{m}) \to \mathbb{C}^*$, where $I_K(\mathfrak{m})$ stands for the group of fractional ideals coprime to $\mathfrak{m}$,

$$\psi(\alpha\mathcal{O}) = \prod_{\sigma\in\Phi^*} {}^{\sigma}\alpha \quad \text{if } \alpha \equiv 1 \,(\mathrm{mod}^* \,\mathfrak{m}),$$

such that $L(\psi, s) = L(C, s)$. The infinite type Φ^* is the reflex of the CM-type Φ of $\mathrm{Jac}(C)$. Up to a finite number of Euler factors, one has

$$L(\psi, s) = \prod_{\mathfrak{p}} \left(1 - \psi(\mathfrak{p})\,\mathrm{N}(\mathfrak{p})^{-s}\right)^{-1} \quad \text{and} \quad L(C, s) = \prod_{p} L_p(C, p^{-s})^{-1}.$$

Hence, the local factor $L_p(C,T)$ can be obtained from the (monic) irreducible polynomial of $\psi(\mathfrak{p})$ over $\mathbb{Q}$ according to

$$L_p(C,T) = T^6\,\mathrm{Irr}(\psi(\mathfrak{p}), 1/T^f; \mathbb{Q})^{6/(fd)},$$

where f is the residue field degree of p in K, and $d = [\mathbb{Q}(\psi(\mathfrak{p}))\colon \mathbb{Q}]$.

LEMMA 1.1. *There exists a Grössencharakter $\psi\colon I_K(\mathfrak{m}) \to \mathbb{C}^*$ of conductor $\mathfrak{m} = (1+\zeta+\zeta^4)^4$ and infinite type $\Phi^* = \{\sigma_5, \sigma_7, \sigma_8\} = \{\sigma_2^3, \sigma_2^4, \sigma_2^5\}$.*

PROOF. The following holds

$$\epsilon_0^{18} \equiv 1 \pmod{\mathfrak{m}}, \qquad \epsilon_1^9 \equiv 1 \pmod{\mathfrak{m}}, \qquad \epsilon_2^3 \equiv 1 \pmod{\mathfrak{m}}.$$

Moreover, one readily checks that $\epsilon_0^a\epsilon_1^b\epsilon_2^c \equiv 1\,(\mathrm{mod}\,\mathfrak{m})$ if and only if

$$\begin{aligned}(a,b,c) &\equiv (0,0,0), \quad (2,1,2), \quad (4,2,1),\\ &\equiv (6,3,0), \quad (8,4,2), \quad (10,5,1),\\ &\equiv (12,6,0), \quad (14,7,2), \quad (16,8,1),\end{aligned}$$

mod $(18,9,3)$, respectively. Now an easy computation case-by-case shows that if $\epsilon_0^a\epsilon_1^b\epsilon_2^c \equiv 1 \pmod{\mathfrak{m}}$, then

$$\prod_{\sigma\in\Phi^*} {}^{\sigma}(\epsilon_0^a\epsilon_1^b\epsilon_2^c) = 1.$$

By using that K has class number one, we define $\psi(\mathfrak{p})$ over prime ideals $\mathfrak{p}$ of $\mathcal{O}$ coprime with $\mathfrak{m}$ as follows. First we find a generator of $\mathfrak{p} = (\alpha)$, and then search for

$$\epsilon_0^a\epsilon_1^b\epsilon_2^c\alpha \equiv 1 \pmod{\mathfrak{m}}$$

with $0 \le a < 18$, $0 \le b < 9$, and $0 \le c < 3$. The existence of such triple (a,b,c) is guaranteed by the fact that $(\alpha, \mathfrak{m}) = 1$ and the classes of the 486 possible products $\epsilon_0^a\epsilon_1^b\epsilon_2^c$ exhaust the all the elements in $(\mathcal{O}/\mathfrak{m})^*$. It follows that

$$\psi(\mathfrak{p}) = \prod_{\sigma\in\Phi^*} {}^{\sigma}(\epsilon_0^a\epsilon_1^b\epsilon_2^c\alpha)$$

is well-defined. Finally, one extends ψ over all ideals prime to $\mathfrak{m}$ multiplicatively. An argument along the same lines shows the non existence of a Grössencharakter of K of modulus $(1+\zeta+\zeta^4)^i$ for $i < 4$. Thus, ψ has conductor $\mathfrak{m}$. $\square$

TABLE 2. Values of the Grössencharakter ψ.

p	$\psi(\mathfrak{p})$	$L_p(C,T) = T^6\,\mathrm{Irr}(\psi(\mathfrak{p}),1/T^f;\mathbb{Q})^{6/(fd)}$
5	-125	$(1+5T^2)(1-5T^2+25T^4)$
7	$21\zeta^3+7$	$1+7T^3+343T^6$
11	-1331	$(1+11T^2)(1-11T^2+121T^4)$
13	$-39\zeta^3+13$	$1-65T^3+2197T^6$
17	-17	$(1+17T^2)^3$
19	$-\zeta^5-4\zeta^4+2\zeta^3+\zeta^2-2\zeta+2$	$1-6T-12T^2+169T^3-228T^4-2166T^5+6859T^6$
23	-12167	$(1+23T^2)(1-23T^2+529T^4)$
29	-24389	$(1+29T^2)(1-29T^2+841T^4)$
31	$-186\zeta^3-155$	$1+124T^3+29791T^6$
37	$4\zeta^5+4\zeta^4-2\zeta^3+5\zeta^2+2\zeta$	$1-6T+42T^2-47T^3+1554T^4-8214T^5+50653T^6$

PROPOSITION 1.2. *Let ψ be the above Grössencharakter. Then, one has*

$$L(C,s) = L(\psi,s).$$

PROOF. For every prime $\mathfrak{p}$ in $I_K(\mathfrak{m})$, let $\mathbb{F}_\mathfrak{p} = \mathcal{O}/\mathfrak{p}$ be the residue field of $\mathfrak{p}$ and consider the character $\chi_\mathfrak{p}\colon \mathbb{F}_\mathfrak{p}^* \to K^*$ such that

$$\chi_\mathfrak{p}(x) \equiv x^{(\mathrm{N}(\mathfrak{p})-1)/9} \pmod{\mathfrak{p}},$$

that we extend by $\chi_\mathfrak{p}(0) = 0$. By Hasse [**Has54**], the Jacobi sum

$$\mathrm{J}(\mathfrak{p}) := -\sum_{x\in\mathbb{F}_\mathfrak{p}} \chi_\mathfrak{p}^3(x)\chi_\mathfrak{p}(1-x)$$

is uniquely determined by the three properties:

(i) $|J(\mathfrak{p})| = \sqrt{\mathrm{N}(\mathfrak{p})}$;
(ii) $J(\mathfrak{p}) \equiv 1 \pmod{\mathfrak{m}}$;
(iii) $J(\mathfrak{p})\,\mathcal{O} = \mathfrak{p}\cdot\mathfrak{p}^{\sigma_2}\cdot\mathfrak{p}^{\sigma_2^2}$.

Moreover, Holzapfel and Nicolae [**HN02**] show that for a prime power q such that $q \not\equiv 1 \pmod 9$ one has $|C(\mathbb{F}_q)| = q+1$, while for $q \equiv 1 \pmod 9$ it follows

$$|C(\mathbb{F}_\mathfrak{p})| = \mathrm{N}(\mathfrak{p}) + 1 - \mathrm{Tr}_{K/\mathbb{Q}}(J(\mathfrak{p})),$$

where $\mathfrak{p}$ is any prime ideal of the factorization of $q\mathcal{O}$.

Now it is easy to check that $\psi(\mathfrak{p})$ satisfies (i), (ii) and that $J(\mathfrak{p}) = \overline{\psi(\mathfrak{p})}$. Therefore, $\mathrm{Tr}_{K/\mathbb{Q}}(J(\mathfrak{p})) = \mathrm{Tr}_{K/\mathbb{Q}}(\psi(\mathfrak{p}))$ and the claim follows. □

REMARK 1.3. The proof of the last equality takes 4 pages in the referenced article [**HN02**]. We are grateful to Francesc Fité for a more concise proof included in the appendix of the present paper.

The Grössencharakter ψ satisfies ${}^\sigma\psi(\mathfrak{p}) = \psi({}^\sigma\mathfrak{p})$ for every prime ideal $\mathfrak{p}$ and $\sigma \in \mathrm{Gal}(K/\mathbb{Q})$. The L-function of the curve C over K satisfies

$$L(C_K,s) = \prod_{\sigma\in\mathrm{Gal}(K/\mathbb{Q})} L({}^\sigma\psi,s) = L(C,s)^6.$$

The CM-type of Jac(C) is $\Phi = \{\sigma_2,\sigma_4,\sigma_8\}$, i.e. the reflex of Φ^*.

2. The Sato-Tate group ST(C)

For every prime $p \neq 3$, let us normalize the polynomials

$$L_p^{\mathrm{ST}}(C,T) = L_p\left(C, \frac{T}{\sqrt{p}}\right)$$

and call them *normalized local factors* of C. Since they are monic, palindromic with real coefficients, roots lying in the unit circle and Galois stable, one can think of them as the characteristic polynomials of (conjugacy classes of) matrices in the unitary symplectic group

$$\mathrm{USp}(6,\mathbb{C}) = \{M \in \mathrm{GL}(6,\mathbb{C})\colon M^{-1} = J^{-1}M^tJ = M^*\},$$

where M^* denotes the complex conjugate transpose of M, and J denotes the skew-symmetric matrix

$$J = \begin{pmatrix} 0 & 1 & 0 & 0 & 0 & 0 \\ -1 & 0 & 0 & 0 & 0 & 0 \\ 0 & 0 & 0 & 1 & 0 & 0 \\ 0 & 0 & -1 & 0 & 0 & 0 \\ 0 & 0 & 0 & 0 & 0 & 1 \\ 0 & 0 & 0 & 0 & -1 & 0 \end{pmatrix}.$$

The Sato-Tate group attached to C is defined by a precise recipe of Serre [**Ser12**] to be (a conjugacy class of) a compact subgroup $\mathrm{ST}(C) \subseteq \mathrm{USp}(6,\mathbb{C})$, to the effect of generalizing the Sato-Tate conjecture (originally for elliptic curves). Then, the Sato-Tate group $\mathrm{ST}(C)$ is conjectured to be such that the characteristic polynomials of the matrices in $\mathrm{ST}(C)$ *fit well* with the normalized local factors $L_p^{\mathrm{ST}}(C,T)$, in the sense that the normalized local factors $L_p^{\mathrm{ST}}(C,T)$, as p varies, are equidistributed with respect to the Haar measure of $\mathrm{ST}(C)$ projected on the set of its conjugacy classes.

In analogy with Galois theory, the presence of some extra structure on C gives rise to proper subgroups of the symplectic group; moreover, the distribution of $L_p^{\mathrm{ST}}(C,T)$ can be viewed as a generalization of the classical Chebotarev distribution.

In this section, we calculate the Sato-Tate group $\mathrm{ST}(C)$ for our Picard curve C in Proposition 2.1, and then prove that it satisfies the Sato-Tate conjecture in Proposition 3.1.

PROPOSITION 2.1. *Up to conjugacy in* $\mathrm{USp}(6,\mathbb{C})$, *the Sato-Tate group of* C *is*

$$\mathrm{ST}(C) = \left\langle \begin{pmatrix} u_1 &&&&& \\ & \bar{u}_1 &&&& \\ && u_2 &&& \\ &&& \bar{u}_2 && \\ &&&& u_3 & \\ &&&&& \bar{u}_3 \end{pmatrix}, \begin{pmatrix} 0 & 0 & 1 & 0 & 0 & 0 \\ 0 & 0 & 0 & 1 & 0 & 0 \\ 0 & 0 & 0 & 0 & 1 & 0 \\ 0 & 0 & 0 & 0 & 0 & 1 \\ 0 & -1 & 0 & 0 & 0 & 0 \\ 1 & 0 & 0 & 0 & 0 & 0 \end{pmatrix}, |u_i| = 1 \right\rangle$$

In particular, there is an isomorphism $\mathrm{ST}(C) \simeq U(1)^3 \rtimes (\mathbb{Z}/9\mathbb{Z})^*$.

PROOF. The recipe of Serre in [**Ser12**] is as follows. Fix an auxiliary prime ℓ of good reduction (say $\ell > 3$), and fix an embedding $\iota\colon \mathbb{Q}_\ell \hookrightarrow \mathbb{C}$. Let

$$\rho_\ell\colon \mathrm{Gal}(\overline{\mathbb{Q}}/\mathbb{Q}) \to \mathrm{GL}(V_\ell(\mathrm{Jac}(C))) \simeq \mathrm{GL}(6,\mathbb{Q}_\ell)$$

be the ℓ-adic Galois representation attached to the ℓ-adic Tate module of the Jacobian variety of C. Denote by G the Zariski closure of the image $\rho_\ell(\mathrm{Gal}(\overline{\mathbb{Q}}/\mathbb{Q}))$, and let G_1 be the Zariski closure of $G \cap \mathrm{Sp}_6(\mathbb{Q}_\ell)$, where Sp_6 denotes the symplectic group. By definition, the Sato-Tate group $\mathrm{ST}(C)$ is a maximal compact subgroup of $G_1 \otimes_\iota \mathbb{C}$. In general, one hopes that this construction does not depend on ℓ and ι, and this is the case for our Picard curve C. Indeed, since the CM-type of $\mathrm{Jac}(C)$ is non-degenerate then the twisted Lefschetz group $\mathrm{TL}(C)$ satisfies $G_1 = \mathrm{TL}(C) \otimes \mathbb{Q}_\ell$ for all primes ℓ (see [**FGL14**, Lemma 3.5]). Recall that the twisted Lefschetz group is defined as

$$\mathrm{TL}(C) = \bigcup_{\tau \in \mathrm{Gal}(\overline{\mathbb{Q}}/\mathbb{Q})} \mathrm{L}(C)(\tau),$$

where $\mathrm{L}(C)(\tau) = \{\gamma \in \mathrm{Sp}_6(\mathbb{Q})\colon \gamma\alpha\gamma^{-1} = \tau(\alpha) \text{ for all } \alpha \in \mathrm{End}(\mathrm{Jac}(C)_{\overline{\mathbb{Q}}}) \otimes \mathbb{Q}\}$, where $\mathrm{Jac}(C)_{\overline{\mathbb{Q}}}$ denotes the base change to $\overline{\mathbb{Q}}$. Here, α is seen as an endomorphism of $H_1(\mathrm{Jac}(C)_{\mathbb{C}}, \mathbb{Q})$. The reason why the CM-type of $\mathrm{Jac}(C)$ is non-degenerate is due to the fact that Φ^* is simple and $\dim \mathrm{Jac}(C) = 3$ (see [**Kub65**, **Rib81**]); alternatively, one checks that the $\mathbb{Z}$-linear map:

$$\mathbb{Z}[\mathrm{Gal}(K/\mathbb{Q})] \to \mathbb{Z}[\mathrm{Gal}(K/\mathbb{Q})], \qquad \sigma_a \mapsto \sum_{\sigma_b \in \Phi} \sigma_b^{-1}\sigma_a$$

has maximal rank $1 + \dim(\mathrm{Jac}(C)) = 4$. Then, by combining [**BGK03**] and [**FKRS12**, Thm.2.16(a)], it follows that the connected component of the identity $\mathrm{TL}(C)^0$ satisfies

$$G_1^0 = \mathrm{TL}(C)^0 \otimes \mathbb{Q}_\ell = \{\mathrm{diag}(x_1, y_1, x_2, y_2, x_3, y_3) \mid x_i, y_i \in \mathbb{Q}_\ell^*, x_i y_i = 1\}.$$

Thus, the connected component of the Sato-Tate group for C is equal to

$$\mathrm{ST}(C)^0 = \{\mathrm{diag}(u_1, \overline{u}_1, u_2, \overline{u}_2, u_3, \overline{u}_3)\colon u_i \in U(1)\} \simeq U(1)^3.$$

According to [**FKRS12**, Prop. 2.17], it also follows that the group of components of $\mathrm{ST}(C)$ is isomorphic to $\mathrm{Gal}(K/\mathbb{Q})$. We claim that $\mathrm{ST}(C) = \mathrm{ST}(C)^0 \rtimes \langle\gamma\rangle$, where

$$\gamma = \begin{pmatrix} 0 & 0 & 1 & 0 & 0 & 0 \\ 0 & 0 & 0 & 1 & 0 & 0 \\ 0 & 0 & 0 & 0 & 1 & 0 \\ 0 & 0 & 0 & 0 & 0 & 1 \\ 0 & -1 & 0 & 0 & 0 & 0 \\ 1 & 0 & 0 & 0 & 0 & 0 \end{pmatrix}.$$

To this end, we consider the automorphism of the Picard curve C determined by $\alpha(x, y) = (\zeta^6 x, \zeta^2 y)$. We still denote by α the induced endomorphism of $\mathrm{Jac}(C)$. Under the basis of regular differentials of $\Omega^1(C)$:

$$\omega_1 = \frac{dx}{y^2}, \quad \omega_2 = \frac{dx}{y}, \quad \omega_3 = \frac{x\,dx}{y^2},$$

the action induced is given by $\alpha^*(\omega_1) = \zeta^2\omega_1$, $\alpha^*(\omega_2) = \zeta^4\omega_2$, $\alpha^*(\omega_3) = \zeta^8\omega_3$. By taking the symplectic basis of $H_1(\mathrm{Jac}(C)_{\mathbb{C}}, \mathbb{C})$ corresponding to the above basis (with respect to the skew-symmetric matrix J), we get the matrix

$$\alpha = \begin{pmatrix} \zeta^2 & 0 & 0 & 0 & 0 & 0 \\ 0 & \overline{\zeta}^2 & 0 & 0 & 0 & 0 \\ 0 & 0 & \zeta^4 & 0 & 0 & 0 \\ 0 & 0 & 0 & \overline{\zeta}^4 & 0 & 0 \\ 0 & 0 & 0 & 0 & \zeta^8 & 0 \\ 0 & 0 & 0 & 0 & 0 & \overline{\zeta}^8 \end{pmatrix}.$$

One checks that the matrix γ satisfies

$$\gamma\alpha\gamma^{-1} = {}^{\sigma_2}\alpha\,,$$

which implies that $\gamma \in \mathrm{TL}(\sigma_2)$. Hence, γ belongs to $\mathrm{ST}(C)$; finally, a short computations shows that $\gamma^6 = -\,\mathrm{Id} \in \mathrm{ST}(C)^0$, but γ^i is not in $\mathrm{ST}(C)^0$ for $1 \leq i < 6$. □

REMARK 2.2. For future use, we compute the shape of the characteristic polynomials in each component of the Sato-Tate group. To this end, we take a random matrix $\mathrm{diag}(u_1, \overline{u}_1, u_2, \overline{u}_2, u_3, \overline{u}_3)$ in the connected component $\mathrm{ST}^0(C)$, and we get:

$$\begin{aligned} \mathrm{ST}^0(C)\cdot \mathrm{Id}: &\quad \textstyle\prod_{i=1}^{3}(T-u_i)(T-\overline{u}_i) \\ \mathrm{ST}^0(C)\cdot\gamma\ : &\quad T^6+1 \\ \mathrm{ST}^0(C)\cdot\gamma^2: &\quad T^6+(u_1\overline{u}_2u_3+\overline{u}_1u_2\overline{u}_3)\,T^3+1 \\ \mathrm{ST}^0(C)\cdot\gamma^3: &\quad (T^2+1)^3 \\ \mathrm{ST}^0(C)\cdot\gamma^4: &\quad T^6-(u_1\overline{u}_2u_3+\overline{u}_1u_2\overline{u}_3)\,T^3+1 \\ \mathrm{ST}^0(C)\cdot\gamma^5: &\quad T^6+1\,. \end{aligned}$$

REMARK 2.3. As a consequence of [**FKRS12**, Prop. 2.17], we also obtain that, for every subextension $K/K'/\mathbb{Q}$, one has $\mathrm{ST}(C_{K'}) = \mathrm{ST}(C)^0 \rtimes \langle\gamma^{[K':\mathbb{Q}]}\rangle$, where $C_{K'}$ denotes the base change $C \times_{\mathbb{Q}} K'$.

3. Sato-Tate distribution

A general strategy to prove the expected distribution is due to Serre [**Ser98**]. For every non-trivial irreducible representation $\phi\colon \mathrm{ST}(C) \to \mathrm{GL}_m(\mathbb{C})$, one needs to consider the L-function

$$L(\phi,s) = \prod_{p\neq 3} \det(1-\phi(x_p)p^{-s})^{-1}\,,$$

where $x_p = \frac{1}{\sqrt{p}}\rho_\ell(\mathrm{Frob}_p) \in \mathrm{ST}(C)$, and then show that $L(\phi,s)$ is invertible, in the sense that it has meromorphic continuation to $\mathrm{Re}(s) \geq 1$ and it holds

$$L(\phi,1) \neq 0\,.$$

PROPOSITION 3.1. *The Picard curve $C\colon y^3 = x^4 - x$ satisfies the generalized Sato-Tate conjecture. More explicitly, the sequence*

$$\left\{\left(\frac{{}^{\sigma_2}\psi(\mathfrak{p})}{\sqrt{N(\mathfrak{p})}}, \frac{{}^{\sigma_4}\psi(\mathfrak{p})}{\sqrt{N(\mathfrak{p})}}, \frac{{}^{\sigma_8}\psi(\mathfrak{p})}{\sqrt{N(\mathfrak{p})}}, p\right)\right\}_{p\neq 3} \subseteq U(1)^3 \rtimes (\mathbb{Z}/9\mathbb{Z})^* \simeq \mathrm{ST}(C)\,,$$

where $\mathfrak{p}$ is any prime ideal of the factorization of $p\mathcal{O}$, is equidistributed over $U(1)^3 \rtimes (\mathbb{Z}/9\mathbb{Z})^*$ with respect to the Haar measure.

PROOF. The irreducible representations of $\mathrm{ST}(C) \simeq U(1)^3 \rtimes (\mathbb{Z}/9\mathbb{Z})^*$ can be described as follows (see [**Ser77**, §8.2]). For every triple $\underline{b} = (b_1, b_2, b_3)$ in $\mathbb{Z}^3$, we consider the irreducible character of $U(1)^3$ given by

$$\phi_{\underline{b}} \colon U(1)^3 \to \mathbb{C}^*, \quad \phi_{\underline{b}}(u_1, u_2, u_3) = \prod_{i=1}^{3} u_i^{b_i},$$

and let

$$H_{\underline{b}} = \{h \in (\mathbb{Z}/9\mathbb{Z})^* : \phi_{\underline{b}}(u_1, u_2, u_3) = \phi_{\underline{b}}({}^h(u_1, u_2, u_3))\}.$$

The action of $(\mathbb{Z}/9\mathbb{Z})^*$ on $U(1)^3$ is given by conjugation through powers of the matrix γ; more precisely, for the generator $g = 2$ of $(\mathbb{Z}/9\mathbb{Z})^*$ we have ${}^g(u_1, u_2, u_3) = (u_2, u_3, \overline{u}_1)$ since

$$\gamma \begin{pmatrix} u_1 &&&&& \\ & \bar{u}_1 &&&& \\ && u_2 &&& \\ &&& \bar{u}_2 && \\ &&&& u_3 & \\ &&&&& \bar{u}_3 \end{pmatrix} \gamma^{-1} = \begin{pmatrix} u_2 &&&&& \\ & \bar{u}_2 &&&& \\ && u_3 &&& \\ &&& \bar{u}_3 && \\ &&&& \bar{u}_1 & \\ &&&&& u_1 \end{pmatrix}.$$

An easy computation shows that $H_{\underline{b}} = \langle 2 \rangle$ or $\langle 2^3 \rangle$ if and only if $\underline{b} = (0, 0, 0)$, while $H_{\underline{b}} = \langle 2^2 \rangle$ for $\underline{b} = (b_1, -b_1, b_1)$ with $b_1 \neq 0$, and $H_{\underline{b}}$ is trivial otherwise. Then, one has that

$$\phi_{\underline{b}}(u_1, u_2, u_3, h) = \prod_{i=1}^{3} u_i^{b_i}$$

is a character of $H := U(1)^3 \rtimes H_{\underline{b}}$. By [**Ser77**, Prop. 25] every irreducible representation of $G := U(1)^3 \rtimes (\mathbb{Z}/9\mathbb{Z})^*$ is of the form $\theta := \mathrm{Ind}_H^G(\phi_{\underline{b}} \otimes \chi)$, where χ is a character of $H_{\underline{b}}$ that may be viewed as a character of H by composing with the projection $H \to H_{\underline{b}}$.

Let $\theta = \mathrm{Ind}_H^G(\phi_{\underline{b}} \otimes \chi)$ be an irreducible representation of $U(1)^3 \rtimes (\mathbb{Z}/9\mathbb{Z})^*$ as above. If we denote the sequence by

$$x_p = \left(\frac{{}^{\sigma_2}\psi(\mathfrak{p})}{\sqrt{N(\mathfrak{p})}}, \frac{{}^{\sigma_4}\psi(\mathfrak{p})}{\sqrt{N(\mathfrak{p})}}, \frac{{}^{\sigma_8}\psi(\mathfrak{p})}{\sqrt{N(\mathfrak{p})}}, p \right) \in U(1)^3 \rtimes (\mathbb{Z}/9\mathbb{Z})^*$$

where $\mathfrak{p}$ is any prime ideal of the factorization of $p\mathcal{O}$, our claim is equivalent to showing that the corresponding L-function

$$L(\theta, s) = \prod_{p \neq 3} (1 - \det(\theta(x_p)) p^{-s})^{-1}$$

is invertible provided that $(b_1, b_2, b_3) \neq (0, 0, 0)$. Assume first that $H_{\underline{b}}$ is trivial. Then, also χ is trivial and one has

$$L(\theta, s) = L(\phi_{\underline{b}}, s) = \prod_{p \neq 3} \left(1 - \frac{{}^{\sigma_2}\psi(\mathfrak{p})^{b_1}\, {}^{\sigma_4}\psi(\mathfrak{p})^{b_2}\, {}^{\sigma_8}\psi(\mathfrak{p})^{b_3}}{\sqrt{\mathrm{N}(\mathfrak{p})}^{b_1+b_2+b_3}} \right).$$

This can be seen as the L-function of the unitarized Grössencharakter

$$\Psi := \frac{{}^{\sigma_2}\psi(\cdot)^{b_1}\, {}^{\sigma_4}\psi(\cdot)^{b_2}\, {}^{\sigma_8}\psi(\cdot)^{b_3}}{\mathrm{N}(\cdot)^{(b_1+b_2+b_3)/2}}$$

Under our assumption $(b_1, b_2, b_3) \neq (0,0,0)$ and by using the factorization of $\psi(\mathfrak{p})\mathcal{O}$ into prime ideals (see property (iii) in the proof of Proposition 1.2), an easy computation shows that Ψ is non-trivial. Hecke showed [**Hec20**] that the L-function of a non-trivial unitarized Grössencharkter is holomorphic and non-vanishing for $\mathrm{Re}(s) \geq 1$. In the remaining case, that is for $H_{\underline{b}}$ of order 3, one gets $L(\Psi, s) = L(\theta, s)^3$ and the claim also follows by the same argument. □

4. The moment sequences

In this section we will compute the moment sequences of the Sato-Tate group and compare them with the moment statistics obtained by computing the local factors of our curve up to some bound, with the goal of restating that they are in accordance with the proven assymptotic distribution in Proposition 3.1.

Let μ be a positive measure on $I = [-d, d]$. Then, on the one hand, for every integer $n \geq 0$, the nth moment $M_n[\mu]$ is by definition $\mu(\varphi_n)$, where φ_n is the function $z \mapsto z^n$. That is, we have

$$M_n[\mu] = \int_I z^n \mu(z).$$

In our setting, the measure μ is *uniquely* determined by its moment sequence $M_n[\mu]$ (see Proposition 1 in [**KS09**]).

On the other hand, if a sequence $\{a(p)\}_p$ is μ-equidistributed, then the following equality holds:

$$M_n[\mu] = \lim_{x\to\infty} \frac{1}{\pi(x)} \sum_{p\leq x} a_p^n .$$

From now on, we shall denote by $a_1(p)$, $a_2(p)$, $a_3(p)$ the *higher* traces according to

$$L_p^{\mathrm{ST}}(C,T) = 1 + a_1(p)T + a_2(p)T^2 + a_3(p)T^3 + a_2(p)T^4 + a_1(p)T^5 + T^6 .$$

Recall that due to the Weil's conjectures, we know that

$$a_1(p) \in [-6,6], \quad a_2(p) \in [-15,15], \quad a_3(p) \in [-20,20] .$$

4.1. The distribution of $ST(C)$. For each i in $\{1,2,3\}$, let μ_i denote the projection on the interval $I_i = [-\binom{6}{i}, +\binom{6}{i}]$ obtained from the Haar measure of the Sato-Tate group $\mathrm{ST(C)} \simeq U(1)^3 \rtimes (\mathbb{Z}/9\mathbb{Z})^*$.

In general it is difficult to obtain the explicit distribution function, but because of the isomorphism stated in Proposition 2.1, we can easily compute the moment sequence of the Sato-Tate measure.

As in [**FGL14**] we shall split each measure as a sum of its restrictions to each component of $\mathrm{ST}(C)^0 \cdot \gamma^k$, where $0 \leq k \leq 5$.

Therefore one has

$$\mu_i = \frac{1}{6} \sum_{0\leq k\leq 5} {}^k\mu_i , \quad M_n[\mu_i] = \frac{1}{6} \sum_{0\leq k \leq 5} M_n[{}^k\mu_i],$$

so we can compute the moments $M_n[{}^k\mu_i]$ separately for every k and then get the *total* moments $M_n[\mu_i]$. To ease notation, we shall denote the moment sequences by

$$M[\mu_i] := (M_0[\mu_i], M_1[\mu_i], M_1[\mu_i], \dots, M_n[\mu_i], \dots) ,$$

and similarly for every $M[{}^k\mu_i]$.

In what follows, the characteristic polynomial of a matrix in USp(6) will be denoted by

$$P(T) = 1 + a_1T + a_2T^2 + a_3T^3 + a_2T^4 + a_1T^5 + T^6\,.$$

Case $k = 1, 5$: In these components, according to Remark 2.2 one has that $P(T) = T^6 + 1$, so that

$$a_1 = a_2 = a_3 = 0\,.$$

Hence,

$$M_n[{}^k\mu_1] = M_n[{}^k\mu_2] = M_n[{}^k\mu_3] = 0 \text{ for all } n \geq 1\,.$$

Case $k = 2, 4$: In these components, we have

$$P(T) = T^6 \pm (u_1\overline{u}_2u_3 + \overline{u}_1u_2\overline{u}_3)T^3 + 1\,.$$

So that $a_1 = a_2 = 0$. Hence, it follows that

$$M_n[{}^k\mu_1] = M_n[{}^k\mu_2] = 0 \text{ for all } n \geq 1\,.$$

To get the distribution of the third trace, since u_1, u_2, and u_3 are independent elements of U(1), the distribution of $a_3(p)$ will correspond to the distribution of $\alpha := u + \overline{u}$ for $u \in \mathrm{U}(1)$, and hence its associated moment sequence is

$$M[{}^k\mu_3] = (1, 0, 2, 0, 6, 0, 20, 0, \dots)\,.$$

Case $k = 3$: In this case, one has $P(T) = (1 + T^2)^3$, so that we have $a_1 = a_3 = 0$, while $a_2 = 3$. Hence, we obtain

$$M_n[{}^3\mu_1] = M_n[{}^3\mu_3] = 0\,,\ M_n[{}^3\mu_2] = 3^n \quad \text{for all } n \geq 1\,.$$

Case $k = 0$: In this case one has that $P(T) = \prod_{i=1}^3 (T - u_i)(T - \overline{u}_i)$. If we develop this expression we get the following coefficients, where as above α_i stands for the sum of u_i and its complex conjugate:

$$\begin{aligned} a_1 &= \alpha_1 + \alpha_2 + \alpha_3\,,\\ a_2 &= 3 + \alpha_1\alpha_2 + \alpha_2\alpha_3 + \alpha_1\alpha_3\,,\\ a_3 &= 2\,\alpha_1 + 2\,\alpha_2 + 2\,\alpha_3 + \alpha_1\alpha_2\alpha_3. \end{aligned}$$

To get the sequences we proceed as follows. Recall that if X and Y denote independent random variables, then $M_n[X] = E(X^n)$, $E(X + Y) = E(X) + E(Y)$, and $E(XY) = E(X)E(Y)$. Hence, one has

$$\begin{aligned} M_n[X + Y] = E((X + Y)^n) &= E\left(\sum_{k=0}^{n} \binom{n}{k} X^kY^{n-k}\right)\\ &= \sum_{k=0}^{n} \binom{n}{k} E(X^k)E(Y^{n-k})\\ &= \sum_{k=0}^{n} \binom{n}{k} M_k[X]M_{n-k}[Y]\,. \end{aligned}$$

Since we know that $M[\alpha] := M[\alpha_i] = (1, 0, 2, 0, 6, 0, 20, 0, \dots)$ for $i = 1, 2, 3$, one gets:

$$M_n[{}^0\mu_1] = \sum_{a+b+c=n} \binom{n}{a,b,c} M_a[\alpha]M_b[\alpha]M_c[\alpha],$$

$$M_n[{}^0\mu_2] = \sum_{a+b+c+d=n} \binom{n}{a,b,c,d} 3^a M_{b+d}[\alpha]M_{b+c}[\alpha]M_{c+d}[\alpha],$$

$$M_n[{}^0\mu_3] = \sum_{a+b+c+d=n} \binom{n}{a,b,c,d} 2^{a+b+c} M_{a+d}[\alpha]M_{b+d}[\alpha]M_{c+d}[\alpha].$$

Therefore we obtain the sequences:

$$\begin{aligned} M[{}^0\mu_1] &= (1, 0, 6, 0, 90, 0, 1860, \dots), \\ M[{}^0\mu_2] &= (1, 3, 21, 183, 1845, \dots), \\ M[{}^0\mu_3] &= (1, 0, 32, 0, 4920, 0, 1109120, \dots). \end{aligned}$$

We can summarize the above results in the following proposition.

PROPOSITION 4.1. *With the above notations, the first moments of the measures of* ${}^k\mu_i$ *and* μ_i *are as follows:*

(i) *The moments of the first trace are:*

$$M[{}^k\mu_1] = \begin{cases} (1, 0, 0, \dots) & \textit{if } k = 1, \dots, 5; \\ (1, 0, 6, 0, 90, 0, 1860, \dots) & \textit{if } k = 0. \end{cases}$$

Hence, $M[\mu_1] = (1, 0, 1, 0, 15, 0, 310, \dots)$.

(ii) *The moments of the second trace are:*

$$M[{}^k\mu_2] = \begin{cases} (1, 0, 0, \dots) & \textit{if } k = 1, 2, 4, 5; \\ (1, 3, 9, 27, \dots) & \textit{if } k = 3; \\ (1, 3, 21, 183, 1845, \dots) & \textit{if } k = 0. \end{cases}$$

Hence, $M[\mu_2] = (1, 1, 5, 35, 321, \dots)$.

(iii) *The moments of the third trace are:*

$$M[{}^k\mu_3] = \begin{cases} (1, 0, 0, \dots) & \textit{if } k = 1, 3, 5; \\ (1, 0, 2, 0, 6, 0, 20, 0, \dots) & \textit{if } k = 2, 4; \\ (1, 0, 32, 0, 4920, 0, 1109120, \dots) & \textit{if } k = 0. \end{cases}$$

Hence, $M[\mu_3] = (1, 0, 6, 0, 822, 0, 184860, 0 \dots)$.

4.2. The numerical sequences for C. Once we have computed the theoretical moment sequences from the Sato-Tate group $\mathrm{ST}(C)$, we wish to compute for every prime (up to some bound) its associated normalized local factor $L_p^{\mathrm{ST}}(A, T)$ to get the corresponding traces $a_1(p)$, $a_2(p)$ and $a_3(p)$ and do the experimental equidistribution matching.

The Grössencharakter ψ attached to the Picard curve C permits us to perform this numerical experimentation within a reasonable time, in this case $p \leq 2^{26}$ (about two hours of a standard laptop). We display the data obtained in Table 3.

We include figures 1-4 to display the histograms (for primes up to $p \leq 2^{26}$) showing the nondiscret components of the three distributions μ_i.

TABLE 3. Numerical moment sequences computed for p up to 2^{26}.

	a_1		a_2		a_3	
n	$M_n[\mu_1]$	$M_n[\mu_1]_{\leq 2^{26}}$	$M_n[\mu_2]$	$M_n[\mu_2]_{\leq 2^{26}}$	$M_n[\mu_3]$	$M_n[\mu_3]_{\leq 2^{26}}$
0	1	1	1	1	1	1
1	0	−0.000	1	0.999	0	−0.000
2	1	0.998	5	4.991	6	5.984
3	0	−0.005	35	34.868	0	−0.147
4	15	14.946	321	319.058	822	815.937
5	0	−0.151				
6	310	308.160				

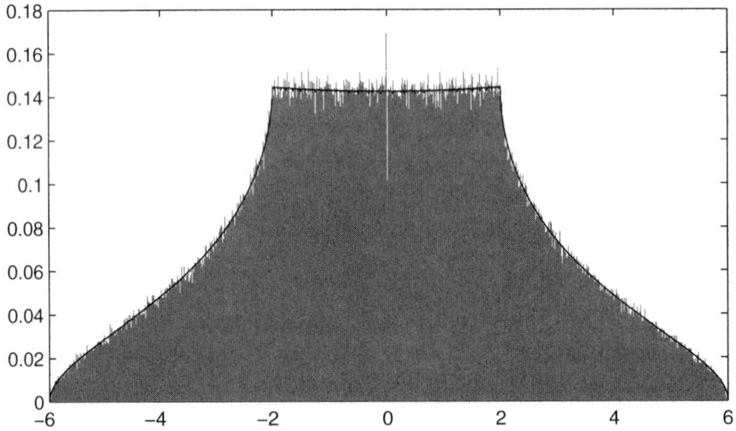

FIGURE 1. Histogram of the first trace for primes $p \equiv 1 \pmod 9$.

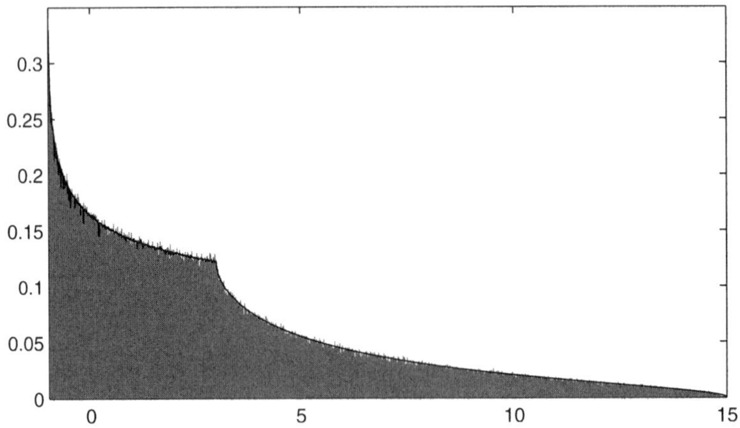

FIGURE 2. Histogram of the second trace for primes $p \equiv 1 \pmod 9$.

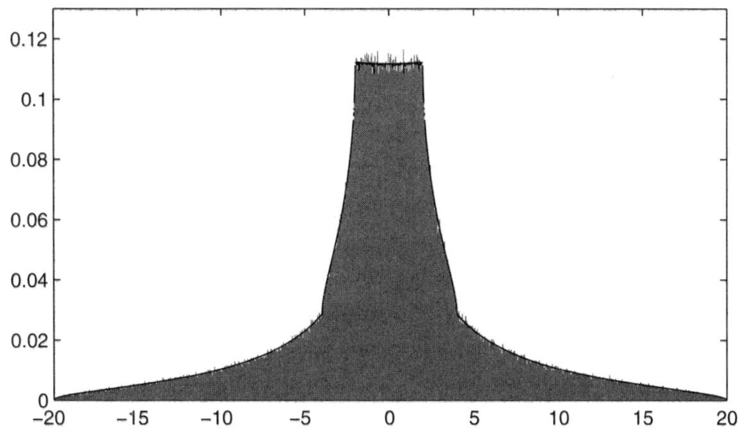

FIGURE 3. Histogram of the third trace for primes $p \equiv 1 \pmod 9$.

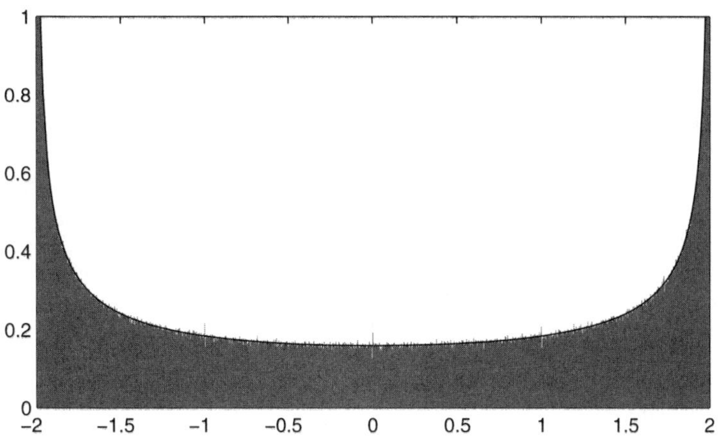

FIGURE 4. Histogram of the third trace for primes $p \equiv 4, 7 \pmod 9$.

References

[BGK03] G. Banaszak, W. Gajda, and P. Krasoń, *On Galois representations for abelian varieties with complex and real multiplications*, J. Number Theory **100** (2003), no. 1, 117–132, DOI 10.1016/S0022-314X(02)00121-X. MR1971250

[FGL14] F. Fité, J. González, and J.-C. Lario, *Frobenius distribution for quotients of Fermat curves of prime exponent*, arXiv:1403.0807.

[FKRS12] Francesc Fité, Kiran S. Kedlaya, Víctor Rotger, and Andrew V. Sutherland, *Sato-Tate distributions and Galois endomorphism modules in genus 2*, Compos. Math. **148** (2012), no. 5, 1390–1442, DOI 10.1112/S0010437X12000279. MR2982436

[Has54] Helmut Hasse, *Zetafunktion und L-Funktionen zu einem arithmetischen Funktionenkörper vom Fermatschen Typus* (German), Abh. Deutsch. Akad. Wiss. Berlin. Kl. Math. Nat. **1954** (1954), no. 4, 70 pp. (1955). MR0076807

[Hec20] E. Hecke, *Eine neue Art von Zetafunktionen und ihre Beziehungen zur Verteilung der Primzahlen* (German), Math. Z. **6** (1920), no. 1-2, 11–51, DOI 10.1007/BF01202991. MR1544392

[HN02] Rolf-Peter Holzapfel and Florin Nicolae, *Arithmetic on a family of Picard curves*, Finite fields with applications to coding theory, cryptography and related areas (Oaxaca, 2001), Springer, Berlin, 2002, pp. 187–208. MR1995336

[Joh13] C. Johansson, *On the sato-tate conjecture for non-generic abelian surfaces.*

[KS09] Kiran S. Kedlaya and Andrew V. Sutherland, *Hyperelliptic curves, L-polynomials, and random matrices*, Arithmetic, geometry, cryptography and coding theory, Contemp. Math., vol. 487, Amer. Math. Soc., Providence, RI, 2009, pp. 119–162, DOI 10.1090/conm/487/09529. MR2555991

[Kub65] Tomio Kubota, *On the field extension by complex multiplication*, Trans. Amer. Math. Soc. **118** (1965), 113–122, DOI 10.2307/1993947. MR0190144

[Rib81] K. A. Ribet, *Division fields of abelian varieties with complex multiplication*, Mém. Soc. Math. France (N.S.) **2** (1980/81), 75–94. Abelian functions and transcendental numbers (Colloq., Étole Polytech., Palaiseau, 1979). MR608640

[Ser77] Jean-Pierre Serre, *Linear representations of finite groups*, Springer-Verlag, New York-Heidelberg, 1977. Translated from the second French edition by Leonard L. Scott; Graduate Texts in Mathematics, Vol. 42. MR0450380

[Ser98] Jean-Pierre Serre, *Abelian l-adic representations and elliptic curves*, Research Notes in Mathematics, vol. 7, A K Peters, Ltd., Wellesley, MA, 1998. With the collaboration of Willem Kuyk and John Labute; Revised reprint of the 1968 original. MR1484415

[Ser12] Jean-Pierre Serre, *Lectures on $N_X(p)$*, Chapman & Hall/CRC Research Notes in Mathematics, vol. 11, CRC Press, Boca Raton, FL, 2012. MR2920749

[ST61] Goro Shimura and Yutaka Taniyama, *Complex multiplication of abelian varieties and its applications to number theory*, Publications of the Mathematical Society of Japan, vol. 6, The Mathematical Society of Japan, Tokyo, 1961. MR0125113

[Wei49] André Weil, *Numbers of solutions of equations in finite fields*, Bull. Amer. Math. Soc. **55** (1949), 497–508, DOI 10.1090/S0002-9904-1949-09219-4. MR0029393

Appendix (by Francesc Fité)

We keep the notation of the article. Let $C\colon y^3 = x^4 - x$. Let K denote the cyclotomic field $\mathbb{Q}(\zeta)$, where ζ is a 9th root of unity. For every prime $\mathfrak{p}$ of K coprime to 3, consider the character $\chi_{\mathfrak{p}}\colon \mathbb{F}_{\mathfrak{p}}^* \to K^*$ such that $\chi_{\mathfrak{p}}(x)$ is the only 9th root of unity satisfying

$$\chi_{\mathfrak{p}}(x) \equiv x^{(N(\mathfrak{p})-1)/9} \pmod{\mathfrak{p}}\,.$$

For $a, b \in \mathbb{Z}/9\mathbb{Z}$, define

$$J_{(a,b)}(\mathfrak{p}) := \sum_{x\in\mathbb{F}_{\mathfrak{p}}} \chi_{\mathfrak{p}}^a(x)\chi_{\mathfrak{p}}^b(1-x)\,.$$

PROPOSITION A.1. *The number of points of C defined over the finite field $\mathbb{F}_{\mathfrak{p}}$ is*

$$|C(\mathbb{F}_{\mathfrak{p}})| = \begin{cases} 1 + N(\mathfrak{p}) & \text{if } N(\mathfrak{p}) \not\equiv 1 \pmod 9, \quad (i) \\ 1 + N(\mathfrak{p}) + \mathrm{Tr}_{K/\mathbb{Q}}(J_{(6,1)}(\mathfrak{p})) & \text{if } N(\mathfrak{p}) \equiv 1 \pmod 9. \quad (ii) \end{cases}$$

PROOF. Case (i) is considered in Proposition 1 and Proposition 2 of [**HN02**]. We now show case (ii), by giving an alternative and shorter proof of Proposition 3 of [**HN02**]. Let $C'\colon v^9 = u(u+1)^6$. There is an isomorphism between C and C' given by

$$\phi\colon C \to C'\,, \qquad \phi(x,y) = \left(-\frac{1}{x^3}, -\frac{y^2}{x^3}\right)\,.$$

One easily sees that the inverse of ϕ is given by

$$\phi^{-1}\colon C' \to C\,, \qquad \phi^{-1}(u,v) = \left(-\frac{(u+1)^2}{v^3}, -\frac{(u+1)^3}{v^4}\right)\,.$$

Note that if $N(\mathfrak{p}) \not\equiv 1 \pmod 3$, then exponentiation by 9 is an isomorphism of $\mathbb{F}_\mathfrak{p}$. Thus C' has $N(\mathfrak{p})$ affine points plus one point at infinity. Assume now that $N(\mathfrak{p}) \equiv 1 \pmod 9$. By [**IR90**, Prop. 8.1.5], we have that

$$\begin{aligned}|C'(\mathbb{F}_\mathfrak{p})| &= 1 + \sum_{u\in\mathbb{F}_\mathfrak{p}} \sum_{a\in\mathbb{Z}/9\mathbb{Z}} \chi_\mathfrak{p}^a(u)\chi_\mathfrak{p}^{6a}(u+1)\\ &= 1 + N(\mathfrak{p}) + \sum_{a\in(\mathbb{Z}/9\mathbb{Z})^*} \sum_{u\in\mathbb{F}_\mathfrak{p}} \chi_\mathfrak{p}^a(u)\chi_\mathfrak{p}^{6a}(u+1)\,,\end{aligned}$$

where for the second equality we have used [**IR90**, Thm. 1 (b), p. 93]. But writing $x = u+1$, we obtain

$$\sum_{u\in\mathbb{F}_\mathfrak{p}} \chi_\mathfrak{p}(u)\chi_\mathfrak{p}^6(u+1) = \chi_\mathfrak{p}(-1)\sum_{x\in\mathbb{F}_\mathfrak{p}} \chi_\mathfrak{p}^6(x)\chi_\mathfrak{p}(1-x)\,.$$

Case (ii) of the proposition is a consequence of the equality $\chi_\mathfrak{p}(-1) = 1$ (this follows form the fact that the order of $\chi_\mathfrak{p}$ is odd). ☐

To show that our result agrees with Proposition 3 of [**HN02**] it remains to show that $\mathrm{Tr}_{K/\mathbb{Q}}(J_{(6,1)}(\mathfrak{p})) = \mathrm{Tr}_{K/\mathbb{Q}}(J_{(3,1)}(\mathfrak{p}))$. Indeed, by [**BEW98**, Thm. 2.1.5], one has

$$J_{(6,1)}(\mathfrak{p}) = J_{(2,1)}(\mathfrak{p})\,, \qquad J_{(3,1)}(\mathfrak{p}) = J_{(5,1)}(\mathfrak{p})\,.$$

Since $5\cdot 2 \equiv 1 \pmod 9$, we deduce that

$$\mathrm{Tr}_{K/\mathbb{Q}}(J_{(2,1)}(\mathfrak{p})) = \mathrm{Tr}_{K/\mathbb{Q}}(J_{(5,1)}(\mathfrak{p}))\,.$$

Finally, note that $J(\mathfrak{p}) = -J_{(3,1)}(\mathfrak{p})$ in the notation of the article.

References

[BEW98] Bruce C. Berndt, Ronald J. Evans, and Kenneth S. Williams, *Gauss and Jacobi sums*, Canadian Mathematical Society Series of Monographs and Advanced Texts, John Wiley & Sons, Inc., New York, 1998. A Wiley-Interscience Publication. MR1625181

[HN02] Rolf-Peter Holzapfel and Florin Nicolae, *Arithmetic on a family of Picard curves*, Finite fields with applications to coding theory, cryptography and related areas (Oaxaca, 2001), Springer, Berlin, 2002, pp. 187–208. MR1995336

[IR90] Kenneth Ireland and Michael Rosen, *A classical introduction to modern number theory*, 2nd ed., Graduate Texts in Mathematics, vol. 84, Springer-Verlag, New York, 1990. MR1070716

Departament de Matemàtiques, Universitat Politècnica de Catalunya, Barcelona, Spain

Email address: `joan.carles.lario@upc.edu`

Departament de Matemàtiques, Universitat Politècnica de Catalunya, Barcelona, Spain

Current address: Mathematisch Instituut, Universiteit Leiden, Leiden, The Netherlands

Email address: `anna.somoza@upc.edu`

Contemporary Mathematics
Volume **701**, 2018
http://dx.doi.org/10.1090/conm/701/14149

Arithmetic twists and Abelian extensions

V. Kumar Murty

To the memory of my friend, Fumiyuki Momose

ABSTRACT. Hilbert's twelfth problem asks for the explicit construction of abelian extensions of general number fields. This is still an open problem even for a real quadratic field. In some cases, Shimura succeeded in the 1970s to construct some abelian extensions of real quadratic fields using points of finite order on abelian varieties associated to modular forms of weight 2 with real Nebentypus. Using results of Momose and Ribet, we shall generalize Shimura's construction to forms with non-real Nebentypus.

CONTENTS

1. Introduction

It is an elementary fact that the field $\mathbb{Q}(\zeta_n)$ obtained by adjoining an n-th root of unity is Galois over $\mathbb{Q}$ with group $(\mathbb{Z}/n)^\times$. We may think of this extension as the one generated by the coordinates of points of order dividing n on the multiplicative group $\mathbb{G}_m$. The classical theorem of Kronecker and Weber asserts that any abelian extension of the rational number field $\mathbb{Q}$ is contained in a cyclotomic extension. It is also known that the abelian extensions of an imaginary quadratic field K are contained in the extensions generated by points of finite order on an elliptic curve

2010 *Mathematics Subject Classification.* Primary 11F11, 11F80; Secondary 11R11, 11R29, 14K15.

Key words and phrases. Abelian extensions, real quadratic field, modular forms, arithmetic twists, Galois representations.

Research partially supported by a Discovery grant from NSERC..

with multiplication by (an order in) K. Kronecker's Jugendtraum, or Hilbert's twelfth problem, ask for such constructions for any number field K. This is still an open problem. In particular, even the case of real quadratic fields is not understood.

In some cases, Shimura succeeded in the 1970s to construct ([**22**], Chapter 7 and [**23**]) some abelian extensions of real quadratic fields using points of finite order on abelian varieties associated to modular forms of weight 2. We shall generalize Shimura's construction to forms with non-real Nebentypus.

The results of this paper essentially formed Chapter 3 of my thesis [**10**] which was written in 1982, and this work was directly influenced by the paper of Momose [**12**]. I am grateful to him for patiently explaining his work to me, and for listening to my ideas as they were evolving. I remember spending many pleasant hours of conversation with Momose discussing mathematics.

Since my work (which has not been published until now), several other authors have considered Shimura's construction and generalized it in various ways. In particular, there are some results of Hida [**7**], Brown and Ghate [**1**] and Darmon and Green [**4**]. We give a brief description of this work and its relation to our work in the final section.

2. Abelian varieties of type (T')

Let K be a number field. We consider an Abelian variety $A_{/K}$ defined over K. If M is any extension of K, we denote by $\mathrm{End}_M(A)$ the algebra of M-rational endomorphisms of A. Write $G_K = \mathrm{Gal}(\overline{K}/K)$. If E is a subfield of $\mathrm{End}_K(A) \otimes \mathbb{Q}$ there is a natural action

$$\rho_\ell : G_K \longrightarrow \mathrm{Aut}_{E\otimes\mathbb{Q}_\ell} V_\ell(A)$$

Moreover, writing

$$E \otimes \mathbb{Q}_\ell = \prod_{\lambda|\ell} E_\lambda,$$

we have a K-rational decomposition

$$V_\ell(A) = \prod_{\lambda|\ell} V_\lambda$$

and the action of G_K on V_λ is denoted ρ_λ.

We say that $A_{/K}$ is of type (T') if there is a subfield E of $\mathrm{End}_K(A) \otimes \mathbb{Q}$ such that for all primes ℓ, the triple (K, A, E) satisfies

- $(T1')$ $V_\ell(A)$ is a free $E \otimes \mathbb{Q}_\ell$ module of rank 2.
- $(T2')$ for each prime λ of E dividing ℓ, the λ-adic representation ρ_λ does not have an abelian semi-simplification, and this remains true even if ρ_λ is restricted to an open subgroup of G_K.
- $(T3')$ $\det_{E\otimes\mathbb{Q}_\ell} \rho_\ell = \epsilon\chi_\ell$ where ϵ is a finite order E-valued character independent of ℓ.

Let Π_ℓ denote the set of primes of K which either divide ℓ or at which A has bad reduction. Then, for $v \notin \Pi_\ell$, set

$$a_{v,\ell} = \mathrm{Tr}_{E\otimes\mathbb{Q}_\ell} \rho_\ell(\mathrm{Frob}_v).$$

Then $a_{v,\ell}$ is an element of E and is in fact independent of ℓ (so can be designated a_v). The subfield of E generated by these traces will be denoted P and we call it the trace subfield.

3. The arithmetic twisting group

Let K be a number field and A an Abelian variety defined over K. The construction described below is modelled on the work of Momose [**11**] and Ribet [**18**], [**19**] who worked with the Abelian varieties associated to quotients of the Jacobian of a modular curve.

Suppose we are given a subfield E of $\mathrm{End}_K(A)\otimes\mathbb{Q}$ such that the triple (K, A, E) is of type (T'). Let ℓ denote a fixed rational prime and denote by Π_ℓ the set of primes of K which divide ℓ or at which A has bad reduction.

The decomposition

$$E \otimes_{\mathbb{Q}} \overline{\mathbb{Q}}_\ell \;=\; \prod \overline{\mathbb{Q}}_\ell$$

induces a corresponding decomposition

$$\overline{V_\ell} \;=\; V_\ell(A) \otimes_{\mathbb{Q}_\ell} \overline{\mathbb{Q}}_\ell \;=\; \prod V_\sigma$$

indexed by the various embeddings $\sigma : E \hookrightarrow \overline{\mathbb{Q}}_\ell$. By a result of Ribet ([**16**], Lemma 4.4.4), the hypothesis $(T2')$ implies that the V_σ are simple $\overline{\mathbb{Q}}_\ell[H]$ modules for any open subgroup H of G_K.

Throughout this section, we view E as a subfield of $\overline{\mathbb{Q}}_\ell$ and we shall write $1 : E \hookrightarrow \overline{\mathbb{Q}}_\ell$ for this distinguished embedding.

LEMMA 3.1. *We have* $\mathbb{Q}(\epsilon) \subseteq P$.

PROOF. Let $\sigma : \overline{\mathbb{Q}}_\ell \longrightarrow \overline{\mathbb{Q}}_\ell$ be an automorphism of $\overline{\mathbb{Q}}_\ell$ which leaves P fixed. Let $\sigma 1$ denote the composition of σ and 1. We consider the representation spaces V_1 and $V_{\sigma 1}$ of G_K. For $v \notin \Pi_\ell$, they have equal traces. Hence, by the Chebotarev density theorem, they have equal traces. As they are simple $\overline{\mathbb{Q}}_\ell[G_K]$-modules (in particular, semisimple), it follows that they are isomorphic. Thus, their determinants are equal, so $\epsilon^\sigma \chi_\ell \;=\; \epsilon\chi_\ell$. The result follows. $\square$

We now describe the construction of the twisting group of A. Consider the set Γ of embeddings

$$\gamma : P \hookrightarrow \overline{\mathbb{Q}}_\ell$$

for which there exists a character

$$\chi_\gamma : G_K \longrightarrow \overline{\mathbb{Q}}_\ell^\times$$

such that for almost all v (that is, for all but finitely many v), we have

$$a_v^\gamma \;=\; a_v \chi_\gamma(\mathrm{Frob}_v). \tag{1}$$

LEMMA 3.2. *The character* χ_γ *is of finite order and* $\mathbb{Q}(\chi_\gamma) \subseteq \mathbb{Q}(\epsilon)$.

PROOF. Choose an extension of γ to an embedding $E \hookrightarrow \overline{\mathbb{Q}}_\ell$ and denote it again by γ. The relation (1) implies that as $\overline{\mathbb{Q}}_\ell[G_K]$-modules, we have

$$V_\gamma \;\simeq\; V_1 \otimes \chi_\gamma.$$

In particular, the determinants are equal and so we have

$$\epsilon^\gamma \;=\; \epsilon\chi_\gamma^2. \tag{2}$$

Since ϵ is of finite order, so is χ_γ. Furthermore, χ_γ must be of the form $\epsilon^i\omega_\gamma$ where i is an integer and ω_γ is a character satisfying $\omega_\gamma^2 = 1$. This shows that $\mathbb{Q}(\chi_\gamma) \subseteq \mathbb{Q}(\epsilon)$. $\square$

The following lemma is modelled on Ribet ([**17**], p. 40).

LEMMA 3.3. *If $\gamma \in \Gamma$, there is a unique character χ_γ satisfying* (1).

PROOF. If χ_γ is not unique, then there is a non-trivial character χ of G_K taking values in P such that

$$a_v = a_v \chi(\mathrm{Frob}_v)$$

for almost all v. This implies that there is an automorphism M of the vector space V_1 such that for all $g \in G_K$, we have

$$M^{-1}\rho_{1,\ell}(g)M = \chi(g)\rho_{1,\ell}(g).$$

It follows that M is not a scalar and that it lies in $\mathrm{End}_{\overline{\mathbb{Q}}_\ell[H]}V_1$ where $H = \mathrm{Ker}\ \chi$. This contradicts the simplicity of the $\overline{\mathbb{Q}}_\ell[H]$-module V_1. □

PROPOSITION 3.4. *Γ is a finite Abelian group. It consists of the automorphisms γ of P for which there exists a character χ_γ such that*

$$a_v^\gamma = a_v \chi_\gamma(\mathrm{Frob}_v)$$

for almost all v.

PROOF. The second statement is implied by Lemma 3.1 and Lemma 3.2. It is clearly a finite group and it remains only to check that it is abelian. If $\gamma_1, \gamma_2 \in \Gamma$, we see that for almost all v, we have

$$\gamma_1\gamma_2(a_v) = a_v(\chi_{\gamma_1}\chi_{\gamma_2}^{\gamma_1})(\mathrm{Frob}_v).$$

It follows from Lemma 3.3 that

$$\chi_{\gamma_1\gamma_2} = \chi_{\gamma_1}\chi_{\gamma_2}^{\gamma_1}$$

and similarly that

$$\chi_{\gamma_2\gamma_1} = \chi_{\gamma_2}\chi_{\gamma_1}^{\gamma_2}.$$

Again by Lemma 3.3, we have

$$\gamma_1\gamma_2 = \gamma_2\gamma_1$$

if and only if

$$\chi_{\gamma_1\gamma_2} = \chi_{\gamma_2\gamma_1}.$$

For $i = 1, 2$, write

$$\chi_{\gamma_i} = \epsilon^{m_i}\omega_i$$

where ω_i is a character such that $\omega_i^2 = 1$ and m_i is an integer such that

$$\epsilon^{\gamma_i - 1} = \chi_{\gamma_i}^2 = \epsilon^{2m_i},$$

the first equality following from (2). Thus,

$$\chi_{\gamma_1\gamma_2}\chi_{\gamma_2\gamma_1}^{-1} = \epsilon^{[m_2(\gamma_1-1)-m_1(\gamma_2-1)]} = 1.$$

□

Now, let M be a finite extension of K and set

$$\Gamma_M = \{\gamma \in \Gamma,\ \ker \chi_\gamma \supseteq \mathrm{Gal}(\overline{K}/M)\}.$$

4. A general construction

We describe an abstract method of constructing abelian extensions. Let p be an odd prime and $\mathbb{F}$ a finite field of characteristic p. Let X be a 2-dimensional $\mathbb{F}$ vector space and let G be an abstract group. Suppose we are given a normal subgroup H of G such that G/H is finite and cyclic. Moreover, suppose that the following condition holds.

HYPOTHESIS 4.1. *We have a representation*

$$\rho : G \longrightarrow \mathrm{Aut}_{\mathbb{F}}(X) \simeq \mathrm{GL}_2(\mathbb{F})$$

such that $\mathrm{Tr}\ \rho(g) = 0$ *whenever* gH *generates* G/H.

LEMMA 4.2. *With the above hypothesis,* $[G : H]$ *is even.*

PROOF. We shall say that $g \in G$ is *special* if gH is a generator of G/H. For special g, we see by Hypothesis 4.1 that

$$\mathrm{Tr}\ \rho(g) = 0.$$

Hence,

$$\rho(g^2) = -\det \rho(g).$$

But if $[G : H]$ is odd, then g^2 is also special. Again, by Hypothesis 4.1, this would imply that

$$\mathrm{Tr}\ \rho(g^2) = 0$$

which is a contradiction. □

Let J denote the unique subgroup of G which contains H and is of index 2 in G, its existence and uniqueness being assured by Lemma 4.2 and the cyclicity of G/H.

LEMMA 4.3. *We have* $\rho(J)$ *is abelian, and contained in a Borel subgroup of* $\mathrm{GL}_2(\mathbb{F})$.

PROOF. We observe that for any special $g \in G$, the subgroup J is generated by H and g^2. We claim that for any element $g' \in G \setminus J$, we have $\mathrm{Tr}\ \rho(g') = 0$. Indeed, write $g' = gx$, for some special g and some $x \in J$. Since we can write $x = g^{2a}h$ for some $h \in H$ and some integer $a \geq 0$, we find that

$$\mathrm{Tr}\ \rho(g') = \mathrm{Tr}\ \rho((g^2)^a gh) = (-\det \rho(g))^a\ \mathrm{Tr}\ \rho(gh) = 0.$$

Now consider the representation

$$\overline{\rho} : G \longrightarrow \mathrm{Aut}_{\overline{\mathbb{F}}}(\overline{X}) \simeq \mathrm{GL}_2(\overline{\mathbb{F}})$$

where $\overline{\mathbb{F}}$ denotes an algebraic closure of $\mathbb{F}$ and $\overline{X} = X \otimes_{\mathbb{F}} \overline{\mathbb{F}}$. Consider the $\overline{\mathbb{F}}$ algebra $R = \overline{\mathbb{F}}[\overline{\rho}(J)]$. Fix a special g. By the observation of the previous paragraph, we have

$$\mathrm{Tr}\ R\rho(g) = 0. \tag{3}$$

Moreover,

$$\rho(g^2) \text{ is a scalar.} \tag{4}$$

These conditions imply that R is commutative. Indeed, by (4), there is a basis of X such that $\rho(g)$ has the form

$$\begin{pmatrix} \alpha & 0 \\ 0 & -\alpha \end{pmatrix}$$

for some $\alpha \in \overline{\mathbb{F}}$. If

$$x = \begin{pmatrix} a & b \\ c & d \end{pmatrix} \in M_2(\overline{\mathbb{F}})$$

is a general element of R, then (3) implies that $a = d$. Then, there are several possibilities:

(a) R consists of scalars ($\dim R = 1$)

(b) for all $x \in R$, we have $c = 0$ but $\dim R > 1$:

$$R \subseteq \left\{ \begin{pmatrix} a & b \\ 0 & a \end{pmatrix}, a, b \in \overline{\mathbb{F}} \right\}$$

(c) for all $x \in R$, we have $b = 0$ but $\dim R > 1$:

$$R \subseteq \left\{ \begin{pmatrix} a & 0 \\ c & a \end{pmatrix}, a, c \in \overline{\mathbb{F}} \right\}$$

(d) there exists an $x \in R$ for which $b \neq 0$ and $c \neq 0$. In this case $\dim R > 1$ and

$$R \subseteq \left\{ \begin{pmatrix} a & c\nu \\ c & a \end{pmatrix}, a, c \in \overline{\mathbb{F}} \right\}$$

for some element $\nu \in \overline{\mathbb{F}}^{\times}$.

In the last three cases, $\dim R = 2$. In all cases, R is commutative and $\rho(J)$ is contained in a Borel subgroup. $\square$

Next, we give a criterion for $\rho(J)$ to be contained in a Cartan subgroup (case (a) or (d) above). Let ω denote the non-trivial character of G/J. We shall also think of it as a character of G.

PROPOSITION 4.4. *Suppose that there does not exist an $\overline{\mathbb{F}}^{\times}$ valued character η of G such that $\det \rho = \omega\eta^2$. Then $\rho(J)$ is contained in a Cartan subgroup C of $\mathrm{Aut}_{\mathbb{F}}(X)$. Furthermore, $\rho(G)$ is contained in the normalizer of C but not in C itself.*

PROOF. This is modelled on (Momose [**12**], Lemma 1.2). By Lemma 4.3, we know that there is a line $\overline{V}$ left stable by $\overline{\rho}(J)$. Suppose that

$$\overline{\rho}(G)\overline{V} = \overline{V}.$$

Then, the semisimplification of $\overline{\rho}$ is given by two characters ν and η such that

$$\nu\eta = \det\ \overline{\rho}$$

$$(\nu + \eta)(g) = 0 \text{ if } g \in G \setminus J.$$

From the second relation, it follows that

$$\nu/\eta = \omega.$$

Then, the first relation implies that

$$\omega\eta^2 = \det\ \rho$$

contrary to hypothesis. Thus,

$$\overline{\rho}(G)\overline{V} \neq \overline{V}.$$

It follows that for any $g \in G \setminus J$, we have a non-trivial decomposition

$$\overline{X} = \overline{V} \oplus \overline{\rho}(g)\overline{V}.$$

If we choose another element $g' \in G \setminus J$, then

$$\overline{\rho}(g')\overline{V} \;=\; \overline{\rho}(g)\overline{V}.$$

It follows that $\overline{\rho}(J)$ is contained in the Cartan subgroup

$$C \;=\; \mathrm{Aut}_{\overline{F}}(\overline{V}) \;\times\; \mathrm{Aut}_{\overline{F}}(\overline{\rho}(g)\overline{V}).$$

Since any $g' \in G \setminus J$ interchanges $\overline{V}$ and $\overline{\rho}(g)\overline{V}$, it follows that $\overline{\rho}(G)$ is contained in the normalizer of C but not in C itself. This proves the result. □

Throughout the rest of this section, we shall assume that the hypothesis of Proposition 4.4 holds. Thus, we get two characters

$$\phi_1, \phi_2 : J \;\longrightarrow\; \overline{\mathbb{F}}^{\times}$$

and an $\overline{\mathbb{F}}$ basis of $\overline{X}$ such that ρ restricted to J has the form

$$\begin{pmatrix} \phi_1 & 0 \\ 0 & \phi_2 \end{pmatrix}.$$

Moreover, for any $\delta \in G \setminus J$, we have $\rho(\delta)$ has the form

$$\begin{pmatrix} 0 & x \\ y & 0 \end{pmatrix}$$

for some $x, y \in \overline{\mathbb{F}}^{\times}$. Now let us fix one $\delta \in G \setminus J$.

COROLLARY 4.5. *For any $j \in J$, we have*

$$\phi_1(j)\phi_1(\delta^{-1}j\delta) \;=\; \phi_2(j)\phi_2(\delta^{-1}j\delta) \;=\; \det\ \rho(j).$$

PROOF. We have

$$\det \rho(j) \;=\; \det\ \overline{\rho}(j)$$

and

$$\begin{pmatrix} \phi_1(\delta^{-1}j\delta) & 0 \\ 0 & \phi_2(\delta^{-1}j\delta) \end{pmatrix} \;=\; \overline{\rho}(\delta^{-1}j\delta) \;=\; \overline{\rho}(\delta)^{-1}\overline{\rho}(j)\overline{\rho}(\delta) \;=\; \begin{pmatrix} \phi_2(j) & 0 \\ 0 & \phi_1(j) \end{pmatrix}.$$

Thus,

$$\det\ \rho(j) \;=\; \phi_1(j)\phi_2(j) \;=\; \phi_1(j)\phi_1(\delta^{-1}j\delta) \;=\; \phi_2(j)\phi_2(\delta^{-1}j\delta).$$

□

Let G' (resp. J') denote the commutator subgroup of G (resp. J). We view ϕ_1 and ϕ_2 as characters of the quotient J/J'. Consider the composite homomorphisms

$$\Phi_i : G \;\longrightarrow\; G/G' \;\longrightarrow J/J' \;\longrightarrow \overline{\mathbb{F}}^{\times}$$

for $i = 1, 2$. Here, the rightmost map is ϕ_i and the middle map is the transfer homomorphism Ver (see Serre [**20**], p. 120) which in our case can be made quite explicit:

$$\mathrm{Ver}(g \bmod G') \;=\; \begin{cases} g^2 \bmod J' & \text{if } g \notin J \\ g\delta g\delta^{-1} \bmod J' & \text{if } g \in J. \end{cases}$$

COROLLARY 4.6. *For any $g \in G$, we have*

$$\Phi_1(g) \;=\; \Phi_2(g) \;=\; \omega(g)\det\ \rho(g).$$

PROOF. If $g \notin J$, then

$$\mathrm{Tr}\rho(g) = 0.$$

Thus,

$$\rho(g^2) = -\det\rho(g).$$

If $g \in J$, we have by Corollary 4.5,

$$\phi_i(\mathrm{Ver}(g \bmod G')) = \det\ \rho(g)$$

for $i = 1, 2$. This proves the result. □

COROLLARY 4.7. *$\rho(G)$ is non-abelian.*

PROOF. Since conjugation by $\rho(\delta)$ interchanges the eigenspaces of C, we see that $\rho(G)$ is abelian if and only if $\phi_1 = \phi_2$. Suppose $\rho(G)$ is abelian. Then ϕ_1 can be extended to a character ϕ (say) of G, This character satisfies

$$\Phi_1 = \Phi_2 = \phi^2.$$

Thus, by Corollary 4.6, we have

$$\det\rho = \omega\phi^2.$$

This contradicts the hypothesis of Proposition 4.4. □

PROPOSITION 4.8. *Let M denote the quadratic field corresponding to ω. The splitting field of $\rho_{\mathrm{Gal}(\overline{K}/M)}$ is an abelian extension of M which is at most tamely ramified at p.*

PROOF. We have already seen that the splitting field is abelian as $\rho(\mathrm{Gal}(\overline{K}/M))$ is contained in a Cartan subgroup of $\mathrm{GL}_2(\mathbb{F})$. Such a subgroup has order $(q-1)^2$ or q^2-1 depending on whether it is split or not, where q is the cardinality of $\mathbb{F}$. In particular, the order of $\rho(\mathrm{Gal}(\overline{K}/M))$ is prime to p. □

REMARK 4.9. In section 6 we shall find examples of representations ρ satisfying the hypotheses in this section, as follows. We consider a triple (K, A, E) of type (T')which has a non-trivial twist γ and we associate an ideal S_γ of E. Then, the representation ρ is obtained by studying the Galois action on the S_γ-division points. In the next section, we abstractly define and develop the properties of S_γ.

5. The ideal $S_n(E/F)$

Let E/F be a cyclic Galois extension of (local or global) fields. Let γ be a generator of $\mathrm{Gal}(E/F)$, and let ζ be a root of unity in E such that $\mathrm{Norm}_{E/F}\zeta = 1$. (Note that ζ may or may not be in F). Let $\mathfrak{b}(\gamma, \zeta)$ be the ideal of E generated by the set

$$\mathfrak{b}_0(\gamma, \zeta) = \{x \in \mathcal{O}_E : x^\gamma = \zeta x\}.$$

Our assumption on ζ implies that this set contains non-zero elements.

DEFINITION 5.1. *$S(\gamma, \zeta)$ is the radical of $\mathfrak{b}(\gamma, \zeta)$.*

REMARK 5.2. Whenever we write $S(\gamma, \zeta)$ or $\mathfrak{b}(\gamma, \zeta)$, with some root of unity ζ, we shall be tacitly assuming that $\mathrm{Norm}_{E/F}\zeta = 1$.

REMARK 5.3. In the case that E is a CM-field, γ is complex conjugation and $\zeta = -1$, this ideal was considered by Shimura [**22**], [**23**] and it is this case that is the motivation for this work.

We begin by developing some of the elementary properties of this ideal.

PROPOSITION 5.4. (1) *If ζ and η are two primitive n-th roots of unity, then*

$$S(\gamma,\zeta) = S(\gamma,\eta).$$

(2) *If $\tau \in \mathrm{Aut}(E/\mathbb{Q})$, then*

$$S(\gamma,\zeta)^\tau = S(\tau^{-1}\gamma\tau,\zeta^\tau).$$

PROOF. We can write $\eta = \zeta^i$ for some positive integer i satisfying $(i,n) = 1$. Then, we clearly have

$$\mathfrak{b}_0(\gamma,\zeta)^i \subseteq \mathfrak{b}_0(\gamma,\eta).$$

By symmetry, for some j, we have

$$\mathfrak{b}_0(\gamma,\eta)^j \subseteq \mathfrak{b}_0(\gamma,\zeta).$$

Thus, the radicals of $\mathfrak{b}(\gamma,\zeta)$ and $\mathfrak{b}(\gamma,\eta)$ are equal, proving the first assertion. If $x \in \mathfrak{b}_0(\gamma,\zeta)$, then

$$(x^\tau)^{\tau^{-1}\gamma\tau} = x^{\gamma\tau} = \zeta^\tau x^\tau$$

and so

$$x^\tau \in \mathfrak{b}_0(\tau^{-1}\gamma\tau,\zeta^\tau).$$

Conversely, if $x \in \mathfrak{b}_0(\tau^{-1}\gamma\tau,\zeta^\tau)$, then

$$(x^{\tau^{-1}})^\gamma = (x^{\tau^{-1}\gamma\tau})^{\tau^{-1}} = (\zeta^\tau x)^{\tau^{-1}} = \zeta x^{\tau^{-1}}.$$

Thus, $x^{\tau^{-1}} \in \mathfrak{b}_0(\gamma,\zeta)$. This shows that

$$\mathfrak{b}_0(\gamma,\zeta)^\tau = \mathfrak{b}_0(\tau^{-1}\gamma\tau,\zeta^\tau).$$

It follows that the corresponding equality also holds for $\mathfrak{b}$ and its radical S, proving the second assertion. □

PROPOSITION 5.5. *Let m be the order of γ, and r a positive integer with $(r,m) = 1$. Let ζ be a primitive n-th root of 1 in E. Then*

$$S(\gamma^r,\zeta) = S(\gamma,\zeta).$$

PROOF. Let i be the unique integer satisfying $0 \le i < n$ with $\zeta^\gamma = \zeta^i$. Let

$$j = 1 + i + i^2 + \cdots + i^{r-1}.$$

If for an $x \in \mathcal{O}_E$, we have

$$x^\gamma = \zeta x,$$

then

$$x^{\gamma^r} = \zeta^j x.$$

Thus

$$S(\gamma,\zeta) \subseteq S(\gamma^r,\zeta^j).$$

We claim that $(j,n) = 1$. If this were so, then

$$S(\gamma^r,\zeta^j) = S(\gamma^r,\zeta)$$

by Proposition 5.4. Thus

$$S(\gamma,\zeta) \subseteq S(\gamma^r,\zeta).$$

And then, by symmetry, it follows that equality holds. To complete the proof, it only remains to check that $(j,n) = 1$. For this, we note that

$$i^r - 1 = j(i-1)$$

and

$$(i^m - 1, i^r - 1) \;=\; i^{(m,r)} - 1 \;=\; i - 1.$$

But

$$1 + i + i^2 + \cdots + i^{m-1} \;\equiv\; 0 \pmod{n}$$

since $\gamma^m = 1$. Thus, (n, j) divides $((i^m - 1)/(i - 1), j) = 1$. □

DEFINITION 5.6. *We write $S_n(E/F)$ for $S(\gamma, \zeta)$ where γ is* any *generator of* $\mathrm{Gal}(E/F)$ *and ζ is* any *primitive n-th root of unity such that $Norm_{E/F}(\zeta) = 1$.*

By Propositions 5.4 and 5.5 this is well-defined.

In the next few results, we relate $S_n(E/F)$ to the different of E/F.

PROPOSITION 5.7. *Let $\alpha \in \mathcal{O}_E$ be such that $\alpha^\gamma \;=\; \zeta\alpha$. Then*

$$S_n(E/F) \;=\; \sqrt{S_n(F(\alpha)/F)\mathcal{O}_E}.$$

PROOF. If $\beta \in \mathcal{O}_E$ and $\beta^\gamma \;=\; \zeta\beta$, then $\beta\alpha^{-1} \in F$. Thus,

$$\mathfrak{b}_0(\gamma, \zeta) \;\subseteq\; F(\alpha)$$

and $\mathfrak{b}(\gamma, \zeta)$ is the lifting to E of the ideal in $F(\alpha)$ generated by $\mathfrak{b}_0(\gamma, \zeta)$. □

REMARK 5.8. The field $F(\alpha)$ above depends only on n and the extension E/F, but not on the choice of γ, ζ or the choice of α in $\mathfrak{b}_0(\gamma, \zeta)$. We shall denote this field E' for short, when no ambiguity can arise. In general, it is a proper subfield of E. For example, let $E \;=\; \mathbb{Q}(\zeta_5)$ and $F = \mathbb{Q}$, where ζ_5 is a primitive 5-th root of unity. Then $\mathrm{Gal}(E/F)$ is generated by the automorphism which maps ζ_5 to ζ_5^2. Now, $\sqrt{5} \in E$ and $(\sqrt{5})^\gamma \;=\; -\sqrt{5}$. Hence, $E' \;=\; \mathbb{Q}(\sqrt{5})$. In this case,

$$S_2(E/F) = \sqrt{S_2(\mathbb{Q}(\sqrt{5})/\mathbb{Q})\mathcal{O}_E} \;=\; \sqrt{\sqrt{5}\mathcal{O}_E} \;=\; (1 - \zeta_5)\mathcal{O}_E.$$

DEFINITION 5.9. (1) *If T is a number field and R is a finite extension of T, we denote by $\mathfrak{d}(R/T)$ the different of R/T. It is an ideal of $\mathcal{O}_R$.*

(2) *If I is an ideal of T and r is an integer, $I^{\text{prime to } r}$ denotes the ideal $\prod \mathfrak{p}^{v_\mathfrak{p}}$ where the product is over prime divisors of I which do not divide $r\mathcal{O}_E$ and $v_\mathfrak{p}$ is the exact exponent of $\mathfrak{p}$ in I. We shall also write I^{odd} when $r = 2$.*

(3) *If r is a positive integer, ζ_r denotes a primitive r-th root of unity.*

(4) *The degree $[E : F]$ will be denoted by m.*

PROPOSITION 5.10. *We have*

$$S_n(E/F) \;\supseteq\; \sqrt{\mathfrak{d}(E'/F)\mathcal{O}_E}.$$

PROOF. By Proposition 5.7, we may assume that $E \neq E'$. Let $\mathfrak{p}$ be a prime of E and suppose that $\mathfrak{p} \nmid \mathfrak{d}(E/F)$. We want to show that $\mathfrak{p} \nmid S_n(E/F)$. For this, it is enough to produce an element $y \in \mathcal{O}_E$ such that $y^\gamma \;=\; \zeta_n y$ and $y \notin \mathfrak{p}$. Since the inertia group at $\mathfrak{p}$ is trivial, the ring homomorphisms

$$\chi_i : \mathcal{O}_E \;\longrightarrow^{\gamma^i}\; \mathcal{O}_E \;\longrightarrow\; \mathcal{O}_E/\mathfrak{p}$$

are distinct. Suppose we are given $a_0, a_1, \cdots, a_{m-1} \in \mathcal{O}_E$, not all zero modulo $\mathfrak{p}$. By mimicing the proof of independence of characters, we can show that there is an $x \in \mathcal{O}_E$ such that

$$(a_0\chi_0 \;+\; a_1\chi_1 \;+\; \cdots \;+\; a_{m-1}\chi_{m-1})(x) \not\equiv 0 \pmod{\mathfrak{p}}.$$

In particular, there is an integer $x \in E$ such that

$$y \;=\; \sum_{j=0}^{m-1} \zeta_n^{-(1+\gamma+\gamma^2\;+\;\cdots\;+\;\gamma^{j-1})}\gamma^j x \;\not\equiv\; 0 \pmod{\mathfrak{p}}.$$

Here, the exponent of ζ_n is interpreted as 1 for $j = 0$. Now,

$$y^\gamma \;=\; \zeta_n y$$

as is readily verified. □

PROPOSITION 5.11. *We have*

$$S_n(E/F)^{\text{prime to } n} \;=\; \sqrt{\mathfrak{d}(E'/F)^{\text{prime to } n}}\mathcal{O}_E.$$

PROOF. By Proposition 5.10, it is enough to prove that

$$S_n(E/F)^{\text{prime to } n} \;\subseteq\; \sqrt{\mathfrak{d}(E'/F)^{\text{prime to } n}}\mathcal{O}_E.$$

By Proposition 5.7, we may suppose that $E = E'$. Choose an element

$$\alpha \in \mathfrak{b}_0(\gamma, \zeta_n).$$

Then α is an integral generator of E/F. Thus,

$$(\alpha - \alpha^\gamma)\cdots(\alpha - \alpha^{\gamma^{m-1}}) \;\in \mathfrak{d}(E/F).$$

This expression simplifies to

$$\alpha^{m-1}(1-\zeta_n)(1-\zeta_n^{1+\gamma})\cdots(1-\zeta_n^{1+\gamma+\gamma^2+\cdots+\gamma^{m-2}}) \;\in\; \mathfrak{d}(E//F).$$

Thus,

$$\alpha \in \sqrt{\mathfrak{d}(E/F)^{\text{prime to } n}}\mathcal{O}_E.$$

□

PROPOSITION 5.12. *We have*

$$S_n(E/F) \;\subseteq\; \sqrt{\mathfrak{d}(E'/F(\zeta_n))^{\text{prime to } m}}\mathcal{O}_E.$$

PROOF. let $e = [E' : F]$. Suppose we show that

$$S_n(E'/F) \;\subseteq\; \sqrt{\mathfrak{d}(E'/F(\zeta_n))^{\text{prime to } e}}\mathcal{O}_E.$$

Then, since e divides $m = [E : F]$, we get the stated result. Write

$$d = [F(\zeta_n) : F].$$

Then $\mathrm{Gal}(E'/F(\zeta_n))$ is generated by $\tau = \gamma^d$. In particular, for any

$$\alpha \in \mathfrak{b}_0(\gamma, \zeta_n),$$

we have

$$\alpha^\tau \;=\; \zeta_n^{1+\gamma+\cdots\;+\;\gamma^{d-1}}\alpha$$

and let us write the right hand side as $\eta\alpha$. It is easily checked that η is a primitive f-th root of unity, where $f = e/d$. Since α is an integral generator of $E'/F(\zeta_n)$, we have

$$(\alpha - \alpha^\tau)\cdots(\alpha - \alpha^{\tau^{f-1}}) \;\in \mathfrak{d}(E'/F(\zeta_n)).$$

But this element is equal to

$$\alpha^{f-1}(1-\eta)(1-\eta^{1+\tau})\cdots(1-\eta^{1+\tau+\tau^2+\cdots+\tau^{f-2}}).$$

Hence,

$$\alpha \in \sqrt{\mathfrak{d}(E'/F(\zeta_n))^{\text{prime to } e}}\mathcal{O}_E.$$

proving the result. □

REMARK 5.13. Without a further restriction, it is *not* true that

$$S_n(E/F) \subseteq \sqrt{\mathfrak{d}(E'/F)^{\text{prime to } m}}\mathcal{O}_E.$$

For example, take $E = \mathbb{Q}(\zeta_3)$, $F = \mathbb{Q}$ *and* γ *to be complex conjugation. Then* $S_3(E/F) = 1$ *since*

$$\zeta_3^\gamma = \zeta_3^2 = \zeta_3\zeta_3.$$

But

$$\mathfrak{d}(E/F)^{\text{prime to } 2} = \sqrt{-3}\mathcal{O}_E.$$

Next, we give some necessary conditions for $S_n(E/F)$ to be non-trivial.

PROPOSITION 5.14. *Let* i *be an integer such that* $0 < i < n$ *and* $\zeta^\gamma = \zeta^i$. *Let* $d = (i-1, n)$. *If* $(d, n/d) = 1$, *then*

$$S_n(E/F) = S_d(E/F).$$

In particular, if $d = 1$, $S_n(E/F) = 1$.

PROOF. Write $n = dn_0$. Then

$$\mathfrak{b}_0(\gamma, \zeta_n)^{n_0} \subseteq \mathfrak{b}_0(\gamma, \zeta_d),$$

and so

$$S_n(E/F) \subseteq S_d(E/F).$$

Similarly,

$$\mathfrak{b}_0(\gamma, \zeta_d)\mathfrak{b}_0(\gamma, \zeta_{n_0}) \subseteq \mathfrak{b}_0(\gamma, \zeta_{n/(n_0,d)}),$$

and so

$$S_d(E/F)S_{n_0}(E/F) \subseteq S_{n/(n_0,d)}(E/F) = S_n(E/F).$$

Now let $i - 1 = da$. Then

$$\zeta_n^\gamma = \zeta_n^{da}\zeta_n = \zeta_{n_0}^a\zeta_n.$$

Since $(a, n_0) = 1$, we have

$$\zeta_n \in S(\gamma, \zeta_{n_0}^a) = S(\gamma, \zeta_{n_0}) = S_{n_0}(E/F).$$

Hence, $S_{n_0}(E/F) = 1$ and the result follows. □

COROLLARY 5.15. *If* $(m, n) = 1$, *then* $S_n(E/F) = 1$.

PROOF. Let i be as in Proposition 5.14. Then

$$1 + i + \cdots + i^{m-1} \equiv 0 \pmod{n}.$$

But also

$$1 + i + \cdots + i^{m-1} \equiv m \pmod{(i-1)}.$$

Thus, $(i-1, n)$ divides (m, n). Now the result follows from Proposition 5.14. □

Finally, we determine the ideal $S_n(E/F)$ when $m = 2$.

PROPOSITION 5.16. *Suppose that* γ *has order* 2 *(that is,* $m = 2$*). Then, we have the following:*

(1) *If* n *is odd,* $S_n(E/F) = 1$.
(2) *If* $n = 2n_0$ *with* n_0 *odd, then* $S_n(E/F) = S_2(E/F)$.

(3) *If* $4|n$ *and* n *is not a power of* 2, *then* $S_n(E/F) = S_2(E/F) = 1$.
(4) *If* n *is a power of* 2 *but* $n \neq 2$, *then* $S_2(E/F) = 1$ *and* $S_n(E/F)$ *divides* $(1-\zeta_n)$.

PROOF. Let i be as above, that is $\zeta_n^\gamma = \zeta_n^i$. Since $\gamma^2 = 1$, we have

$$\zeta^{1+i} = 1.$$

If $i = 1$, then $n = 2$ and there is nothing to prove. So assume from now on that $i \equiv -1 \pmod{n}$ and $i \neq 1$. For the first assertion, we have $(m,n) = (2,n) = 1$ and so by Corollary 5.15, $S_n(E/F) = 1$. For the second assertion, we have $(i-1,n) = 2$. Then by Proposition 5.14, we have

$$S_n(E/F) = S_2(E/F).$$

For any even n, we see that

$$\mathfrak{b}(\gamma,\zeta_n)^{n/2} \subseteq \mathfrak{b}(\gamma,-1).$$

Thus,

$$S_n(E/F) \subseteq S_2(E/F).$$

On the other hand, $S_{n/2}(E/F) = 1$ since $\zeta_n \in \mathfrak{b}_0(\gamma,\zeta_{n/2}^{-1})$ from the relation

$$\zeta_n^\gamma = \zeta_n^{-2}\zeta_n = \zeta_{n/2}^{-1}\zeta_n.$$

Thus, if $4|n$, we find that $S_2(E/F) = 1$. In particular, this proves part of the third and part of the fourth assertions. To complete the proof, write $n_0 = n/4$ and consider

$$y = \zeta_n^{n_0-1}(1-\zeta_n).$$

Then

$$y^\gamma = (\zeta_n^{n_0-1})^i - \zeta_n^{n_0 i} = \zeta_n^{1-n_0} - \zeta_n^{-n_0} = -\zeta_n^{n_0+1} + \zeta_n^{n_0} = \zeta_n y.$$

Thus,

$$y \in \mathfrak{b}(\gamma,\zeta_n)$$

and

$$(1-\zeta_n) \in S_n(E/F).$$

If n is a power of 2 but not equal to 2, then $1-\zeta_n$ is a unit. This completes the proof of the third assertion and also proves the fourth assertion. □

COROLLARY 5.17. *Suppose that* γ *has order* 2. *Then*

$$S_n(E/F) = 1 \text{ unless } n = 2q \text{ with } q = 1 \text{ or a prime power.}$$

If q *is odd, then*

$$S_n(E/F) = S_2(E/F)$$

and

$$S_2(E/F)^{\text{odd}} = \mathfrak{d}(E/F)^{\text{odd}}$$

where $\mathfrak{d}(E/F)$ *is the different of* E/F. *Moreover, this different divides* $(1-\zeta_q)\mathcal{O}_E$.

6. Applications to Abelian varieties

We now use the results of the first three sections to construct abelian extensions of number fields. We preserve the notation of section 3. Thus (K, A, E) is a triple of type (T'). We suppose that $\Gamma(A) \neq \{1\}$. We have the following associated notation:

- Π : the set of places of K where A has bad reduction
- P: the trace field
- γ: a non-trivial element of $\Gamma(A)$
- R: the subfield of P fixed by γ
- χ: the twisting character corresponding to γ.

The following elementary lemma is crucial in what follows.

LEMMA 6.1. *If the order n of χ is odd, then $S_n(P/R) = \{1\}$.*

PROOF. If n is odd, then χ^2 also has order n. Let v be a prime of K, not dividing the conductor of χ or ϵ, such that $\chi^2(\mathrm{Frob}_v)$ is a primitive n-th root of unity. Then, from the relation

$$\epsilon^\gamma = \epsilon\chi^2,$$

we deduce that

$$\epsilon(\mathrm{Frob}_v) \in S(\gamma, \chi^2(\mathrm{Frob}_v)).$$

□

Now we introduce the rest of our notation and assumptions:

- n: the order of χ; we assume n is even, say $n = 2q$
- S: the ideal $(S_n(P/R)\mathcal{O}_E)^{\mathrm{odd}}$ which we assume is not the unit ideal
- $\mathfrak{p}$: a prime divisor of S
- $\mathbb{F}$: the residue field $\mathcal{O}_E/\mathfrak{p}$ of $\mathfrak{p}$
- p: the characteristic of $\mathbb{F}$; it is necessarily odd
- $\mathcal{L}$: an $\mathcal{O}_\mathfrak{p}[G_K]$-stable lattice in $V_\mathfrak{p}(A) = V_p(A) \otimes_{E\otimes\mathbb{Q}_p} E_\mathfrak{p}$, such lattices exist (see below).
- X: the two dimensional $\mathbb{F}$ vector space $\mathcal{L}/\mathfrak{p}\mathcal{L}$
- ω: the character χ^q
- M: the fixed field of ker ω
- M_ϵ: the fixed field of ker ϵ
- $\tilde{\omega}$, $\tilde{\epsilon}$, $\tilde{\chi_p}$: the reduction of the appropriate character modulo $\mathfrak{p}$.

REMARK 6.2. *As G_K is compact and $\rho_\mathfrak{p}$ is continuous, there always exist $\mathcal{O}_\mathfrak{p}[G_K]$ stable lattices.*

REMARK 6.3. *The fact that X is two dimensional over $\mathbb{F}$ follows from $(T1')$.*

Since E acts K-rationally on A, the action of G_K on X is $\mathbb{F}$-linear. Thus, we get a representation

$$\rho : G_K \longrightarrow \mathrm{Aut}_\mathbb{F}(X).$$

LEMMA 6.4. *Let $K(X)$ be the fixed field of ker ρ. Then $M \subseteq K(X)$ and $K(X)/M$ is an abelian extension.*

PROOF. If we let $G = G_K$, $H = \ker\chi$, and ρ as above, then it is clear that H is normal and G/H is finite and cyclic. Moreover, the Hypothesis 4.1 is satisfied. Indeed, let $g \in G$ be such that gH generates G/H. There are infinitely many primes v of K so that $g = \mathrm{Frob}_v$. For such v, we see that $\chi(\mathrm{Frob}_v)$ is a primitive n-th root of unity. The relation

$$a_v^\gamma = a_v\chi(\mathrm{Frob}_v)$$

implies that

$$a_v \equiv 0 \bmod \mathfrak{p}.$$

Thus, $\mathrm{Tr}\ \rho(g) = 0$. It follows from Lemma 4.3 that $\rho(J)$ is abelian, where $J = \ker(\omega)$. Furthermore, if $g \in G\backslash J$ then $\mathrm{Tr}\ \rho(g) = 0$. If $g \in \ker\rho$, then $\mathrm{Tr}\ \rho(g) = 2 \neq 0$ as p is odd. Therefore,

$$\ker\ \rho \subseteq J$$

and the Lemma follows. □

THEOREM 6.5. *The extension $K(X)/M$ is unramified outside of the primes of M which divide p or an element of Π. Suppose that at least one of the following holds:*

(i) *K has a real place ∞ and ω and ϵ agree on the inertia group at that place*
(ii) *K has a place $\mathfrak{q}$ over p of odd local degree $[K_\mathfrak{q} : \mathbb{Q}_p]$ such that ω and ϵ agree on an inertia group over $\mathfrak{q}$.*

Then,

(1) *$\mathrm{Gal}(K(X)/M)$ is contained in a Cartan subgroup C of $\mathrm{GL}_2(\mathbb{F})$.*
(2) *$K(X)/K$ is Galois and non-abelian. Its group is contained in the normalizer of C.*
(3) *If (i) above holds, then in fact ω and ϵ are unramified at the place ∞.*
(4) *If (ii) above holds, then $\omega(-1) = \epsilon(-1) = 1$ where -1 denotes an element of order 2 in any inertia group over $\mathfrak{q}$.*

PROOF. The $\mathfrak{p}$-adic representation of G_K on $V_\mathfrak{p}(A)$ is unramified outside of the primes of K which divide p or lie in Π. The first assertion follows from this.

Suppose there exists a character

$$\eta : G_K \longrightarrow \overline{\mathbb{F}}^\times$$

such that

$$\tilde{\omega}\eta^2 = \tilde{\epsilon}\tilde{\chi}_p. \tag{5}$$

It remains to show that either of the assumptions (i) or (ii) contradict this. Then (1) and (2) will follow from Proposition 4.4 and Corollary 4.7.

If (i) holds, denote by c the non-trivial element in the inertia group at ∞. Then Lemma 6.1 implies that

$$\tilde{\omega}(c) = \tilde{\omega}(c)\eta^2(c) = \tilde{\epsilon}(c)\tilde{\chi}_p(c) = \tilde{\omega}(c)\tilde{\chi}_p(c).$$

This implies that $\tilde{\chi}_p(c) = 1$, which is a contradiction.

If (ii) holds, let $I_\mathfrak{q}$ denote any inertial subgroup (of G_K) over $\mathfrak{q}$. Now ω and ϵ agree on on e, and hence on any, such subgroup. Let θ_{p-1} denote the "fundamental character of level 1", that is the composite map

$$I_\mathfrak{q} \simeq \mathcal{O}_\mathfrak{q}^\times \longrightarrow \mathbb{Z}_p^\times \longrightarrow \mathbb{F}_p^\times \hookrightarrow \mathbb{F}^\times,$$

where the first arrow is the Norm map and the second is reduction modulo p. By Serre ([**21**], Proposition 8), we have

$$\tilde{\chi}_p = \theta_{p-1}^e$$

where $e = e(K_{\mathfrak{q}}/\mathbb{Q}_p)$. Let $p^n = |\mathcal{O}_K/\mathfrak{q}|$. Then, by hypothesis, n and e are odd. Thus,

$$\chi_p(-1) = (-1)^{(1+p+\cdots+p^{n-1})e} = (-1)^{ne} = -1.$$

This contradicts (5) and proves (1) and (2).

Now let a be any element of order 2 in G_K. Suppose that $\tilde{\chi}_p(a) = -1$ and $\tilde{\omega}(a) = \tilde{\epsilon}(a)$. Then in fact

$$\tilde{\omega}(a) = \tilde{\epsilon}(a) = +1.$$

Indeed, otherwise, $a \notin \mathrm{Gal}(\overline{K}/M)$. Then by Corollary 4.6, we have

$$1 = \rho(a^2) = \tilde{\omega}(a)\tilde{\epsilon}(a)\tilde{\chi}_p(a) = -1.$$

This is a contradiction as $p \neq 2$. This proves (3) and (4). □

REMARK 6.6. It follows from the Brauer-Nesbitt theorem, that the semisimplification of $\rho|_J$ is independent of the choice of lattice. In particular, this applies if $\rho(J)$ is contained in a Cartan subgroup. Thus in this case, the extension $K(X)/M$ is independent of the choice of lattice. We call X the G_K-module associated to $\mathfrak{p}$.

7. The p-part of the conductor

In this section, we consider the ramification over p in the abelian extensions constructed in Section 6. We preserve the notation and hypotheses of that section. In addition, we shall assume throughout this section that the following hypothesis holds.

HYPOTHESIS 7.1. *$\mathrm{Gal}(K(X)/M)$ is contained in a Cartan subgroup C of $\mathrm{GL}_2(\mathbb{F})$ and $\mathrm{Gal}(K(X)/K)$ is non-abelian.*

Thus, we get two characters

$$\phi_1,\ \phi_2 :\ \mathrm{Gal}(K(X)/M) \longrightarrow \overline{\mathbb{F}}^{\times}$$

such that

$$\overline{\phi}|_{\mathrm{Gal}(\overline{K}/M)} = \phi_1 \oplus \phi_2.$$

PROPOSITION 7.2. *(cf. Shimura [**22**], Corollary 7.3.1) Let N be the compositum of M and M_ϵ. Then, $N(\zeta_p) \subsetneq N(X)$. If $([N : M], p-1) = 1$, then $M(\zeta_p) \subsetneq K(X)$.*

PROOF. Take an element $g \in \mathrm{Gal}(\overline{K}/N(X))$. Then $\rho(g) = 1$ and

$$\tilde{\chi}(g) = \det \rho(g) = 1.$$

This implies that $\zeta_p \in N(X)$. If $N(\zeta_p) = N(X)$, then $N(X)$ (and hence also $K(X)$) is abelian over K. This contradicts Hypothesis 7.1. Now, if $g \in \mathrm{Gal}(\overline{K}/K(X))$, then

$$\tilde{\epsilon}(g)\tilde{\chi}(g) = 1.$$

Again, by Hypothesis 7.1, $M(\zeta_p) \neq K(X)$. □

In the next result, we assume that $K = \mathbb{Q}$. We shall calculate the restriction of the characters ϕ_1, ϕ_2 to an inertia group over p using the methods of Ohta [**13**] and Momose [**12**]. By class field theory, ϕ_1 and ϕ_2 give two idèle class characters

$$\psi_1,\ \psi_2\ : M_{\mathbb{A}}^{\times} \longrightarrow \overline{\mathbb{F}}^{\times}.$$

Let v be a place of M unramified in $\mathbb{Q}(X)$.

REMARK 7.3. We assume that the reciprocity map

$$M_{\mathbb{A}}^{\times} \longrightarrow \mathrm{Gal}(\mathbb{Q}(X)/M)$$

sends a uniformizer at v to the *inverse* of a Frobenius element at v (that is, to the geometric Frobenius).

For each place w of M, we get characters

$$M_w^{\times} \hookrightarrow M_{\mathbb{A}}^{\times} \xrightarrow{\psi_i} \overline{\mathbb{F}}^{\times}$$

by composition. Let $\psi_{i,w}$ denote the restriction of the composed character to $\mathcal{O}_w^{\times}$. (Here, $\mathcal{O}_w$ is the w-adic completion of $\mathcal{O}_M$.)

We shall also make use of the "fundamental characters" $\{\theta_{p^d-1}\}$ (see Serre [**21**], Section 1). We view them as characters of $\mathcal{O}_w^{\times}$.

THEOREM 7.4. *Suppose that $K = \mathbb{Q}$, that A has good reduction at p and that p is unramified in M. If p splits in M, say*

$$p\mathcal{O}_M = ww'$$

then

$$(\psi_{1,w}, \psi_{1,w'}, \psi_{2,w}, \psi_{2,w'}) = (\theta_{p-1}, 1, 1, \theta_{p-1}) \text{ or } (1, \theta_{p-1}, \theta_{p-1}, 1).$$

If p remains prime in M, then

$$(\psi_{1,p}, \psi_{2,p}) = (\theta_{p^2-1}, \theta_{p^2-1}^{p}) \text{ or } (\theta_{p^2-1}^{p}, \theta_{p^2-1}).$$

PROOF. Firstly, we may assume that $\mathcal{O}_E \subseteq \mathrm{End}_{\mathbb{Q}}(A)$. Indeed, there is an Abelian variety $B_{/\mathbb{Q}}$ which is $\mathbb{Q}$-isogenous to A and such that $\mathcal{O}_E \subseteq \mathrm{End}_{\mathbb{Q}}(B)$ (cf. Shimura [**22**], p. 199). For such a B, we see that $(\mathbb{Q}, B, E)$ is again a triple of type (T'), and we can identify $\Gamma(B) = \Gamma(A)$. In the notation of section 6, B will correspond to a (possibly) different choice of $\mathcal{O}_B[G_{\mathbb{Q}}]$-lattice in $V_{\mathfrak{p}}(A)$. By our Hypothesis 7.1, and the remark at the end of Section 6, the extension $\mathbb{Q}(X)/M$ is independent of the choice of lattice. Thus, we may as well assume to start with that $\mathcal{O}_E \subseteq \mathrm{End}_{\mathbb{Q}}(A)$.

Now, let w be a place of M over p. Then, $A_{/M}$ has good reduction at w. Since $e(M_w/\mathbb{Q}_p) = 1$, by a theorem of Raynaud ([**15**], 3.3.4) (see also Ribet [**16**], Theorem 2.2.4), the characters $\psi_{i,w}$ can be written, without multiplicity, as a product of fundamental characters of level f where

$$f = [\mathcal{O}_M/w : \mathbb{F}_p].$$

Also, by Serre ([**21**], Proposition 8), we have

$$\tilde{\chi}_p|_{\mathcal{O}_w^{\times}} = \theta_{p-1}.$$

Consider first the case when p splits: $p\mathcal{O}_M = ww'$. Then

$$\psi_{i,w} = \theta_{p-1}^{a_i}, \ \psi_{i,w'} = \theta_{p-1}^{b_i}, \ i = 1,2 \tag{6}$$

where a_i, b_i are integers such that $0 \le a_i, b_i \le 1$. Then, from the relation

$$\psi_1\psi_2 = \tilde{\epsilon}\tilde{\chi}_p,$$

we deduce that

$$\theta_{p-1}^{a_1+a_2} = \psi_{1,w}\psi_{2,w} = \theta_{p-1}$$

since ϵ is unramified at p. Let r be any rational integer prime to p. We view it as an element of $\mathbb{Z}_p^\times = \mathcal{O}_w^\times$. Then

$$r^{a_1+a_2} \equiv r \pmod{p}.$$

Since

$$0 \leq a_1 + a_2 \leq 2$$

we must have

$$a_1 + a_2 = 1.$$

Similarly, $b_1 + b_2 = 1$.

We get another relation using Corollary 4.6. Let r be a rational prime $r \neq p$, congruent to 1 modulo a sufficiently high power of each prime dividing the non-p part of the conductors of ψ_1 and ψ_2. Let $e(r)$ be the idèle of M given by

$$e(r)_v = \begin{cases} r & \text{if } v|r \\ 1 & \text{otherwise.} \end{cases}$$

We view r itself as a constant idèle. Then

$$1 = \psi_i(r) = \psi_i(e(r))\psi_{i,w}(r)\psi_{i,w'}(r).$$

Now by Corollary 4.6,

$$r^{a_i+b_i} \equiv r \pmod{p}.$$

Thus,

$$a_1 + b_1 = a_2 + b_2 = 1.$$

This, together with the previous relation implies that

$$(a_1, b_1, a_2, b_2) = (1, 0, 0, 1) \text{ or } (0, 1, 1, 0).$$

Now suppose that p remains prime in M. Then, we can write

$$\psi_{i,p} = \theta_{p^2-1}^{a_i+b_ip}, \ a_i, b_i \in \mathbb{Z}, \ 0 \leq a_i, b_i \leq 1$$

for $i = 1, 2$. Proceeding in the same way as above, we find that

$$\theta_{p^2-1}^{(a_1+a_2)+(b_1+b_2)p} = \psi_{1,p}\psi_{2,p} = \theta_{p-1}.$$

Let $(\mathbb{Q}/\mathbb{Z})'$ denote the abelian group of rational numbers of order prime to p. By Serre ([**21**], Proposition 5), the map which associates to each ratonal number a/d with $(d, p) = 1$, the character θ_d^a gives an isomorphism of $(\mathbb{Q}/\mathbb{Z})'$ with the character group of the tame inertia group of the local field M_p. Thus,

$$\frac{(a_1 + a_2) + (b_1 + b_2)p}{p^2 - 1} = \frac{1}{p - 1} \pmod{\mathbb{Z}[1/p]}.$$

Since $0 \leq a_i, b_i \leq 1$, this implies that

$$a_1 + a_2 = b_1 + b_2 = 1.$$

We get another relation if we choose r as in the previous paragraph. Then,

$$1 = \psi_i(r) = \psi_i(e(r))\psi_{i,p}(r)$$

and by Proposition 4.6, we have

$$r^{a_i+b_i} \equiv r \pmod{p}.$$

Therefore,

$$a_1 + b_1 = a_2 + b_2 = 1.$$

Thus,

$$(a_1, b_1, a_2, b_2) = (1, 0, 0, 1) \text{ or } (0, 1, 1, 0).$$

□

COROLLARY 7.5. *If p splits in M, then $\mathbb{Q}(X)/M(\zeta_p)$ is unramified over p.*

8. Numerical examples

Given an Abelian variety A of type (T'), one may try to produce explicit abelian extensions using the above theory. To do this, one has to solve two problems. The first, is to decide whether A has a non-trivial twist or not. The second is to check whether the corresponding ideal is non-trivial or not.

If we consider the Abelian variety A_f attached to a non complex-multiplication newform f of weight 2, both problems can be solved, in principle, using the Selberg trace formula. Indeed, the trace formula can be used to explicitly calculate the field generated by the Fourier coefficients of f. However, this usually entails an excessive amount of computation, since the dimension of the space $S_2(\Gamma_0(N), \epsilon)$ of cusp forms of weight 2 and a given Nebentypus character ϵ, is $\gg N$.

Shimura has extensively studied the case when f has real Nebentypus and has produced many examples, by explicit calculation, in the case that f has prime level. In this section, we give a set of examples with *non-real* Nebentypus and prime level. In these examples, the two problems mentioned above can be solved "by inspection".

Let $f \in S_2(\Gamma_0(N), \epsilon)$ be a normalized non-complex multiplication newform of weight 2, level N and character ϵ. Suppose that $\epsilon \neq 1$. We consider the triple $(\mathbb{Q}, A_f, E_f)$, where E_f is the field generated by the Fourier coefficients of f. We have seen that E_f is a CM-field. Let c denote the canonical complex conjugation of E_f, and denote by E_f^+ the maximal totally real subfield of E_f. Then $c \in \Gamma(A_f)$ and $\chi_c = \epsilon^{-1}$. We retain the notation of our previous sections and set

$$S_f = S_n(E_f/E_f^+)^{\mathrm{odd}}$$

where n is the order of ϵ.

PROPOSITION 8.1. *Suppose that N is squarefree, and that ϵ has conductor N and even order $n = 2q$. Suppose that $S_f \neq 1$ and let $\mathfrak{p}$ be a prime divisor of S_f. Let $p\mathbb{Z} = \mathfrak{p} \cap \mathbb{Z}$ and X the associated $G_{\mathbb{Q}}$ module. Then the fixed field M of ϵ^{-q} is $\mathbb{Q}(\sqrt{N})$ and*

1. *$q = p^a$ with some integer $a \geq 0$.*
2. *$\mathrm{Gal}(\mathbb{Q}(X)/M)$ is contained in a* split *Cartan subgroup of* $\mathrm{GL}_2(\mathbb{F})$.
3. *$M \subseteq M(\zeta_p) \subseteq \mathbb{Q}(X)$.*
4. *The conductor of $\mathbb{Q}(X)/M$ is $p\infty_1\infty_2$, where ∞_1, ∞_2 are the infinite places of M.*

Suppose that p does not divide N. Let u denote the fundamental unit of M. Then

5. *$\mathbb{Q}(X)/M(\zeta_p)$ is everywhere unramified if p splits in M.*
6. *If* $\mathrm{Norm}\, u = +1$, *then* $u \equiv 1 \pmod{p\mathcal{O}_M}$.
7. *If* $\mathrm{Norm}\, u = -1$, *then* $u^2 \equiv 1 \pmod{p\mathcal{O}_M}$ *but* $u \not\equiv 1 \pmod{p\mathcal{O}_M}$.

REMARK 8.2. Conditions (1), (6) and (7) are necessary for $S_f \neq 1$. If f has real Nebentypus, then (1) is vacuous ($a = 0$) but (6) and (7) are still necessary. The extra condition (1) in the case of non-real Nebentypus makes it more difficult for S_f

to be non-trivial. For example, consider the case of prime level. Since $\epsilon(-1) = 1$, we must have $N \equiv 1 \pmod 4$. In this case, Norm $u = -1$ and so we have

$$u^2 - 1 = (\mathrm{Tr}\ u)u.$$

Thus, $u^2 - 1$ and Tr u have the same prime factors.Then (1) and (6), (7) would require that p divides $(N-1, \mathrm{Tr}\ u)$. For $N < 100$, this happens only for $N \equiv 1 \bmod 12$, i.e. $N = 13, 37, 61, 73, 97$. In all these cases $p = 3$. The first case when a prime larger than 3 divides $(N-1, \mathrm{Tr}\ u)$ is $N = 101$. In this case, $u = 10 + \sqrt{101}$ and $p = 5$. If we choose any character $\epsilon \pmod{101}$ of order 10, we find that $\dim S_2(\Gamma_0(N), \epsilon) = 8$. Without explicitly calculating the characteristic polynomial of some Hecke operators, I see no a-priori way of deciding whether there is an eigenform f in this space for which $S_f \neq 1$.

REMARK 8.3. Shimura ([**23**], p. 148) has conjectured that (for real Nebentypus and prime level), condition (6) (i.e. p divides Tr u) is also sufficient to ensure that $S_f \neq 1$ provided $p \geq 5$. Momose pointed out to me that Koike [**8**] has shown the following result. Denote by S_f^* the ideal of E_f generated by the Fourier coefficients a_n such that $a_n^c = -a_n$. Let N be squarefree with $N \equiv 1 \pmod 4$. Suppose that the norm of the fundamental unit u of $\mathbb{Q}(\sqrt{N})$ is -1. Let $p \geq 5$ be any prime which divides the trace of u. Then, there exists a cusp form $h \in S_2(\Gamma_0(N), (\frac{N}{\cdot}))$ which is an eigenform for the Hecke operators such that p divides the norm of S_h^*.

REMARK 8.4. We may conjecture that if $p \neq 2$ is a common divisor of $\phi(N)$ and the trace of u, then there exists an eigenform $h \in S_2(\Gamma_0(N), \epsilon)$ with ϵ a character of order $2p^a$ for some a, such that p divides the norm of S_h. We shall partly prove this in the case $p = 3$. By this method, we shall produce some abelian extensions which it does not seem possible to obtain using forms with real Nebentypus.

REMARK 8.5. The analogue of Koike's theorem for $p = 3$ is false. Indeed, consider $N = 37$. In this case, the space $S_2(\Gamma_0(37), (\frac{37}{\cdot}))$ is 2 dimensional and is spanned by an eigenform f and its conjugate. We have $E_f = \mathbb{Q}(\sqrt{-1})$ and $S_f^{*\mathrm{odd}} = 1$ although 3 divides the trace of the fundamental unit $6 + \sqrt{37}$.

8.1. Proof of Proposition 8.1.

PROOF. The first assertion (1) follows from Corollary 5.17. Next, since condition (1) of Theorem 6.5 is satisfied, $\mathrm{Gal}(\mathbb{Q}(X)/M)$ is contained in a Cartan subgroup C (say) of $\mathrm{GL}_2(\mathbb{F})$. Now assertion (3) follows from Proposition 7.2.

To see that we can take C to be split, let w_N denote the Atkin-Lehner involution of level N. It induces an automorphism of A_f which is defined over M_ϵ, and which is of order 2. Furthermore, for all $g \in G_{\mathbb{Q}}$, we have

$$w_N^g = \bar{\epsilon}(g) w_N \tag{7}$$

and w_N commutes with the action of E_f^+. Thus, if we let

$$X_\pm = (1 \pm w_N)X$$

then we see that

(a) $X = X_+ \oplus X_-$
(b) there exists a $\delta \in G_{\mathbb{Q}}$ such that $X_\pm^\delta = X_\mp$.
(c) $X_\pm$ is an $\mathbb{F}$-vector space of dimension 1
(d) the decomposition of (a) is rational over M.

Indeed, (a) is clear. For (b), we note that since ϵ has even order, there exists a $\delta \in G_{\mathbb{Q}}$ such that $\epsilon(\delta) = -1$. Then, we have from (7)

$$w_N^{\delta} \;=\; -w_N$$

and (b) follows. In particular, both $X_{\pm}$ are non-zero.

Let $\mathcal{O}_+$ denote the ring of integers of E_f^+, and let $\mathfrak{q} = \mathfrak{p} \cap \mathcal{O}_+$. Then $\mathcal{O}_+$ acts on $X_{\pm}$ and this action factors through $\mathfrak{q}$. This gives $X_{\pm}$ the structure of an $\mathcal{O}_+/\mathfrak{q} \simeq \mathbb{F}$ vector space. Moreover, from (a) and (b) it follows that it is of dimension 1. Finally, (d) follows from (7) and the fact that $\zeta_q \equiv 1 \pmod{\mathfrak{p}}$ (cf. (1) and Corollary 5.17). Putting all of these observations together, we see that $\mathrm{Gal}(\mathbb{Q}(X)/M)$ is contained in the split Cartan subgroup

$$C \;=\; \mathrm{Aut}_{\mathbb{F}}(X_+) \;\times\; \mathrm{Aut}_{\mathbb{F}}(X_-).$$

This proves (2).

By Proposition 4.8, we know that the ramification over p is tame, and so (4) will follow from (3) if we show that $\mathbb{Q}(X)/M$ is unramified over $N/(N,p)$. To show this, we consider the diagram of fields below.

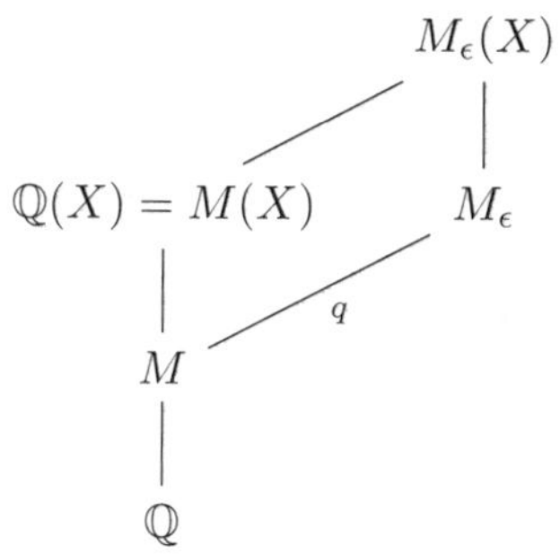

We have $[M_\epsilon : M] = q$ whereas $[\mathbb{Q}(X) : M]$ divides $(q-1)^2$ and is, in particular, prime to q. Therefore, $\mathbb{Q}(X) \cap M_\epsilon = M$. On the other hand, the hypotheses on N and ϵ ensure that A_f acquires everywhere good reduction over M (by Deligne and Rapaport [**5**]). Thus, $M_\epsilon(X)/M_\epsilon$ is unramified at the primes dividing $N/(N,p)$. Combining these two observations, we deduce that $\mathbb{Q}(X)/M$ is unramified at primes dividing $N/(N,p)$ also. Thus (4) is proved. We note that (5) is just (4) combined with Corollary 7.5. To prove (6), let α be any totally positive unit of M. Let ϕ_1, ϕ_2 be the characters

$$\mathrm{Gal}(\mathbb{Q}(X)/M) \;\longrightarrow\; \overline{\mathbb{F}}^{\times}$$

obtained from the Cartan subgroup C. Let ψ_1, ψ_2 be the associated Hecke characters. Then

$$\psi_1(\alpha) \;=\; \psi_2(\alpha) \;=\; 1.$$

Suppose p splits in M, say $p\mathcal{O}_M \;=\; ww'$. By Theorem 7.4, we may assume that

$$(\psi_{1,w}, \psi_{1,w'}, \psi_{2,w}, \psi_{2,w'}) \;=\; (\theta_{p-1}, 1, 1, \theta_{p-1}).$$

Then

$$1 \;=\; \psi_1(\alpha) \;=\; \psi_{1,w}(\alpha) \;=\; \alpha \pmod{w}.$$

Similarly,

$$\alpha \pmod{w'} \;=\; 1.$$

Thus, $\alpha \equiv 1 \pmod{p\mathcal{O}_M}$. If p does not split in M, we find in a similar manner that $\alpha \equiv 1 \pmod{p\mathcal{O}_M}$.

Now if Norm $u = +1$, then u is totally positive so $u \equiv 1 \pmod p$. On the other hand, if Norm $u = -1$, then u^2 is totally positive, so $u^2 \equiv 1 \pmod p$. Moreover, $u \not\equiv 1 \pmod p$ for otherwise we would have

$$-1 = uu' \equiv 1 \pmod p$$

contradicting the fact that p is odd. (Here, u' denotes the conjugate of u.) This proves the Proposition. □

8.2. Numerical Examples. Now we find a set of examples which satisfy the hypotheses of Proposition 8.1.

LEMMA 8.6. *Let $M = \mathbb{Q}(\sqrt{N})$ with a positive squarefree integer $N \equiv 1 \pmod 4$. Suppose that* $\mathrm{Norm}_{M/\mathbb{Q}}u = -1$ *where u is the fundamental unit of M. Then* $3|\mathrm{Tr}_{M/\mathbb{Q}}u$ *if and only if* $N \equiv 1 \pmod 3$.

PROOF. This is due to Shimura ([**23**], p. 186). □

LEMMA 8.7. *Let N be a positive integer and ϵ an even character modulo N of conductor r. Then* $\dim S_2(\Gamma_0(N), \epsilon)$ *is equal to*

$$\frac{1}{12}N\prod_{p|N}\left(1+\frac{1}{p}\right) - \frac{1}{2}\sum_{\substack{d|N\\(d,N/d)|(N/r)}}\phi((d,N/d)) - \frac{1}{4}{\sum}'\,\epsilon(x) - \frac{1}{3}{\sum}''\,\epsilon(x)$$

where the first sum is over $x \pmod N$ satisfying $x^2+1 \equiv 0 \pmod N$ and the second sum is over $x \pmod N$ satisfying $x^2+x+1 \equiv 0 \pmod N$.

PROOF. See Cohen and Oesterlé [**3**]. □

LEMMA 8.8. *Let N be a prime* $\equiv 13 \bmod 24$. *Let a be the largest power of 3 that divides $N-1$. Let ϵ be the even character modulo N of order $2\cdot 3^a$. Then* $\dim S_2(\Gamma_0(N), \epsilon)$ *is $(N-1)/12$. In particular, it is odd.*

PROOF. Let g be a primitive root modulo N and ζ a primitive 3^a-th root of unity, such that $\epsilon(g) = -\zeta$. If $x^2+1 \equiv 0 \pmod N$, then

$$x = \pm g^{(N-1)/4}.$$

Thus,

$$\epsilon(x) = (-1)^{(N-1)/4} = -1.$$

Similarly, if $x^2+x+1 \equiv 0 \pmod N$, then

$$x = g^{(N-1)/3} \text{ or } g^{2(N-1)/3}.$$

Thus,

$$\epsilon(x) = \zeta_3 \text{ or } \zeta_3^2$$

where ζ_3 is a primitive cube root of 1. Thus, by Lemma 8.7,

$$\dim S_2(\Gamma_0(N), \epsilon) = \frac{1}{12}(N+1) - 1 + \frac{1}{2} + \frac{1}{3} = (N-1)/12.$$

□

LEMMA 8.9. *Let N be an arbitrary positive integer and ϵ an even character modulo N. Suppose that the dimension of $S_2(\Gamma_0(N), \epsilon)$ is odd. Then, there exists an eigenform $f \in S_2(\Gamma_0(N), \epsilon)$ such that the degree $[E_f : \mathbb{Q}(\epsilon)]$ is odd.*

PROOF. Choose a basis X of $S_2(\Gamma_0(N), \epsilon)$ consisting of eigenforms for all the Hecke operators T_n with $(n, N) = 1$. Choose a maximal subset Y of X such that no two elements of Y are Galois conjugates of one another. Then

$$\dim S_2(\Gamma_0(N), \epsilon) \;=\; \sum_{h \in Y} [E_h : \mathbb{Q}(\epsilon)].$$

Indeed, viewing the E_h as subfields of $\mathbb{C}$, we have

$$X \;\subseteq\; Y' \;=\; \{h^\sigma \mid h \in Y,\ \sigma \;\in \operatorname{Aut}(\mathbb{C}/\mathbb{Q}(\epsilon))\}$$

and the forms in Y' are independent. Now, since $S_2(\Gamma_0(N), \epsilon)$ has odd dimension, one at least of the $[E_h : \mathbb{Q}(\epsilon)]$ must be odd. This proves the lemma. □

THEOREM 8.10. *Let N be a prime $\equiv 13 \bmod 24$ and let a be the highest power of 3 that divides $N-1$. Let ϵ be the even character $\pmod N$ of order $2 \cdot 3^a$. Then, there exists a normalized eigenform $f \in S_2(\Gamma_0(N), \epsilon)$ and a prime $\mathfrak{p}$ of E_f which lies over 3, such that $\mathfrak{p}$ divides S_f. Let X be the associated $G_{\mathbb{Q}}$-module, and $M = \mathbb{Q}(\sqrt{N})$. Then, $\mathbb{Q}(X)$ contains the unique class field of conductor $3\infty_1\infty_2$ and degree 4 over M.*

PROOF. By Lemma 8.8, the dimension of $S_2(\Gamma_0(N), \epsilon)$ is odd. Then by Lemma 8.9, we can choose a normalized eigenform f such that the degree $[E_f : \mathbb{Q}(\epsilon)]$ is odd. Let $n = 2q$ with $q = 3^a$. Then, by Corollary 5.17, we have

$$S_f \;=\; \mathfrak{d}(E_f/E_f^+)^{\mathrm{odd}} \;\supseteq\; (1 - \zeta_q)\mathcal{O}_{E_f}.$$

Now by the diagram of fields below, it follows that $\mathfrak{d}(E_f/E_f^+)^{\mathrm{odd}} \;\neq\; 1$. We thus may apply Proposition 8.1 to deduce that $\mathbb{Q}(X)/M$ is abelian of conductor $3\infty_1\infty_2$. Let h denote the class number of M.

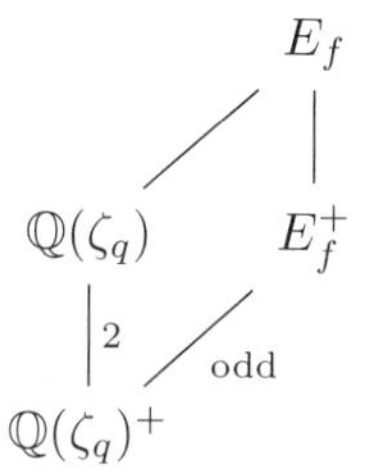

We claim that the $3\infty_1\infty_2$ ray class field H of M is of degree $4h$ over M. Indeed, let u be the fundamental unit of M. Since N is a prime $\equiv 1 \pmod 4$, we have $Norm\ u \;=\; -1$. By Proposition 8.1 (4), we deduce that $u^2 \;\equiv 1 \pmod 3$ but $u \not\equiv 1 \pmod 3$, Since $N \equiv 1 \pmod 3$, it follows that 3 splits in M. Putting these two facts together, we see that H has degree $2 \cdot h \cdot (3-1)^2/2 \;=\; 4h$ over M (cf. Lang ([**9**], p. 127)).

Since N is prime, h is odd. Therefore, H contains a unique field H' (say) of degree 4 over M and of conductor $3\infty_1\infty_2$. To show that $\mathbb{Q}(X) \supseteq H'$ it is enough to show that 4 divides the degree $[\mathbb{Q}(X) : M]$. Since 3 splits in M, this follows, for example, from Theorem 7.4. □

COROLLARY 8.11. *If the class number $h(N)$ of $\mathbb{Q}(\sqrt{N})$ is prime to* Norm $\mathfrak{p}$ -1, *then $\mathbb{Q}(X)/M$ is the unique class field of conductor $3\infty_1\infty_2$ and degree 4. In particular, this is the case if $h(N) = 1$.*

PROOF. Since the order of $\mathrm{Gal}(\mathbb{Q}(X)/M)$ divides $(Norm\ \mathfrak{p}\ -1)^2$, and also $4h$, it follows that this order is in fact 4. The rest of the assertion follows from Theorem 8.10. □

REMARK 8.12. From the tables in Cohen and Roblot [**2**], we see that if D is a prime such that $D \equiv 1 \pmod 4$ and $D < 573$, then $h(D) \leq 5$ and is odd. It is easily checked that 3 and 5 are coprime to any number of the form $3^b - 1$ with b not divisible by 4. Thus, if $N \equiv 13 \pmod{24}$ is a prime with $N < 573$, then $\mathbb{Q}(X)/M$ is the unique class field of conductor $3\infty_1\infty_2$ and degree 4.

9. Remarks on other work

Brown and Ghate [**1**] consider a quadratic field $F = \mathbb{Q}(\sqrt{D})$. Let χ_D denote the associated quadratic Dirichlet character modulo $|D|$. Consider a cusp form f of some weight $k \geq 2$ for the congruence subgroup $\Gamma_0(|D|)$ with Nebentypus χ_D and assume that f is a normalized Hecke eigenform. Factor $D = D_1 D_2$ and assume that $F_1 = \mathbb{Q}(\sqrt{D_1})$ is a real quadratic field. Assume that f has a twist by (γ, χ_{D_1}) for some automorphism γ of the field E_f of Fourier coefficients of f. Let $\mathfrak{p}$ be a prime that divides the different of E_f/E_f^{γ} and p the rational prime below it. Assume that $(p, 2D) = 1$ and that f is ordinary at $\mathfrak{p}$. Assume also that the mod $\mathfrak{p}$ representation associated to f is absolutely irreducible. Then, they use Shimura's method to construct some abelian extensions of F_1.

The case $F_1 = F$ is considered by Hida [**7**]. However, Hida allows forms of any weight $k \geq 2$ and he restricts his attention to identifying and characterizing 'dihedral primes' (that is primes such as $\mathfrak{p}$ above at which the mod $\mathfrak{p}$ representation is dihedral) and does not explicitly discuss the construction of class fields. Of course, for weight larger than 2, one does not have a corresponding Abelian variety so it is not at first clear how one might generate class fields. However, Brown and Ghate remark that one can work in a Hida family which contains a form of weight 2 congruent to the form of higher weight, and this form can then be used to generate some class fields.

The work of Darmon and Green [**4**] takes a completely different approach to the construction of class fields of real quadratic fields. Their attempt is to generalize the construction of Heegner points and they propose a conjectural construction of Stark-Heegner points. The construction is local, and it is conjectured that the points are actually global and defined over a ring class field of a real quadratic field. This theory is very intriguing given that there are very few general ways of producing rational points on curves or higher dimensional varieties.

References

[1] Alexander F. Brown and Eknath P. Ghate, *Dihedral congruence primes and class fields of real quadratic fields*, J. Number Theory **95** (2002), no. 1, 14–37, DOI 10.1006/jnth.2001.2753. MR1916078

[2] Henri Cohen and Xavier-François Roblot, *Computing the Hilbert class field of real quadratic fields*, Math. Comp. **69** (2000), no. 231, 1229–1244, DOI 10.1090/S0025-5718-99-01111-4. MR1651747

[3] Henri Cohen and Joseph Oesterlé, *Dimensions des espaces de formes modulaires* (French), Modular functions of one variable, VI (Proc. Second Internat. Conf., Univ. Bonn, Bonn, 1976), Springer, Berlin, 1977, pp. 69–78. Lecture Notes in Math., Vol. 627. MR0472703

[4] Henri Darmon and Peter Green, *Elliptic curves and class fields of real quadratic fields: algorithms and evidence*, Experiment. Math. **11** (2002), no. 1, 37–55. MR1960299

[5] Pierre Deligne, *Courbes elliptiques: formulaire d'après J. Tate* (French), Modular functions of one variable, IV (Proc. Internat. Summer School, Univ. Antwerp, Antwerp, 1972), Springer, Berlin, 1975, pp. 53–73. Lecture Notes in Math., Vol. 476. MR0387292

[6] Koji Doi and Masatoshi Yamauchi, *On the Hecke operators for $\Gamma_0(N)$ and class fields over quadratic number fields*, J. Math. Soc. Japan **25** (1973), 629–643, DOI 10.2969/jmsj/02540629. MR0344226

[7] Haruzo Hida, *Global quadratic units and Hecke algebras*, Doc. Math. **3** (1998), 273–284. MR1650571

[8] Masao Koike, *Congruences between cusp forms and linear representations of the Galois group*, Algebraic number theory (Kyoto Internat. Sympos., Res. Inst. Math. Sci., Univ. Kyoto, Kyoto, 1976), Japan Soc. Promotion Sci., Tokyo, 1977, pp. 109–116. MR0466024

[9] Serge Lang, *Algebraic number theory*, Addison-Wesley Publishing Co., Inc., Reading, Mass.-London-Don Mills, Ont., 1970. MR0282947

[10] Vijayakumar Pedaprolu Murty, *ALGEBRAIC CYCLES ON ABELIAN VARIETIES*, ProQuest LLC, Ann Arbor, MI, 1982. Thesis (Ph.D.)–Harvard University. MR2632244

[11] Fumiyuki Momose, *On the l-adic representations attached to modular forms*, J. Fac. Sci. Univ. Tokyo Sect. IA Math. **28** (1981), no. 1, 89–109. MR617867

[12] Fumiyuki Momose, *Galois action on some ideal section points of the abelian variety associated with a modular form and its application*, Nagoya Math. J. **91** (1983), 19–36. MR716785

[13] Masami Ohta, *On l-adic representations of Galois groups obtained from certain two-dimensional abelian varieties*, J. Fac. Sci. Univ. Tokyo Sect. IA Math. **21** (1974), 299–308. MR0419368

[14] Masami Ohta, *The representation of Galois group attached to certain finite group schemes, and its application to Shimura's theory*, Algebraic number theory (Kyoto Internat. Sympos., Res. Inst. Math. Sci., Univ. Kyoto, Kyoto, 1976), Japan Soc. Promotion Sci., Tokyo, 1977, pp. 149–156. MR0457445

[15] Michel Raynaud, *Schémas en groupes de type $(p,\dots,p)$* (French), Bull. Soc. Math. France **102** (1974), 241–280. MR0419467

[16] Kenneth A. Ribet, *Galois action on division points of Abelian varieties with real multiplications*, Amer. J. Math. **98** (1976), no. 3, 751–804, DOI 10.2307/2373815. MR0457455

[17] Kenneth A. Ribet, *On l-adic representations attached to modular forms*, Invent. Math. **28** (1975), 245–275, DOI 10.1007/BF01425561. MR0419358

[18] Kenneth A. Ribet, *Twists of modular forms and endomorphisms of abelian varieties*, Math. Ann. **253** (1980), no. 1, 43–62, DOI 10.1007/BF01457819. MR594532

[19] Kenneth A. Ribet, *Endomorphism algebras of abelian varieties attached to newforms of weight* 2, Seminar on Number Theory, Paris 1979–80, Progr. Math., vol. 12, Birkhäuser, Boston, Mass., 1981, pp. 263–276. MR633903

[20] Jean-Pierre Serre, *Local fields*, Graduate Texts in Mathematics, vol. 67, Springer-Verlag, New York-Berlin, 1979. Translated from the French by Marvin Jay Greenberg. MR554237

[21] Jean-Pierre Serre, *Propriétés galoisiennes des points d'ordre fini des courbes elliptiques* (French), Invent. Math. **15** (1972), no. 4, 259–331, DOI 10.1007/BF01405086. MR0387283

[22] Goro Shimura, *Introduction to the arithmetic theory of automorphic functions*, Publications of the Mathematical Society of Japan, No. 11. Iwanami Shoten, Publishers, Tokyo; Princeton University Press, Princeton, N.J., 1971. Kanô Memorial Lectures, No. 1. MR0314766

[23] Goro Shimura, *Class fields over real quadratic fields and Hecke operators*, Ann. of Math. (2) **95** (1972), 130–190, DOI 10.2307/1970859. MR0314801

Department of Mathematics, University of Toronto, 40 St. George Street, Toronto, Canada M5S 2E4

Email address: `murty@math.toronto.edu`

Contemporary Mathematics
Volume **701**, 2018
http://dx.doi.org/10.1090/conm/701/14148

Transcendental numbers and special values of Dirichlet series

M. Ram Murty

To the memory of F. Momose, with respect and admiration

Abstract. We give a short survey of results and conjectures regarding special values of certain Dirichlet series

Contents

1. Introduction

In this paper, we are concerned with special values of Dirichlet series of the form

$$\sum_{n=1}^{\infty} \frac{a_n}{n^s},$$

assuming that the series is convergent. Most of the time, these Dirichlet series will be zeta and L-functions that arise out of number theory. Sometimes, they

2010 *Mathematics Subject Classification.* Primary 1102, 11M06, 11M32, 11M41.
Key words and phrases. special values, L-functions, Dirichlet series.
Research partially supported by an NSERC Discovery grant.

will be general series arising from other considerations (as we will see below as in the case of the Chowla and Erdös problems or the Chowla-Milnor conjectures). Several different perspectives are available to study the former case of zeta functions. The cohomological approach of Beilinson and his predictions regarding the special values in terms of generalized regulators is an active area of current research and we refer the reader to several good expositions such as Soulé [**51**] and Ramakrishnan [**44**]. What we want to highlight in this exposition is a more analytic and classical approach.

Beginning with Euler's work on the explicit evaluation of the Riemann zeta function at even arguments and the role the cotangent function plays in this evaluation, we amplify the role of special functions that emerge in our understanding of these special values. The logarithm function, the dilogarithm function and the polylogarithm functions assume a central place in this study, as well as the gamma function, the digamma function and the polygamma functions. Special values of modular forms are also closely connected to special values of certain Dirichlet series and L-functions. Other zeta functions, such as the Hurwitz zeta function, the Lerch zeta function, the multiple zeta functions and the multiple Hurwitz zeta functions also make an appearance.

Often the evaluation of these Dirichlet series and the determination of their algebraic or transcendental nature require the marriage of several disparate branches of number theory. The former is an analytic-arithmetic viewpoint, and the latter being transcendental number theory. It is hoped that this survey will serve to highlight the beauty of all these viewpoints and propose some new questions for further research. The reader may find additional exposition in the monograph [**36**].

2. The discovery of transcendental numbers

An *algebraic number* is a complex number which is a root of a non-trivial polynomial with integer coefficients. It is a beautiful theorem of algebra that the totality of algebraic numbers forms a field. If a complex number is not algebraic, we call it *transcendental*. The notion of a transcendental number may be traced back to Euler but the first use of the term "transcendental" occurs in a 1682 paper of Leibnitz where he showed that $\sin x$ is not an algebraic function of x. In 1844, using the theory of continued fractions, Joseph Liouville proved that transcendental numbers exist. It was not until 1851 that he realized that there was a simple way to construct some examples. His construction was based on the following elementary idea which had a profound impact on the development of transcendental number theory.

Liouville observed that if α is algebraic then we may consider a polynomial $f(x) \in \mathbb{Z}[x]$ of minimal degree for which it is a root. This polynomial is unique up to an integral factor. The *degree of* α is defined to be the degree of f. Thus, the algebraic numbers which have degree 1 are precisely the rational numbers. If α has degree ≥ 2, and $f(x)$ is an irreducible polynomial with integer coefficients such that $f(\alpha) = 0$, Liouville proved that there is a positive constant C (depending on f) such that for any rational number p/q, we have

$$\left|\alpha - \frac{p}{q}\right| \geq \frac{C}{q^n}.$$

To prove this, let us note that if $|\alpha - p/q| > 1$, we are done since $1 \geq 1/q^n$ and we can choose $C = 1$. So, let us suppose that $|\alpha - p/q| \leq 1$. If $\alpha = \alpha_1, ..., \alpha_n$ are the roots of f, and A is the leading coefficient of $f(x)$, then,

$$A(\alpha - p/q)(\alpha_2 - p/q) \cdots (\alpha_n - p/q) = f(p/q).$$

Observe that for any rational number p/q,

$$|f(\alpha) - f(p/q)| = |f(p/q)| \geq 1/q^n$$

because the numerator is a non-zero integer. Since

$$\left|\alpha_i - \frac{p}{q}\right| \leq |\alpha_i - \alpha_1| + \left|\alpha_1 - \frac{p}{q}\right| \leq |\alpha_i - \alpha_1| + 1,$$

we see immediately on choosing

$$M = |A| \prod_{i=2}^{n} (|\alpha_i - \alpha| + 1),$$

that

$$\left|\alpha - \frac{p}{q}\right| \geq \frac{1}{Mq^n}.$$

Setting $C = \max(1, 1/M)$ gives the result.

What Liouville's theorem says is that algebraic numbers are not too well approximable by rational numbers. Consequently, if a number is too well approximable by rational numbers, in the above sense, it must be transcendental. Applying his theorem to numbers of the form

$$\sum_{n=0}^{\infty} \frac{1}{2^{n!}},$$

Liouville deduced that these must be transcendental numbers since partial sums of these numbers are rational numbers that approximate the sum too well for the sum to be algebraic. These were perhaps the first class of infinite series shown to be transcendental.

Liouville's numbers were exotic constructions. It wasn't clear at that time whether numbers like e or π were transcendental. This had already been conjectured by Johann Heinrich Lambert in his 1761 paper where he proved that π is irrational. In 1873, Charles Hermite proved that e is transcendental. A year later, in 1874, Cantor gave his famous diagonal argument to show that transcendental numbers are uncountable. It was as late as 1882 when Ferdinand von Lindemann proved that π is transcendental using methods initiated by Hermite. In his paper, Lindemann stated many results without proof. For example, he stated that if α is algebraic and non-zero, then e^α is transcendental. Since $e^{\pi i} = -1$, we deduce that π is transcendental. This more general result was later proved rigorously by Hermite. Lindemann also stated that if $\alpha_1, ..., \alpha_n$ are algebraic numbers which are linearly independent over $\mathbb{Q}$, then $e^{\alpha_1}, ..., e^{\alpha_n}$ are algebraically independent. The proof of this statement was given by Weierstrass. (For some reason, Lindemann did not give much attention to the line of research he initiated. Instead, he turned his gaze to Fermat's Last Theorem and published a book on it with a general "proof", which was unfortunately wrong.)

Perhaps the oldest explicit evaluation of an infinite series was first carried out by Madhava in 14th century India, when he proved that

$$\frac{\pi}{4} = 1 - \frac{1}{3} + \frac{1}{5} - \frac{1}{7} + \cdots.$$

This evaluation is often attributed to Leibnitz, but there is now well-documented evidence to the contrary. Recent research has uncovered the contributions of the Kerala school of mathematicians, led by Madhava. Their writings show that much of what we would now call "pre-calculus" was well developed in 14th century India by the Kerala school and we refer the reader to [**26**] for an account of this fascinating history.

Several centuries later, the famous Basel problem asking for an explicit evaluation of

$$1 + \frac{1}{4} + \frac{1}{9} + \frac{1}{16} + \cdots$$

was solved by Euler in 1735 almost a century after Pietro Mengoli proposed the problem in 1644. Euler went on to show, after a decade of work, that

$$\sum_{n=1}^{\infty} \frac{1}{n^{2k}}$$

is a rational multiple of π^{2k}. The underlying reason for the explicit evaluation of both the Madhava-Leibnitz series and Euler's determination of the special values of the Riemann zeta function at even arguments is due to the properties of the cotangent function, though this is not overtly clear from the study of their works.

The Madhava-Leibnitz formula is an explicit evaluation of a Dirichlet L-series. In fact, it is $L(1, \chi)$ with χ being the non-trivial Dirichlet character (mod 4). From this, together with Lindemann's result that π is transcendental and Euler's explicit evaluation of special values of the Riemann zeta function, there emerges a new theme of determining the nature of special values of general zeta and L-functions and more generally special values of Dirichlet series. This determination requires a two-fold (perhaps three-fold) development of number theory. On the one hand, one needs a general method to evaluate these series and this is often difficult. It involves an analytic and arithmetic study of special functions. One then needs results from transcendental number theory regarding the precise nature of the special values. These special values often factor as a product of an algebraic number and a transcendental number (a "period" in modern parlance), and the algebraic number is often pregnant with arithmetic meaning. These determinations stimulate a three-fold development of number theory. At the moment, some of these strands are developing faster than others. The slowest seems to be transcendental number theory where it is difficult to determine whether a given number (or "period") is transcendental. Such results require tremendous advances in our understanding the nature of special functions. This is best highlighted by relating the story of some of the Hilbert problems.

In his famous list of problems at the International Congress of Mathematicians held in 1900, Hilbert asked: if α is algebraic $\neq 0, 1$ and β is any irrational algebraic number, then is α^{β} transcendental? If true, it would imply that e^{π} is transcendental since $e^{\pi} = (-1)^{-i}$. Apparently, Hilbert considered this problem to be notoriously difficult. He predicted that the Riemann hypothesis and Fermat's Last Theorem would be solved before this problem (see p. 84 of [**49**]). History proved otherwise! It is dangerous to make predictions!

In 1934, Gelfond and Schneider independently resolved this problem. This represented a major advance in the field. Thus, not only are e and π transcendental numbers but so is e^π by the Gelfond-Schneider theorem. But what about $e+\pi$ or πe? Are these transcendental? (It is easy to see that at least one of the two is transcendental.) Is there an algebraic relation between e and π? The answers to these questions are unknown at the moment. However, in the 1960's, Schanuel (as cited in Lang's monograph on Transcendental Numbers) conjectured that if $x_1, ..., x_n$ are linearly independent over $\mathbb{Q}$, then

$$\operatorname{tr}\deg \mathbb{Q}(x_1, ..., x_n, e^{x_1}, ..., e^{x_n}) \geq n.$$

The Lindemann- Weierstrass theorem states that if $x_1, ..., x_n$ are linearly independent *algebraic* numbers, then this conjecture is true. For $n = 1$, Schanuel's conjecture follows from the Hermite - Lindemann theorem. Already for $n = 2$, we do not know the truth of the conjecture. However, the general case is still a major unsolved problem and many interesting consequences emerge from it. For example, we immediately deduce that e and π are algebraically independent. To see this, consider $x_1 = 1, x_2 = \pi i$. These are linearly independent over $\mathbb{Q}$ and so the

$$\operatorname{tr}\deg \mathbb{Q}(\pi, e) = \operatorname{tr}\deg \mathbb{Q}(1, \pi i, e, e^{\pi i})$$

is at least 2. Thus, e and π are algebraically independent. Therefore Schanuel implies that both $e+\pi$ and $e\pi$ are transcendental and algebraically independent. Also, $1, \log\pi$ are linearly independent over $\mathbb{Q}$ for otherwise, we have $a + b\log\pi = 0$, for some integers a, b from which we get $e^a\pi^b = 1$, contradicting the algebraic independence of e and π. To deduce that $\log\pi$ is transcendental, we consider $\pi i, \log\pi$ which are linearly independent over $\mathbb{Q}$ since $\pi \neq \pm 1$. Thus, π and $\log\pi$ are algebraically independent. In particular, $\log\pi$ is transcendental (modulo Schanuel).

There has been some progress on Schanuel's conjecture but not much. A special case of the conjecture is the following conjecture of Gelfond and Schneider: if α is an algebraic number and $\alpha \neq 0, 1$, and if β is an irrational algebraic number of degree d, then the $d-1$ numbers

$$\alpha^\beta, \alpha^{\beta^2}, ..., \alpha^{\beta^{d-1}}$$

are algebraically independent. We call this the Gelfond-Schneider conjecture. What is know is a result of Diaz [**14**] which states that the transcendence degree of the field generated by the $d-1$ numbers above is at least $[(d+1)/2]$.

Schanuel's conjecture will help us later to determine the transcendental nature of special values of many Dirichlet series (more specifically Artin L-series at $s = 1$).

3. An overview of problems and results

The prototypical example of an L-series is the Riemann zeta function, denoted by $\zeta(s)$ and defined by the Dirichlet series

$$\sum_{n=1}^{\infty} \frac{1}{n^s},$$

for $\Re(s) > 1$ and then extended to the entire complex plane via a classical method of Riemann [**46**]. Related to this, perhaps the most celebrated and most beautiful of results on special values of L-series is Euler's theorem [**15**] that

$$\zeta(2k) = \frac{(2\pi i)^{2k} B_{2k}}{2(2k)!},$$

for every positive integer k. Here, B_k denotes the k-th Bernoulli number, given via the generating function:

$$\frac{t}{e^t - 1} = \sum_{k=0}^{\infty} \frac{B_k t^k}{k!}.$$

By the functional equation for $\zeta(s)$, this can be written equivalently as $\zeta(1-k) = -B_k/k$. It is easy to see that $B_k = 0$ for k odd and greater than 1. Thus, the Riemann zeta function vanishes at $s = -2, -4, \ldots$ and these are referred to as the trivial zeros. Since the Bernoulli numbers are rational numbers, we see that $\zeta(2k) \in \pi^{2k}\mathbb{Q}^*$, and hence the special value is a transcendental number by a famous theorem of Lindemann [**30**]. It is conjectured that all of the special values $\zeta(2k+1)$ with k a natural number are transcendental and that in fact, $1, \pi, \zeta(3), \zeta(5), \zeta(7), \ldots$ are algebraically independent (see [**8**]). Though this conjecture is still open, some spectacular progress has been made in the recent past. Beginning with the work of Apéry [**1**] in 1978 that $\zeta(3)$ is irrational, we have the theorem of Rivoal [**47**] that infinitely many of the values $\zeta(2k+1)$ are irrational. There are even some stronger results giving a lower bound for the dimension of the $\mathbb{Q}$-vector space spanned by the values $\zeta(2k+1)$ with $k \leq a$.

There are several different directions in which these results can be extended. Firstly, Euler's theorem on the special values of the Riemann zeta function was first extended by Hecke [**24**] to the case of a real quadratic field. He showed that if F is a real quadratic field and $\zeta_F(s)$ is the Dedekind zeta function of F, then $\zeta_F(2k)$ is an algebraic multiple of π^{4k}. This led him to conjecture that if F is a totally real algebraic number field, then $\zeta_F(2k)$ is an algebraic multiple of π^{2dk} where $d = [F : \mathbb{Q}]$. This conjecture was later proved by Siegel and Klingen [**28**]. These results raise further questions. What happens if F is not totally real? If F is totally real, what about $\zeta_F(2k+1)$? Are these transcendental numbers? The answer is most likely "yes" but we are far from knowing this. We discuss what is known and unknown in the larger context of Artin L-series, in a later section. In this connection, it is worth noting that Euler's theorem was first extended to Dirichlet L-series by Hecke as late as 1940, though the ideas needed for this work were already there at the time of Euler in the 18th century. We give the details of this development in a section below.

Another direction worthy of study is to fix a value of s and study special values of a family of L-series at s. The most celebrated example of this phenomenon is Dirichlet's class number formula. This formula states that if

$$L(s, \chi) = \sum_{n=1}^{\infty} \frac{\chi(n)}{n^s}$$

is the Dirichlet L-series attached to a Dirichlet character $\chi \pmod q$, then for χ a non-trivial quadratic character,

$$L(1, \chi) = \begin{cases} \frac{2\pi h}{w\sqrt{|d|}}, & \chi(-1) = -1 \\ \frac{h \log \epsilon}{\sqrt{|d|}} & \chi(-1) = 1, \end{cases}$$

where h denotes the class number of the quadratic field cut out by χ and ϵ is the fundamental unit in the real quadratic case, and w is the number of roots of unity in the associated quadratic field.

If χ is not a quadratic character, one can also write down a precise formula for $L(1, \chi)$. In all cases, Dirichlet's celebrated result is that $L(1, \chi) \neq 0$ and this is equivalent to the infinitude of primes in a given arithmetic progression (mod q). Inspired by this formula, Chowla **[12]** proposed the following problem. Let $f : \mathbb{Z}/q\mathbb{Z} \to \mathbb{C}$ and consider the Dirichlet series

$$L(s, f) = \sum_{n=1}^{\infty} \frac{f(n)}{n^s}.$$

Under what general conditions can we assert that $L(1, f) \neq 0$? Is it possible to evaluate $L(1, f)$? If f is algebraic valued, can we say that if $L(1, f) \neq 0$, then $L(1, f)$ is transcendental? More generally, what can we say about $L(k, f)$? These questions led Chowla and his daughter, Paromita Chowla, **[13]** to a variety of conjectures. These conjectures were generalized by Milnor **[31]**. One can ask for analogues of Chowla's question to number fields. But this investigation is still in its infancy. V.K. Murty and M.R. Murty **[34]** initiated this study by considering first imaginary quadratic fields. We describe this below.

Yet another direction of study is via multiple zeta values. These are defined as:

$$\zeta(k_1, ..., k_r) = \sum_{n_1 > n_2 > ... > n_r} n_1^{-k_1} n_2^{-k_2} \cdots n_r^{-k_r},$$

with $k_1 \geq 2$, and $k_2, ..., k_{r-1} \geq 1$ and the n_i run over all positive integers.. The *weight* of the multiple zeta value $\zeta(k_1, ..., k_r)$ is defined as the sum $k_1 + \cdots + k_r$ and its *depth* as r. A recent remarkable theorem of Brown **[7]** is that **all** multiple zeta values of weight n are $\mathbb{Q}$-linear combinations of

$$\{\zeta(a_1, ..., a_r) : \text{where} \quad a_i = 2 \ \text{ or } \ 3, \ \text{ and } \ a_1 + \cdots + a_r = n\}.$$

Using this theorem, we can see that the dimension of the $\mathbb{Q}$-vector space V_n spanned by multiple zeta values of weight n is bounded by d_n where d_n satisfies the recurrence relation $d_n = d_{n-2} + d_{n-3}$, with $d_0 = 1$, $d_1 = 0$, and $d_2 = 1$. It is conjectured that the dimension of V_n is exactly d_n but this has not yet been proved. The recursion shows that d_n grows exponentially as a function of n and yet, not a single value of n is known for which dim V_n is at least 2. Gun, Murty and Rath **[21]** showed that if the Chowla-Milnor conjecture is true, then there are infinitely many values of n for which the dimension is at least 2. This goes to indicate that the Chowla-Milnor conjecture is quite difficult.

A fourth direction of study is the special values of L-series attached to modular forms and more generally automorphic forms. Already, in the modular forms case, there are quite a number of results and conjectures. The full extent of these conjectures is beyond the scope of this survey. We relegate this to a future occasion. For the time being, we refer the reader to the excellent survey by Raghuram and Shahidi **[43]** where more references on this theme can be found.

4. Euler's theorem revisited

The analytic continuation of the Riemann zeta function to the entire complex plane was first proved by Riemann **[46]**. He also established the functional equation for the zeta function:

$$\pi^{-\frac{s}{2}}\Gamma\left(\frac{s}{2}\right)\zeta(s) = \pi^{-\frac{1-s}{2}}\Gamma\left(\frac{1-s}{2}\right)\zeta(1-s).$$

If it is only the analytic continuation we desire and not the functional equation, there is an elementary way to derive it. This was noted in the author's paper with Reece [**37**]. For this purpose, it is useful to consider the Hurwitz zeta function:

$$\zeta(s,x) = \sum_{n=0}^{\infty} \frac{1}{(n+x)^s},$$

for $0 < x \leq 1$. Thus, $\zeta(s,1) = \zeta(s)$. This series converges absolutely for $\Re(s) > 1$ and it is surprising that one can derive an analytic continuation of the Hurwitz zeta function by a simple induction argument as follows. Let us observe that

$$-\frac{1}{x^s} + \zeta(s,x) - \zeta(s) = \sum_{n=1}^{\infty} \left\{ \frac{1}{(n+x)^s} - \frac{1}{n^s} \right\}.$$

Writing the summand as

$$\frac{1}{n^s}\left(\left(1+\frac{x}{n}\right)^{-s} - 1\right),$$

we can apply the binomial theorem for $0 < x < 1$ and get

$$-\frac{1}{x^s} + \zeta(s,x) - \zeta(s) = \sum_{r=1}^{\infty} \binom{-s}{r} \zeta(s+r)x^r. \tag{4.1}$$

Several observations can now be made. First, if $x = 1/2$, we observe that $\zeta(s,1/2) = (2^s - 1)\zeta(s)$ so that

$$-2^s + (2^s - 2)\zeta(s) = \sum_{r=1}^{\infty} \zeta(s+r)2^{-r},$$

which serves to provide a meromorphic continuation of the Riemann zeta function to the entire complex plane by a simple induction argument. This argument shows that $\zeta(s)$ extends analytically to the entire complex plane except possibly at those s for which $2^s = 2$. Indeed, let us first consider the region $\Re(s) > 0$. In this region, the only poles are at

$$s = \frac{2\pi i m}{\log 2}, \qquad m \in \mathbb{Z}.$$

For any natural number $q > 1$, we also have the identity

$$(q^s - q)\zeta(s) = \sum_{a=1}^{q} \left(\zeta(s, \frac{a}{q}) - \zeta(s)\right)$$

and the right hand side is analytic for $\Re(s) > 0$ by (4.1). This identity serves to imply that if $\zeta(s)$ has any poles in this region, they occur at

$$s = 1 + \frac{2\pi i m}{\log q}, \qquad m \in \mathbb{Z},$$

and they are all simple. Taking $q = 3$ say, and noting that $s = 1$ is the only element in the intersection of

$$\left\{1 + \frac{2\pi i m}{\log 2} : m \in \mathbb{Z}\right\} \cap \left\{1 + \frac{2\pi i n}{\log 3} : n \in \mathbb{Z}\right\},$$

we deduce that $\zeta(s)$ extends to $\Re(s) > 0$, except for a simple pole at $s = 1$. Proceeding inductively, we deduce the analytic continuation of $\zeta(s)$ to the entire complex plane. This then serves to give the analytic continuation of $\zeta(s,x)$ as well.

What is impressive about (4.1) is that it shows that $\zeta(1-k,x)$ is a polynomial in x of degree $k-1$ for any positive value of k. To see this, one need only note that for $s = 1-k$, the infinite series becomes a finite series because the binomial coefficients vanish for $r \geq k$. One can also derive from this that $\zeta(1-k) = -B_k/k$, by an elementary induction argument as follows. The recursion above allows us to deduce that for any positive integer m,

$$m\zeta(1-m) = \frac{(-1)^m}{m+1} - m\sum_{r=1}^{m-1}(-1)^r\binom{m-1}{r}\frac{\zeta(1-m+r)}{r+1}.$$

Recall that the generating function for the Bernoulli numbers is

$$\frac{t}{e^t-1} = \sum_{k=0}^{\infty}\frac{B_k t^k}{k!}.$$

From this we deduce that the B_k are rational numbers and the following recurrence for them:

$$\sum_{k=0}^{n}\binom{n}{k}\frac{B_{n-k}}{k+1} = 0.$$

Moreover,

$$\frac{t}{e^t-1} + \frac{t}{2}$$

is an even function of t so that the Bernoulli numbers for odd subscripts ≥ 3 vanish. We can then prove by induction the formula

$$\zeta(1-k) = (-1)^{k-1}\frac{B_k}{k}$$

using (4.1). One can now use the functional equation to deduce the explicit value of $\zeta(2k)$.

Of course, Euler's approach was completely different. Since this approach will be useful later in our study of multiple zeta values, we indicate briefly his point of view.

In 1735, Euler discovered experimentally that

$$1 + \frac{1}{2^2} + \frac{1}{3^2} + \frac{1}{4^2} + \cdots = \frac{\pi^2}{6}. \tag{4.2}$$

He gave a "rigorous" proof much later, in 1742. Here is a sketch of Euler's proof. The polynomial

$$\left(1-\frac{x}{r_1}\right)\left(1-\frac{x}{r_2}\right)\ldots\left(1-\frac{x}{r_n}\right)$$

has roots equal to $r_1, r_2, ..., r_n$. When we expand the polynomial, the coefficient of x is

$$-\left(\frac{1}{r_1} + \frac{1}{r_2} + \cdots + \frac{1}{r_n}\right).$$

Using this observation, Euler proceeded "by analogy." Supposing that $\sin \pi x$ "behaves" like a polynomial and noting that its roots are at $x = 0, \pm 1, \pm 2, ...$, Euler puts

$$f(x) = \frac{\sin \pi x}{\pi x}.$$

By l'Hôpital's rule, $f(0) = 1$. Now $f(x)$ has roots at $x = \pm 1, \pm 2, \ldots$ and so

$$f(x) = (1-x)(1+x)\left(1-\frac{x}{2}\right)\left(1+\frac{x}{2}\right)\left(1-\frac{x}{3}\right)\left(1+\frac{x}{3}\right)\cdots.$$

That is,

$$f(x) = (1-x^2)\left(1-\frac{x^2}{4}\right)\left(1-\frac{x^2}{9}\right)\cdots.$$

The coefficient of x^2 on the right hand side is

$$-\left(1+\frac{1}{4}+\frac{1}{9}+\cdots+\right).$$

By Taylor's expansion,

$$\sin \pi x = \pi x - \frac{(\pi x)^3}{3!} + \cdots$$

so that comparing the coefficients gives us formula (4.2).

The main question is whether all of this can be justified. Euler certainly didn't have a completely rigorous proof of his argument. To make the above discussion rigorous, one needs either the theory of Weierstrass products discovered in 1876 (see page 79 of [**45**]) or Hadamard's theory of factorization of entire funtions, a theory developed much later in 1892, in Jacques Hadamard's doctoral thesis. Still, we credit Euler for the discovery of this result since the basic idea is sound.

The next question is whether Euler's result can be generalized. For example, can we evaluate

$$\sum_{n=1}^{\infty} \frac{1}{n^3} \quad \text{or} \quad \sum_{n=1}^{\infty} \frac{1}{n^4}.$$

Euler had difficulty with the first question but managed to show, using a similar argument, that

$$\sum_{n=1}^{\infty} \frac{1}{n^4} = \frac{\pi^4}{90}$$

and more generally that

$$\sum_{n=1}^{\infty} \frac{1}{n^{2k}} \in \pi^{2k}\mathbb{Q}.$$

It is not hard to see that Euler's proof can be modified to deduce the above results. Indeed, if $i = \sqrt{-1}$, then observing that

$$f(ix) = (1+x^2)\left(1+\frac{x^2}{4}\right)\cdots$$

we see that

$$f(x)f(ix) = (1-x^4)\left(1-\frac{x^4}{2^4}\right)\left(1-\frac{x^4}{3^4}\right)\cdots$$

But the Taylor expansion of $f(x)f(ix)$ is

$$\left(1-\frac{\pi^2x^2}{3!}+\frac{\pi^4x^4}{5!}-\cdots\right)\left(1+\frac{\pi^2x^2}{3!}+\frac{\pi^4x^4}{5!}+\cdots\right).$$

Computing the coefficient of x^4 yields

$$\sum_{n=1}^{\infty} \frac{1}{n^4} = \frac{\pi^4}{90}.$$

Continuing in this way, it is not difficult to see how Euler arrived at the assertion that

$$\sum_{n=1}^{\infty} \frac{1}{n^{2k}} \in \pi^{2k}\mathbb{Q}.$$

Euler's viewpoint is useful in the study of multiple zeta values. He seemed to suggest that for $\zeta(2k+1)$ the value is a product of π^{2k+1} and "a function of $\log 2$" (see the last line of p. 1078 of [**2**]). This conjecture is probably wrong but no definite disproof has yet been found.

5. Special values of Dirichlet L-series

The logarithmic derivative of the Γ-function is called the *digamma function*. Higher derivatives of the digamma function give rise to the *polygamma functions*. More precisely, the digamma function is defined by

$$\psi(z) = -\gamma - \frac{1}{z} - \sum_{n=1}^{\infty}\left(\frac{1}{n+z} - \frac{1}{n}\right), \qquad z \neq 0, \pm 1, \pm 2, \ldots$$

so that the polygamma functions $\psi_k(z)$ are given by

$$\psi_k(z) = (-1)^{k-1} k! \sum_{n=0}^{\infty} \frac{1}{(n+z)^{k+1}}.$$

It is easily seen from the series expansion for $\psi_k(z)$ that

$$\psi_k(z+1) = \psi_k(z) + \frac{(-1)^k k!}{z^{k+1}}.$$

This allows us to deduce that for every integer $k \geq 0$,

$$-\frac{d^k}{dz^k}(\pi \cot \pi z) = \psi_k(z) + (-1)^{k+1}\psi_k(-z) + (-1)^k \frac{k!}{z^{k+1}}.$$

Indeed, from the partial fraction expansion of the cotangent function, we have

$$\pi \cot \pi z = \frac{1}{z} + \sum_{n=1}^{\infty}\left(\frac{1}{z-n} + \frac{1}{z+n}\right).$$

The result is now easily deduced by taking successive derivatives. This allows us to relate the cotangent function to the polygamma functions. Indeed, it is readily seen that

$$-\frac{d^k}{dz^k}(\pi \cot \pi z) = \psi_k(z) + (-1)^{k+1}\psi_k(1-z). \tag{5.1}$$

These identities are at the heart of Hecke's 1940 generalization of Euler's explicit determination of (see Ayoub [**2**] and [**15**]) $\zeta(2k)$ of 1749 and it is surprising that it took almost two centuries to write them down. This highlights the importance of a survey, when we can look back and see what has been done and what is yet to be done. Here is a brief description of Hecke's theorem.

Let q be a natural number and let $\chi : (\mathbb{Z}/q\mathbb{Z})^* \to \mathbb{C}$ be a Dirichlet character and define the L-series $L(s, \chi)$ by

$$L(s,\chi) := \sum_{n=1}^{\infty} \frac{\chi(n)}{n^s}.$$

Then

$$(k-1)!L(k,\chi) = (-q)^{-k}\sum_{a=1}^{q}\chi(a)\psi_{k-1}\left(\frac{a}{q}\right).$$

Here k is a positive integer and if $k=1$, we assume χ is non-trivial. Indeed, we may write

$$L(k,\chi) = \sum_{a=1}^{q}\chi(a)\sum_{n\equiv a(\operatorname{mod} q)} n^{-k} = \sum_{a=1}^{q}\chi(a)\sum_{j=0}^{\infty}(qj+a)^{-k}$$

from which the desired result is apparent.

If χ is an even character (that is, $\chi(-1)=1$) and k is even, then

$$-2(k-1)!L(k,\chi) = (-q)^{-k}\sum_{a=1}^{q}\chi(a)\frac{d^k}{dz^k}(\pi\cot\pi z)\Big|_{z=a/q}.$$

Then, $L(k,\chi)$ is an algebraic multiple of π^k.

To see this, we can write

$$2(k-1)!L(k,\chi) = (-q)^{-k}\sum_{a=1}^{q}\chi(a)\psi_{k-1}\left(\frac{a}{q}\right) + \chi(q-a)\psi_{k-1}\left(1-\frac{a}{q}\right)$$

which is

$$= (-q)^{-k}\sum_{a=1}^{q}\chi(a)\left(\psi_{k-1}\left(\frac{a}{q}\right) + \psi_{k-1}\left(1-\frac{a}{q}\right)\right).$$

By our earlier observation (5.1), the result follows. The last assertion is immediate upon noting that

$$\frac{d^k}{dz^k}(\pi\cot\pi z)\Big|_{z=a/q}$$

is an algebraic multiple of π^k. It is also important to note that this calculation allows us to determine the algebraic number precisely. An analogous calculation can be made for odd characters.

If χ is an odd character (that is, $\chi(-1)=-1$) and k is odd, then

$$-2(k-1)!L(k,\chi) = (-q)^{-k}\sum_{a=1}^{q}\chi(a)\frac{d^k}{dz^k}(\pi\cot\pi z)\Big|_{z=a/q}.$$

and $L(k,\chi)$ is an algebraic multiple of π^k.

As before we can write

$$2(k-1)!L(k,\chi) = (-q)^{-k}\sum_{a=1}^{q}\chi(a)\psi_{k-1}\left(\frac{a}{q}\right) + \chi(q-a)\psi_{k-1}\left(1-\frac{a}{q}\right)$$

which is now equal to (since χ is odd)

$$= (-q)^{-k}\sum_{a=1}^{q}\chi(a)\left(\psi_{k-1}\left(\frac{a}{q}\right) - \psi_{k-1}\left(1-\frac{a}{q}\right)\right).$$

As before, the result follows by noting the relation of the polygamma function to the cotangent function. The last assertion is also immediate upon noting (as before) that

$$\frac{d^k}{dz^k}(\pi \cot \pi z)\Big|_{z=a/q}$$

is an algebraic multiple of π^k.

When k and χ have opposite parity, the situation is as difficult as the determination of the transcendental nature of the Riemann zeta function at odd arguments. The simplest unknown case concerns the famous Catalan constant:

$$1 - \frac{1}{3^2} + \frac{1}{5^2} - \cdots$$

We do not know if this number is irrational.

6. Summation of infinite series of rational functions

The essential success of explicit evaluations of the special values of Riemann's zeta function and Dirichlet's L-series discussed in the earlier sections is mainly due to our understanding of the cotangent function. More precisely, the identity

$$\pi \cot \pi z = \sum_{n\in\mathbb{Z}} \frac{1}{n+z}, \qquad z \notin \mathbb{Z}$$

and its derivatives allow us a precise knowledge of

$$\sum_{n\in\mathbb{Z}} \frac{1}{(n+z)^k}.$$

This idea also allows us to explicitly evaluate infinite series of the form

$$\sum_{n\in\mathbb{Z}} \frac{A(n)}{B(n)},$$

where $A(x), B(x)$ are polynomials and we assume that the sum converges and that $B(x)$ has no integral roots. We may further allow for $B(x)$ to have integer roots and then restrict the sum so that we exclude these (finite number of) roots in the summation. In any case, one can apply partial fraction expansions and derive very beautiful explicit formulas for these sums. In the case $A(x)$ and $B(x)$ have algebraic coefficients, the transcendental nature of these sums can sometimes be determined thanks to the work of Nesterenko [**42**]. All of these investigations have been carried out in the paper by Murty and Weatherby [**40**]. Among other results proved in [**40**], here is a representative one. If the Gelfond-Schneider conjecture is true, then for any $A(x), B(x) \in \overline{\mathbb{Q}}[x]$, with degree of A less than the degree of B and $B(n) \neq 0$ for any $n \in \mathbb{Z}$, the sum

$$\sum_{n\in\mathbb{Z}} \frac{A(n)}{B(n)}$$

is either zero or transcendental. Perhaps the most striking of formulas derived through these investigations is a superb generalization of Euler's theorem. The sum

$$\sum_{n\in\mathbb{Z}} \frac{1}{An^2+Bn+C} = \frac{2\pi}{\sqrt{D}} \left(\frac{e^{2\pi\sqrt{D}/A} - 1}{e^{2\pi\sqrt{D}/A} - 2(\cos(\pi B/A))e^{\pi\sqrt{D}/A} + 1} \right)$$

is transcendental if $A, B, C \in \mathbb{Z}$ and $-D = B^2 - 4AC < 0$. The explicit evaluation is simply an application of the cotangent expansion. The transcendence is derived by applying a result of Nesterenko which states that π and $e^{\pi\sqrt{D}}$ are algebraically independent. Viewing the right hand side as a function of C and applying successive differentiation with respect to C, we deduce an explicit formula for

$$\sum_{n\in\mathbb{Z}} \frac{1}{(An^2+Bn+C)^k},$$

and this is exposed in [**41**]. More can be done in this direction and one can study sums of the form

$$\sum_{n\in\mathbb{Z}} \frac{f(n)}{(An^2+Bn+C)^k},$$

with f a periodic function (mod q). More generally, we can study

$$\sum_{n\in\mathbb{Z}} \frac{A(n)}{B(n)} f(n).$$

Some partial results have been obtained in [**54**]. However, the full extent of our knowledge of these special values has not yet been determined.

7. Multiple zeta values

To understand the arithmetic nature of special values of the Riemann zeta function, it has become increasingly clear that multiple zeta values (MZV's for short) must be studied. These are defined as follows:

$$\zeta(k_1,\ldots,k_r) = \sum_{n_1>n_2>\cdots>n_r>0} \frac{1}{n_1^{k_1} n_2^{k_2}\cdots n_r^{k_r}},$$

where $k_1, k_2, \ldots, k_r$ are positive integers with the proviso that $k_1 \geq 2$. The last condition is imposed to ensure convergence of the series.

There are several advantages to introducing these multiple zeta functions. First, they have an algebraic structure which we describe. It is easy to see that

$$\zeta(s_1)\zeta(s_2) = \zeta(s_1,s_2) + \zeta(s_2,s_1) + \zeta(s_1+s_2).$$

Indeed, the left hand side can be decomposed as

$$\sum_{n_1,n_2} \frac{1}{n_1^{s_1} n_2^{s_2}} = \sum_{n_1>n_2} \frac{1}{n_1^{s_1} n_2^{s_2}} + \sum_{n_2>n_1} \frac{1}{n_1^{s_1} n_2^{s_2}} + \sum_{n_1=n_2} \frac{1}{n_1^{s_1} n_2^{s_2}}$$

from which the identity becomes evident. In a similar way, one can show that

$$\zeta(s_1)\zeta(s_2,\ldots,s_t)$$

is again an integral linear combination of multiple zeta values. More generally, the product of any two MZV's is an integral linear combination of MZV's. These identities lead to new relations, like

$$\zeta(2,1) = \zeta(3),$$

an identity which appears in Apéry's proof [**1**] of the irrationality of $\zeta(3)$.

If we let V_r be the $\mathbb{Q}$-vector space spanned by

$$\zeta(s_1,s_2,\ldots,s_k)$$

with $s_1+s_2+\cdots+s_k = r$, then the product formula for MZV's shows that

$$V_r V_s \subseteq V_{r+s}.$$

In this way, we obtain a graded algebra of MZV's. Let d_r be the dimension of V_r as a vector space over $\mathbb{Q}$. For convenience, we set $d_0 = 1$ and $d_1 = 0$. Clearly, $d_2 = 1$ since V_2 is spanned by $\pi^2/6$. Zagier [**55**] has made the following conjecture:

$$d_r = d_{r-2} + d_{r-3},$$

for $r \geq 3$. In other words, d_r satisfies a Fibonacci-type recurrence relation. Consequently, d_r is expected to have exponential growth. Given this prediction, it is rather remarkable that not a single value of r is known for which $d_r \geq 2$! We relate this to the Chowla-Milnor conjecture in a later section.

In view of the identity, $\zeta(2,1) = \zeta(3)$, we see that $d_3 = 1$. What about d_4? V_4 is spanned by $\zeta(4), \zeta(3,1), \zeta(2,2), \zeta(2,1,1)$. What are these numbers? Zagier's conjecture predicts that $d_4 = d_2 + d_1 = 1 + 0 = 1$. Is this true? Let us see.

We can adapt Euler's technique to evaluate $\zeta(2,2)$. As noted earlier

$$\left(1 - \frac{x}{r_1}\right)\left(1 - \frac{x}{r_2}\right)\cdots\left(1 - \frac{x}{r_n}\right)$$

has roots equal to $r_1, r_2, ..., r_n$. When we expand the polynomial, the coeffcient of x is

$$-\left(\frac{1}{r_1} + \frac{1}{r_2} + \cdots + \frac{1}{r_n}\right).$$

The coefficient of x^2 is

$$\sum_{i<j} \frac{1}{r_i r_j}.$$

With this observation, we see from the product expansion

$$f(x) = \frac{\sin \pi x}{\pi x} = (1 - x^2)\left(1 - \frac{x^2}{4}\right)\left(1 - \frac{x^2}{9}\right)\cdots$$

that the coefficient of x^4 is precisely $\zeta(2,2)$. An easy computation shows that

$$\zeta(2,2) = \frac{\pi^4}{5!}.$$

It is now clear that this method can be used to evaluate $\zeta(2,2,...,2) = \zeta(\{2\}^m)$ (say). By comparing the coefficient of x^{2m} in our expansion of $f(x)$, we obtain that

$$\zeta(\{2\}^m) = \frac{\pi^{2m}}{(2m+1)!},$$

which can also be viewed as another generalization of Euler's result. We could have also evaluated $\zeta(2,2)$ using the identity

$$\zeta(2)^2 = 2\zeta(2,2) + \zeta(4),$$

but we opted for the method above to indicate its generalization which allows us to also evaluate $\zeta(2,2,...,2)$.

What about $\zeta(3,1)$? What can we say about it? This is a bit more difficult and will not come out of our earlier results. In 1998, Borwein, Bradley, Broadhurst and Lisonek [**6**] showed that $\zeta(3,1) = 2\pi^4/6!$.

What about $\zeta(2,1,1)$? Can we relate it to a known constant? With some work, one can show that this is equal to $\zeta(4)$. Thus, we conclude that $d_4 = 1$ as predicted by Zagier.

What about d_5? With more work, we can show that

$$\zeta(2,1,1,1) = \zeta(5); \quad \zeta(3,1,1) = \zeta(4,1) = 2\zeta(5) - \zeta(2)\zeta(3).$$
$$\zeta(2,1,1) = \zeta(2,3) = 9\zeta(5)/2 - 2\zeta(2)\zeta(3).$$
$$\zeta(2,2,1) = \zeta(3,2) = 3\zeta(2)\zeta(3) - 11\zeta(5)/2.$$

This proves that $d_5 \leq 2$. Zagier conjectures that $d_5 = 2$. In other words, $d_5 = 2$ if and only if $\zeta(2)\zeta(3)/\zeta(5)$ is irrational. Of course, this is not yet known.

Can we prove Zagier's conjecture? To this date, not a single example is known for which $d_n \geq 2$. In the next section, we will show its relation to another conjecture due to Chowla and Milnor. If we write

$$(1 - x^2 - x^3)^{-1} = \sum_{n=1}^{\infty} D_n x^n,$$

then it is easy to see that Zagier's conjecture is equivalent to the assertion that $d_n = D_n$ for all $n \geq 1$. Deligne and Goncharov [**17**] and (independently) Terasoma [**53**] showed that $d_n \leq D_n$. As noted earlier, Brown [**7**] showed that the MZV's are generated by

$$\zeta(k_1, ..., k_r)$$

with the k_i equal to 2 or 3. The number of such values is clearly seen to be D_n since we are looking for the number of ways of writing n as a sum of 2's and 3's. The conjecture therefore is that this collection is a basis for all MZV's.

It is a fruitful line of research to study multiple Hurwitz zeta functions and this was initiated in [**39**]. These are defined as:

$$\zeta(k_1, ..., k_r; x_1, ..., x_r) := \sum_{n_1 > n_2 > \cdots > n_r \geq 1} \frac{1}{(n_1 + x_1)^{k_1} \cdots (n_r + x_r)^{k_r}}.$$

However, many of the elegant evaluations of MZV's have not yet been extended to MHZV's.

8. The Chowla-Milnor conjecture

In the sums considered in the previous section, if one restricts the summation to only positive integers, we are led to study

$$\sum_{n=1}^{\infty} \frac{A(n)}{B(n)}. \tag{8.1}$$

Again, we restrict the sum over those positive n which avoid the zeros of $B(x)$. In these situations, a partial fraction expansion leads us to write the value of the sum in terms of the digamma function and special values of the Hurwitz zeta function. Recall that the digamma function $\psi(x)$ is the logarithmic derivative of the Γ-function and it appears as the constant term in the Laurent expansion of the Hurwitz zeta function:

$$\zeta(s, x) = \frac{1}{s-1} - \psi(x) + \cdots$$

Indeed, the partial fraction expansions lead to sums of the form

$$\sum_{n=1}^{\infty} \frac{1}{(n + \alpha)^k}.$$

If $k \geq 2$, then the sum is the special value of the Hurwitz zeta function $\zeta(k, \alpha)$. If we assume that the series (8.1) converges, the terms corresponding to $k = 1$ disappear. The nature of the emerging sums has not been investigated since there is only meager knowledge about the special values of Hurwitz zeta functions. In fact, there are very few conjectures on what we may expect. Foremost among the sparse set of conjectures is the one due to Paromita and Sarvadaman Chowla [**13**] and its generalization due to Milnor [**31**]. Their conjecture is that the special values for a fixed natural number $k \geq 2$ and $q \geq 1$

$$\zeta\left(k, \frac{a}{q}\right), \qquad (a, q) = 1,$$

are linearly independent over the rational numbers. In particular, this conjecture would imply that if $f : (\mathbb{Z}/q\mathbb{Z})^* \to \mathbb{Q}$ is a rational-valued function and supported on the coprime residue classes (mod q), then

$$\sum_{n=1}^{\infty} \frac{f(n)}{n^k} \neq 0.$$

In [**20**], Gun, Murty and Rath proved that the Chowla-Milnor conjecture implies that

$$\left(\frac{\zeta(2k+1)}{\pi^{2k+1}}\right)^2 \notin \mathbb{Q}$$

for any $k \geq 1$. They also showed that the Chowla-Milnor conjecture for the single modulus $q = 4$ is equivalent to the irrationality of

$$\frac{\zeta(2k+1)}{\pi^{2k+1}}$$

for all $k \geq 1$ (see Proposition 4 in [**20**]). At the end of their paper, the authors formulate a stronger form of the conjecture, namely that,

$$1, \zeta(k, a/q), \qquad (a, q) = 1, \qquad 1 \leq a < q,$$

are linearly independent over the rationals. If this stronger conjecture is true either for $q = 3$ or $q = 4$, then it would imply that $\zeta(2k+1)$ is irrational for every value of k. It would also imply that

$$\sum_{n=1}^{\infty} \frac{f(n)}{n^k}$$

is irrational whenever f is rational-valued. From these implications, we see how difficult the Chowla-Milnor conjecture is.

9. The Riemann zeta function at odd arguments

In the notebooks of Ramanujan published by the Tata Institute in 1957, we find some elegant formulas for the special values of the Riemann zeta function at odd arguments. In particular, on page 171 of Volume 2, we see

(9.1)
$$\alpha^{-k}\left\{\frac{1}{2}\zeta(2k+1) + \sum_{n=1}^{\infty} \frac{n^{-2k-1}}{e^{2\alpha n} - 1}\right\} = (-\beta)^{-k}\left\{\frac{1}{2}\zeta(2k+1) + \sum_{n=1}^{\infty} \frac{n^{-2k-1}}{e^{2\beta n} - 1}\right\}$$

$$- 2^{2k}\sum_{j=0}^{k+1}(-1)^j \frac{B_{2j}B_{2k+2-2j}}{(2j)!(2k+2-2j)!}\alpha^{k+1-j}\beta^j, \tag{9.2}$$

where $\alpha, \beta > 0$ with $\alpha\beta = \pi^2$, k is any nonzero integer and B_j is the jth Bernoulli number. If we put $\alpha = \beta = \pi$ and k is odd, we deduce

$$\zeta(2k+1)+2\sum_{n=1}^{\infty}\frac{1}{n^{2k+1}(e^{2\pi n}-1)} = \pi^{2k+1}2^{2k}\sum_{j=0}^{k+1}(-1)^{j+1}\frac{B_{2j}B_{2k+2-2j}}{(2j)!(2k+2-2j)!}, \tag{9.3}$$

a formula apparently due to Lerch [**29**] and published in an obscure journal (see also[**5**]). It seems that Grosswald [**18**] rediscovered this formula and published its proof in 1970 only to learn later that it was discovered earlier by Ramanujan and even earlier by Lerch.

Since the left hand side of this equation is non-zero (both the terms being positive), the right hand side is a non-zero rational multiple of π^{2k+1}. Consequently, at least one of

$$\zeta(4k+3), \quad \sum_{n=1}^{\infty}\frac{1}{n^{4k+3}(e^{2\pi n}-1)}$$

is transcendental for every integer $k \geq 0$.

Motivated by these identities, the authors of [**22**] introduced the function:

$$F_k(z) = \sum_{n=1}^{\infty}\sigma_{-k}(n)e^{2\pi i n z}.$$

and proved the following theorem:

THEOREM 9.1. *Let k be a non-negative integer and set $\delta = 0, 1, 2, 3$ according as the* gcd$(k, 6)$ *equals 1,2,3 or 6. With at most $2k+2+\delta$ exceptions, the number*

$$F_{2k+1}(\alpha) - \alpha^{2k}F_{2k+1}(-1/\alpha)$$

is transcendental for every algebraic $\alpha \in \mathfrak{H}$. In particular, there are at most $2k+2+\delta$ algebraic numbers $\alpha \in \mathfrak{H}$ such that $F_{2k+1}(\alpha)$ and $F_{2k+1}(-1/\alpha)$ are both algebraic.

Notice the similarity between $F_k(z)$ and the classical Eisenstein series $E_{k+1}(z)$ with k odd. It is therefore not unreasonable to study special values of modular forms and quasi-modular forms at algebraic arguments and this is investigated in [**23**]. The interest in this theorem as it relates to $\zeta(2k+1)$ is highlighted by the following observation. There exist algebraic numbers α in the upper half plane $\mathfrak{H}$ for which the numbers

$$F_{2k+1}(\alpha) - \alpha^{2k}F_{2k+1}(-1/\alpha)$$

are non-zero algebraic multiples of $\zeta(2k+1)$ for all $k \geq 4$. Thus, Theorem 9.1 comes very close to showing transcendence of $\zeta(2k+1)$. The function $F_k(z)$ is closely related to the classical Eisenstein series $E_{k+1}(z)$ and is really an example of an Eichler integral. Clearly, the study of the special values of these Eichler integrals will shed new light on the nature of the Riemann zeta function at odd arguments.

Much of the analysis used to derive (9.2) can be carried out for Dirichlet L-functions $L(2k+1,\chi)$ for χ even (and similarly for $L(2k,\chi)$ for χ odd). But this has not been done in a systematic manner and offers a good program for further research. There are some related papers on this topic by Katayama [**27**].

10. Hecke's conjecture and the Siegel-Klingen theorem

The Dedekind zeta function of an algebraic number field F is defined as

$$\zeta_F(s) = \sum_{\mathfrak{a} \neq 0} N(\mathfrak{a})^{-s},$$

where the summation is over non-zero ideals $\mathfrak{a}$ of the ring of integers of F. The analytic continuation and functional equation of $\zeta_F(s)$ was proved by Hecke [**25**] in 1918. From this functional equation, we can see that the Dedekind zeta function has trivial zeroes at $s = -1, -2, ...$ unless F is totally real, in which case it has trivial zeroes only at $s = -2, -4, ...$ just like the Riemann zeta function. It is for this reason that Hecke was able to surmise and Siegel and Klingen were able to extend Euler's theorem describing the special values of $\zeta_F(2k)$ as algebraic multiples of π^{2kd} with $d = [F : \mathbb{Q}]$ when F is totally real.

The gist of the results of an earlier section is that the value of $L(k, \chi)$ when k and χ have the same parity, is a non-zero algebraic multiple of π^k. As mentioned earlier, this fact was first proved by Hecke in 1940 and he noted that this implies an interesting result for real quadratic fields. Namely, if F is a real quadratic field, then $\zeta_F(2m)$ is an algebraic multiple of π^{4m}. This motivated him to ask if such a result holds generally for any totally real field F. That is, if F is totally real of degree d over the rationals, then is it true that $\zeta_F(2m)$ is an algebraic multiple of π^{2dm}? Hecke's calculation also answers his question in another case, namely the case of the cyclotomic subfield $\mathbb{Q}(\zeta + \zeta^{-1})$ where $\zeta = e^{2\pi i/q}$ is a primitive q-th root of unity. This field is also totally real and its Dedekind zeta function is the product of Dirichlet L-functions $L(s, \chi)$ with χ an even character (mod q). Thus, Hecke's simple extension of Euler's theorem allowed him to prove his conjecture in two important cases.

Hecke's question was anwered in the affirmative by Siegel and Klingen and is now known as the Siegel-Klingen theorem. The proof of the Siegel-Klingen theorem makes use of the theory of classical modular forms and Hilbert modular forms. Since a detailed explanation of the proof is beyond the scope of this survey, we content ourselves with a brief outline and refer the reader to Garrett [**16**] as well as Siegel's exposition [**50**] or the authors [**33**] for further details.

Let F be a totally real number field of degree r and discriminant D. Let $x \mapsto x^{(i)}$ be an indexing of real embeddings of F and let $\mathcal{O}_F$ be the ring of integers of $\mathcal{O}_F$ and $\mathfrak{h}$ denote the upper half-plane. The group $SL_2(\mathcal{O}_F)$ is called the Hilbert modular group and it acts on $\mathfrak{h}^r$ via the map:

$$g := \begin{pmatrix} a & b \\ c & d \end{pmatrix} \in SL_2(\mathcal{O}_F),$$

$$g \cdot (z_1, .., z_r) = \left(\frac{a^{(1)} z_1 + b^{(1)}}{c^{(1)} z_1 + d^{(1)}}, ..., \frac{a^{(r)} z_r + b^{(r)}}{c^{(r)} z_r + d^{(r)}} \right).$$

For $a, b \in F$ and $z \in \mathfrak{h}^r$, it is convenient to use the notation

$$N(az + b) := \prod_{i=1}^{r} (a^{(i)} z_i + b^{(i)}).$$

Let k be a natural number (which is even if K has a unit of norm -1). For an ideal $\mathfrak{a}$, we define the analogue of the Eisenstein series as

$$F_k(\mathfrak{a}, z) = N(\mathfrak{a})^k \sum_{(\lambda,\mu)\neq(0,0)} N(\lambda z + \mu)^{-k},$$

where the sum runs over a complete system of pairs of numbers in the ideal $\mathfrak{a}$ different from $(0, 0)$ and not differing from one another by a factor which is a unit (that is, not associated). One can show that the series converges absolutely for $k > 2$ and for $k = 2$, one can apply a limit process (Hecke's trick). The function $F_k(\mathfrak{a}, z)$ is an example of a Hilbert modular form in the sense that

$$F_k(\mathfrak{a}, gz) = N(cz + d)^k F_k(\mathfrak{a}, z) \qquad \forall \quad g \in SL_2(\mathcal{O}_F).$$

As $\mathcal{O}_F$ is a lattice and its dual is the inverse *different* $\mathfrak{d}^{-1}$ of $\mathcal{O}_F$, one has a Fourier expansion of $F_k(\mathfrak{a}, z)$ of the form

$$\zeta(\mathfrak{a}, k) + \left(\frac{(2\pi i)^k}{(k-1)!}\right)^r D^{1/2-k} \sum_{\nu\in\mathfrak{d}^{-1},\nu\gg 0} \sigma_{k-1}(\mathfrak{a}, \nu) e^{2\pi i Tr(z\nu)},$$

where ν runs over totally positive numbers in $\mathfrak{d}^{-1}$ and

$$\zeta(\mathfrak{a}, k) = N(\mathfrak{a})^k \sum_{\mathfrak{a}|(\mu)} N(\mu)^{-k},$$

$$\sigma_{k-1}(\mathfrak{a}, \nu) = \sum_{\mathfrak{d}^{-1}|(\alpha)\mathfrak{a}|\nu} \operatorname{sign}(N(\alpha)^k) N((\alpha)\mathfrak{a}\mathfrak{d})^{k-1}.$$

The summation is over principal ideals $(\mu), (\alpha)$ under the conditions given. If we set all the variables $z_1, ..., z_r$ to z, then $F_k(\mathfrak{a}, z)$ becomes a classical modular form of weight rk for the full modular group. The final result is then deduced using the classical theory of modular forms. Indeed, let E_k denote the usual normalized Eisenstein series of weight k, Δ denote Ramanujan's normalized cusp form and j the modular invariant. If $M_k(SL_2(\mathbb{Z}))$ is the space of modular forms of weight k for the full modular group, let $t = \dim M_k(SL_2(\mathbb{Z}))$. Put

$$F_k := E_{12r-k+2}\Delta^{-t}.$$

Then F_k has q-expansion

$$C_{kr}q^{-t} + \cdots + C_{k1}q^{-1} + C_{k0} + \cdots$$

with $C_{kr} = 1$ and $C_{k\ell} \in \mathbb{Z}$ for all ℓ. We need to make one further observation. Then for any integer $m \geq 0$,

$$j^m \frac{dj}{dz}$$

has a q-expansion without a constant term. Using these facts, one shows that if f is a modular form of weight k for the full modular group with q-expansion

$$f(z) = a_0 + a_1 q + a_2 q^2 + \cdots$$

and $C_{k\ell}$ are as above, then

$$C_{k0}a_0 + C_{k1}a_1 + \cdots + C_{kr}a_r = 0.$$

The key observation is then that $C_{k0} \neq 0$, so that the constant term a_0 is a rational linear combination of $a_1, ..., a_r$. Applying this to our modular form $F_k(\mathfrak{a}, z)$, we deduce that $\zeta(\mathfrak{a}, k)$ is a an algebraic multiple of π^{kr}. Since the Dedekind zeta function $\zeta_F(k)$ can be written as a rational linear combination of the values $\zeta(\mathfrak{a}, k)$, the Siegel-Klingen theorem follows from this. The reader can find further details in [**50**] or [**33**].

11. Artin L-series

The Riemann zeta function and Dirichlet L-functions, as well as the Dedekind zeta function of a number field F are all special cases of Artin L-series. Let K/F be a finite Galois extension of algebraic number fields with Galois group G. Let (ρ, V) be a complex linear representation of G. For each prime ideal $\mathfrak{p}$ of F, and prime ideal $\wp$ of K lying above $\mathfrak{p}$, we have the usual inertia group $I_\wp$ which is a subgroup of the Galois group G. We then define the Artin L-series attached to ρ as

$$L(s, \rho, K/F) = \prod_{\mathfrak{p}} \det(1 - \rho(\sigma_\wp)N(\mathfrak{p})^{-s}|V^{I_\wp})^{-1},$$

where the product is over all prime ideals $\mathfrak{p}$ of $\mathcal{O}_F$. One can show that the product is well-defined and converges absolutely for $\Re(s) > 1$. If ρ is the trivial representation, then the corresponding Artin L-series is the Dedekind zeta function of F. If F is $\mathbb{Q}$ and K is the cyclotomic field $\mathbb{Q}(\zeta)$ where ζ is primitive q-th root of unity, then the Galois group is isomorphic to $(\mathbb{Z}/q\mathbb{Z})^*$ and the Artin L-series attached to this Galois group coincide with the classical Dirichlet L-series.

The first question that arises about special values of these Artin L-series is if there is an analog of the Siegel-Klingen theorem. Surprsingly, this was first proved as late as 1973 by Coates and Lichtenbaum [**10**]. A readable exposition of this theorem can be found in [**16**]. But here is a quick summary.

Their result can be stated in the following form. Given a Galois representation $\rho : \mathrm{Gal}(\overline{\mathbb{Q}}/\mathbb{Q}) \to GL(n, \mathbb{C})$, the Artin L-function $L(s, \rho)$ has a functional equation in which the gamma factors appearing in the functional equation are of the form

$$\Gamma(s/2)^a \Gamma((s+1)/2)^b$$

and we say ρ is totally real if $b = 0$ and totally complex if $a = 0$. (One could refer to (a, b) as the "Hodge type" of ρ.) In any case, one can show that $L(2k, \rho)$ is an algebraic multiple of $\pi^{2k\chi(1)}$ if ρ is totally real. Similarly, if ρ is totally complex, then $L(2k+1, \rho)$ is an algebraic multiple of $\pi^{(2k+1)\chi(1)}$. The essential idea for the proofs of these theorems is to use Brauer's induction formula and reduce it to the case of evaluation of Hecke's L-series and then to use a form of the Siegel-Klingen theorem as described in the previous section. I believe that there is no gentle exposition of these facts using classical analysis. The work that comes closest to such a goal is Shintani's paper [**48**]. It would be a good program of research to simplify considerably these proofs into a readable exposition.

12. Schanuel's conjecture and special values at $s = 1$.

The situation with respect to special values of Artin L-series at $s = 1$ has been studied extensively by Stark in a series of papers (see for example, [**52**]). Essentially, the conjecture predicts that the value is (up to an algebraic factor) a power of π and a determinant of logarithms of algebraic numbers (more precisely units).

A special case of the Schanuel conjecture is the following. Let $\alpha_1, ..., \alpha_n$ be non-zero algebraic numbers such that $\log \alpha_1, ..., \log \alpha_n$ are linearly independent over $\mathbb{Q}$. Then these numbers are algebraical-ly independent. Following [**21**], we call this the weak Schanuel conjecture. Baker's theorem asserts that these numbers are linearly independent over $\overline{\mathbb{Q}}$.

Assuming the weak form of Schanuel's conjecture, the authors in [**21**] show that these values are all transcendental numbers. Stark's conjectures should be viewed as generalizations of Dirichlet's class number formula. In some cases, Stark's conjecture can be proved without Schanuel's conjecture and we refer the reader to [**21**] for more details.

13. The Chowla and Erdös conjectures

As mentioned in section 3, Chowla [**12**] asked in 1970 the following question. Let $f : \mathbb{Z}/q\mathbb{Z} \to \mathbb{Q}$ be a function defined on the residue classes (mod q), not identically zero. Under what conditions is it true that

$$\sum_{n=1}^{\infty} \frac{f(n)}{n} \neq 0? \tag{13.1}$$

Chowla himself conjectured that this is the case if q is a prime number. This conjecture was then settled by Baker, Birch and Wirsing [**3**] and presumably by Chowla (since Chowla never published his proof and there is a comment by the authors in [**3**] that he had done so along the lines of their paper). The author [**32**] has written an exposition on how Chowla may have proved his theorem with the resources available to him, since the methods of [**3**] involve the theory of linear forms of logarithms and this may not have been the method adopted by Chowla.

Chowla's question is undoubtedly inspired by Dirichlet's theorem regarding the non-vanishing of $L(1, \chi)$. The general programs of special values of L-series have focused on those which admit Euler products and multiplicative structure. It may be fruitful to consider the slightly general framework suggested by Chowla. In this connection, Chatterjee, Murty and Pathak [**9**] have characterized all functions f for which $L(1, f) = 0$ in Chowla's problem. Related to this is a question (conjecture) of Erdös, that (13.1) does not vanish whenever $f(n) = \pm 1$ and $f(q) = 0$. This conjecture is non-trivial only in the case q is odd. Using some algebraic number theory, Murty and Saradha [**38**] settled Erdös's conjecture if $q \equiv 3 \pmod 4$. If $q \equiv 1 \pmod 4$, the conjecture is still open, though it was shown by Chatterjee and Murty [**11**] that the conjecture is true for at least 82 percent of q with $q \equiv 1 \pmod 4$. Most likely, one needs to understand the arithmetic significance of the non-vanishing to settle the conjecture completely.

That this approach has value can be seen in a set of analogous results obtained in the case of an imaginary quadratic field. For instance, if k is an imaginary quadratic field, and f is a function defined on the ideal class group of the ring of integers of k, we may consider

$$L(s, f) = \sum_{0 \neq \mathfrak{a} \in \mathcal{O}_k} \frac{f(\mathfrak{a})}{N(\mathfrak{a})^s}$$

and ask under what conditions this is non-zero at $s = 1$. This has been answered by using the Kronecker limit formula, by the author and V. Kumar Murty in [**34**] and further extended to functions on ray class groups of k in [**35**]. A good problem for further research is to study this in general number fields.

14. Concluding remarks

The appearance of the polylogarithm functions is expected in the evaluation of special values of Artin L-series. Zagier formulated a general conjecture for Dedekind zeta functions and this has been extended for Artin L-series by Zagier and Gangl [**56**]. The polylogarithm function is

$$L_k(z) = \sum_{n=1}^{\infty} \frac{z^n}{n^k},$$

and one expects that the value of $L(s, \rho)$ at s equal to an integer m is a power of π times a determinant of a matrix whose entries are values of the polylogarithm function evaluated at certain algebraic numbers. It would be interesting to make this more precise. This has been done in some cases but there is definitely a need to specify which power of π we expect, what the size of the determinant should be, and what are the algebraic numbers that appear as arguments of the polylogarithm function. A part of this conjecture already emerges in the work of [**34**] where the authors evaluated the special values of Hecke L-series attached to an imaginary quadratic field.

The polylogarithm function $L_k(z)$ is of course a generalization of the classical logarithm function since for $k = 1$, $L_1(z) = -\log(1 - z)$. Much of the success of the work so far on the Chowla problem is due to Baker's theorem about special values of $L_1(z)$ at algebraic arguments. This led Gun, Murty and Rath to make the following polylog conjecture. Suppose that $\alpha_1, ..., \alpha_n$ are algebraic numbers satisfying $|\alpha_i| \leq 1$ such that $L_k(\alpha_1), ..., L_k(\alpha_n)$ are linearly independent over $\mathbb{Q}$. Then they are linearly independent over $\overline{\mathbb{Q}}$. They show [**21**] that if the polylog conjecture is true, then the Chowla-Milnor conjecture is true for all $q > 1$ and $k > 1$. Thus, the program to extend Baker's theory of linear forms in logarithms to linear forms in polylogarithms will have tremendous applications in solving many open problems.

In Zagier's formulation of special values, we see the appearance of the polylogarithm function. By contrast, in [**19**], the authors relate special values of zeta and L-functions to the multiple gamma functions. They also study instances of special values of derivatives of the Riemann zeta function and Dirichlet L-functions. In this connection, there is some similarity with other conjectures such as the Birch and Swinnerton-Dyer conjecture. One could consider more generally, L-series of automorphic L-functions or even linear combinations of these. This short survey cannot exhaust the topics or the possibilities of these lines of investigation. But we hope the reader is inspired to explore further this galaxy of special values and behold its stellar beauty.

Acknowledgments

The author thanks Siddhi Pathak and the referee for their careful reading of an earlier version of this manuscript.

References

[1] R. Apéry, Irrationalité de $\zeta(2)$ et $\zeta(3)$, *Astérisque*, **61** (1979), 11–13.

[2] Raymond Ayoub, *Euler and the zeta function*, Amer. Math. Monthly **81** (1974), 1067–1086, DOI 10.2307/2319041. MR0360116

[3] A. Baker, B. J. Birch, and E. A. Wirsing, *On a problem of Chowla*, J. Number Theory **5** (1973), 224–236, DOI 10.1016/0022-314X(73)90048-6. MR0340203

[4] Keith Ball and Tanguy Rivoal, *Irrationalité d'une infinité de valeurs de la fonction zêta aux entiers impairs* (French), Invent. Math. **146** (2001), no. 1, 193–207, DOI 10.1007/s002220100168. MR1859021

[5] Bruce C. Berndt, *Modular transformations and generalizations of several formulae of Ramanujan*, Rocky Mountain J. Math. **7** (1977), no. 1, 147–189, DOI 10.1216/RMJ-1977-7-1-147. MR0429703

[6] Jonathan M. Borwein, David M. Bradley, David J. Broadhurst, and Petr Lisoněk, *Special values of multiple polylogarithms*, Trans. Amer. Math. Soc. **353** (2001), no. 3, 907–941, DOI 10.1090/S0002-9947-00-02616-7. MR1709772

[7] F. Brown, Motivic periods and $\mathbb{P}\backslash\{0,1,\infty\}$, Proceedings of the ICM 2014, to appear.

[8] Pierre Cartier, *A mad day's work: from Grothendieck to Connes and Kontsevich. The evolution of concepts of space and symmetry [in Les relations entre les mathématiques et la physique théorique, 23–42, Inst. Hautes Études Sci., Bures-sur-Yvette, 1998; MR1667896 (2000c:01028)]*, Bull. Amer. Math. Soc. (N.S.) **38** (2001), no. 4, 389–408, DOI 10.1090/S0273-0979-01-00913-2. Translated from the French by Roger Cooke. MR1848254

[9] T. Chatterjee, M. Ram Murty and Siddhi Pathak, A vanishing criterion for Dirichlet series with periodic coefficients, this volume.

[10] J. Coates and S. Lichtenbaum, *On l-adic zeta functions*, Ann. of Math. (2) **98** (1973), 498–550, DOI 10.2307/1970916. MR0330107

[11] Tapas Chatterjee and M. Ram Murty, *On a conjecture of Erdos and certain Dirichlet series*, Pacific J. Math. **275** (2015), no. 1, 103–113, DOI 10.2140/pjm.2015.275.103. MR3336930

[12] S. Chowla, *The nonexistence of nontrivial linear relations between the roots of a certain irreducible equation*, J. Number Theory **2** (1970), 120–123, DOI 10.1016/0022-314X(70)90012-0. MR0249393

[13] P. Chowla and S. Chowla, On irrational numbers, *Skr. K. Nor. Vidensk. Selsk. (Trondheim)*, (1982), 1–5.

[14] Guy Diaz, *Grands degrés de transcendance pour des familles d'exponentielles* (French, with English summary), J. Number Theory **31** (1989), no. 1, 1–23, DOI 10.1016/0022-314X(89)90049-8. MR978097

[15] L. Euler, De summis serierum reciprocarum, *Commentarii academiae scientiarum Petropolitanae,* **7** (1740), 123–134 (available in English translation online at http://eulerarchive.maa.org/pages/E041.html).

[16] Paul B. Garrett, *Holomorphic Hilbert modular forms*, The Wadsworth & Brooks/Cole Mathematics Series, Wadsworth & Brooks/Cole Advanced Books & Software, Pacific Grove, CA, 1990. MR1008244

[17] Pierre Deligne and Alexander B. Goncharov, *Groupes fondamentaux motiviques de Tate mixte* (French, with English and French summaries), Ann. Sci. École Norm. Sup. (4) **38** (2005), no. 1, 1–56, DOI 10.1016/j.ansens.2004.11.001. MR2136480

[18] Emil Grosswald, *Die Werte der Riemannschen Zetafunktion an ungeraden Argumentstellen.* (German), Nachr. Akad. Wiss. Göttingen Math.-Phys. Kl. II **1970** (1970), 9–13. MR0272725

[19] Sanoli Gun, M. Ram Murty, and Purusottam Rath, *A note on special values of L-functions*, Proc. Amer. Math. Soc. **142** (2014), no. 4, 1147–1156, DOI 10.1090/S0002-9939-2014-11858-2. MR3162237

[20] Sanoli Gun, M. Ram Murty, and Purusottam Rath, *On a conjecture of Chowla and Milnor*, Canad. J. Math. **63** (2011), no. 6, 1328–1344, DOI 10.4153/CJM-2011-034-2. MR2894441

[21] Sanoli Gun, M. Ram Murty, and Purusottam Rath, *Transcendental nature of special values of L-functions*, Canad. J. Math. **63** (2011), no. 1, 136–152, DOI 10.4153/CJM-2010-078-9. MR2779135

[22] Sanoli Gun, M. Ram Murty, and Purusottam Rath, *Transcendental values of certain Eichler integrals*, Bull. Lond. Math. Soc. **43** (2011), no. 5, 939–952, DOI 10.1112/blms/bdr031. MR2854564

[23] Sanoli Gun, M. Ram Murty, and Purusottam Rath, *Algebraic independence of values of modular forms*, Int. J. Number Theory **7** (2011), no. 4, 1065–1074, DOI 10.1142/S1793042111004769. MR2812652

[24] E. Hecke, *Analytische Arithmetik der positiven quadratischen Formen* (German), Danske Vid. Selsk. Math.-Fys. Medd. **17** (1940), no. 12, 134. MR0003665

[25] E. Hecke, *Eine neue Art von Zetafunktionen und ihre Beziehungen zur Verteilung der Primzahlen* (German), Math. Z. **1** (1918), no. 4, 357–376, DOI 10.1007/BF01465095. MR1544302

[26] George Gheverghese Joseph, *A passage to infinity*, Sage Publications, Los Angeles, CA, 2009. Medieval Indian mathematics from Kerala and its impact. MR3052306

[27] Koji Katayama, *Ramanujan's formulas for L-functions*, J. Math. Soc. Japan **26** (1974), 234–240, DOI 10.2969/jmsj/02620234. MR0337825

[28] Helmut Klingen, *Über die Werte der Dedekindschen Zetafunktion* (German), Math. Ann. **145** (1961/1962), 265–272, DOI 10.1007/BF01451369. MR0133304

[29] M. Lerch, *Sur la fonction $\zeta(s)$ pour valeurs impaires de l'argument*, J. Sci. Math. Astron., pub. pelo Dr. F. Gomes Teixeira, Coimbra **14** (1901), 65–69.

[30] F. Lindemann, *Ueber die Zahl π.*)* (German), Math. Ann. **20** (1882), no. 2, 213–225, DOI 10.1007/BF01446522. MR1510165

[31] John Milnor, *On polylogarithms, Hurwitz zeta functions, and the Kubert identities*, Enseign. Math. (2) **29** (1983), no. 3-4, 281–322. MR719313

[32] M. Ram Murty, *Some remarks on a problem of Chowla* (English, with English and French summaries), Ann. Sci. Math. Québec **35** (2011), no. 2, 229–237. MR2917833

[33] M. Ram Murty, Michael Dewar, and Hester Graves, *Problems in the theory of modular forms*, Institute of Mathematical Sciences Lecture Notes, vol. 1, Hindustan Book Agency, New Delhi, 2015. MR3330491

[34] M. Ram Murty and V. Kumar Murty, *Transcendental values of class group L-functions*, Math. Ann. **351** (2011), no. 4, 835–855, DOI 10.1007/s00208-010-0619-y. MR2854115

[35] M. Ram Murty and V. Kumar Murty, *Transcendental values of class group L-functions, II*, Proc. Amer. Math. Soc. **140** (2012), no. 9, 3041–3047, DOI 10.1090/S0002-9939-2012-11201-8. MR2917077

[36] M. Ram Murty and Purusottam Rath, *Transcendental numbers*, Springer, New York, 2014. MR3134556

[37] M. Ram Murty and Marilyn Reece, *A simple derivation of $\zeta(1-K) = -B_K/K$*, Funct. Approx. Comment. Math. **28** (2000), 141–154. Dedicated to Włodzimierz Staś on the occasion of his 75th birthday. MR1824000

[38] M. Ram Murty and N. Saradha, *Euler-Lehmer constants and a conjecture of Erdös*, J. Number Theory **130** (2010), no. 12, 2671–2682, DOI 10.1016/j.jnt.2010.07.004. MR2684489

[39] M. Ram Murty and Kaneenika Sinha, *Multiple Hurwitz zeta functions*, Multiple Dirichlet series, automorphic forms, and analytic number theory, Proc. Sympos. Pure Math., vol. 75, Amer. Math. Soc., Providence, RI, 2006, pp. 135–156, DOI 10.1090/pspum/075/2279934. MR2279934

[40] M. Ram Murty and Chester J. Weatherby, *On the transcendence of certain infinite series*, Int. J. Number Theory **7** (2011), no. 2, 323–339, DOI 10.1142/S1793042111004058. MR2782661

[41] M. Ram Murty and Chester Weatherby, *A generalization of Euler's theorem for $\zeta(2k)$*, Amer. Math. Monthly **123** (2016), no. 1, 53–65, DOI 10.4169/amer.math.monthly.123.1.53. MR3453535

[42] Yu. V. Nesterenko, *Algebraic independence*, Published for the Tata Institute of Fundamental Research, Bombay; by Narosa Publishing House, New Delhi, 2009. MR2554501

[43] A. Raghuram and Freydoon Shahidi, *Functoriality and special values of L-functions*, Eisenstein series and applications, Progr. Math., vol. 258, Birkhäuser Boston, Boston, MA, 2008, pp. 271–293, DOI 10.1007/978-0-8176-4639-4_10. MR2402688

[44] Dinakar Ramakrishnan, *Regulators, algebraic cycles, and values of L-functions*, Algebraic K-theory and algebraic number theory (Honolulu, HI, 1987), Contemp. Math., vol. 83, Amer. Math. Soc., Providence, RI, 1989, pp. 183–310, DOI 10.1090/conm/083/991982. MR991982

[45] Reinhold Remmert, *Classical topics in complex function theory*, Graduate Texts in Mathematics, vol. 172, Springer-Verlag, New York, 1998. Translated from the German by Leslie Kay. MR1483074

[46] G.F.B. Riemann, Über die anzahl der primzahlen unter einer gegebenen grösse, *Monatsberichte der Berliner Akademie* 1859 (available online at http://www.claymath.org/sites/default/files/ezeta.pdf).

[47] Tanguy Rivoal, *La fonction zêta de Riemann prend une infinité de valeurs irrationnelles aux entiers impairs* (French, with English and French summaries), C. R. Acad. Sci. Paris Sér. I Math. **331** (2000), no. 4, 267–270, DOI 10.1016/S0764-4442(00)01624-4. MR1787183

[48] Takuro Shintani, *On evaluation of zeta functions of totally real algebraic number fields at non-positive integers*, J. Fac. Sci. Univ. Tokyo Sect. IA Math. **23** (1976), no. 2, 393–417. MR0427231

[49] Carl Ludwig Siegel, *Transcendental Numbers*, Annals of Mathematics Studies, no. 16, Princeton University Press, Princeton, N. J., 1949. MR0032684

[50] Carl Ludwig Siegel, *Advanced analytic number theory*, 2nd ed., Tata Institute of Fundamental Research Studies in Mathematics, vol. 9, Tata Institute of Fundamental Research, Bombay, 1980. MR659851

[51] Christophe Soulé, *Régulateurs* (French), Astérisque **133-134** (1986), 237–253. Seminar Bourbaki, Vol. 1984/85. MR837223

[52] H. M. Stark, *L-functions at $s = 1$. II. Artin L-functions with rational characters*, Advances in Math. **17** (1975), no. 1, 60–92, DOI 10.1016/0001-8708(75)90087-0. MR0382194

[53] Tomohide Terasoma, *Mixed Tate motives and multiple zeta values*, Invent. Math. **149** (2002), no. 2, 339–369, DOI 10.1007/s002220200218. MR1918675

[54] Chester Weatherby, *Transcendence of series of rational functions and a problem of Bundschuh*, J. Ramanujan Math. Soc. **28** (2013), no. 1, 113–139. MR3060302

[55] Don Zagier, *Values of zeta functions and their applications*, First European Congress of Mathematics, Vol. II (Paris, 1992), Progr. Math., vol. 120, Birkhäuser, Basel, 1994, pp. 497–512. MR1341859

[56] Don Zagier and Herbert Gangl, *Classical and elliptic polylogarithms and special values of L-series*, The arithmetic and geometry of algebraic cycles (Banff, AB, 1998), NATO Sci. Ser. C Math. Phys. Sci., vol. 548, Kluwer Acad. Publ., Dordrecht, 2000, pp. 561–615. MR1744961

Queen's University, Kingston, Ontario, K7L 3N6, Canada
Email address: `murty@mast.queensu.ca`